GLOBAL SUSTAINABILITY
and
INNOVATION

SECOND EDITION

Edited by Anoop Desai
Georgia Southern University

Bassim Hamadeh, CEO and Publisher
Kassie Graves, Director of Acquisitions and Sales
Jamie Giganti, Senior Managing Editor
Miguel Macias, Senior Graphic Designer
Carrie Montoya, Manager, Revisions and Author Care
Alisa Munoz, Licensing Coordinator
Kaela Martin, Associate Editor
Abbey Hastings, Associate Production Editor
Chris Snipes, Interior Designer

Copyright © 2018 by Cognella, Inc. All rights reserved. No part of this publication may be reprinted, reproduced, transmitted, or utilized in any form or by any electronic, mechanical, or other means, now known or hereafter invented, including photocopying, microfilming, and recording, or in any information retrieval system without the written permission of Cognella, Inc. For inquiries regarding permissions, translations, foreign rights, audio rights, and any other forms of reproduction, please contact the Cognella Licensing Department at rights@cognella.com.

Trademark Notice: Product or corporate names may be trademarks or registered trademarks, and are used only for identification and explanation without intent to infringe.

Cover image copyright © Depositphotos/Antartis.
         copyright © Depositphotos/Mopic.
         copyright © Depositphotos/Pics4ads.
         copyright © Depositphotos/weerapat.

Printed in the United States of America

ISBN: 978-1-5165-2081-7 (pbk) / 978-1-5165-2082-4 (br) / 978-1-5165-2232-3 (pf)

Contents

1. An Introduction to Sustainability and Why It Is So Important — 5
Anoop Desai

2. Sustainability in A Global Era — 13
Mark Jayne
    Discussion Questions — 31
    In-Class Assignment: Objective Questions — 33

3. The Invisible Green Hand — 35
Martha Amram and Nalin Kulatilaka
    Discussion Questions — 63
    In-Class Assignment: Objective Questions — 65

4. Note on Energy — 67
Sean Harrington and Dennis M. Rohan
    Discussion Questions — 115
    In-Class Assignment: Objective Questions — 117

5. Fuel Cell Technology and Market Opportunities — 119
Andrea Larson and Stephen Keach
    Discussion Questions — 143
    In-Class Assignment: Objective Questions — 145

6. The Need for Nuclear Power — 147
Richard Rhodes and Denis Beller
    Discussion questions — 159

7. Solar Energy — 161
Travis Bradford
    Discussion Questions — 179
    In-Class Assignment: Objective Questions — 181

8. Wind Power — 183
Andrea Larson and Stephen Keach
    Discussion Questions — 209
    In-Class Assignment: Objective Questions — 211

9. Wind Power and Biofuels: A Green Dilemma for Wildlife Conservation — 213
Gregory D. Johnson and Scott E. Stephens

    Discussion Questions — 235
    In-Class Assignment: Objective Questions — 237

10. Climate Change — 239
Andrea Larson and Mark Meier

    Discussion Questions — 255
    In-Class Assignment: Objective Questions — 257

11. Freshwater Availability — 259
Jill Boberg

    Discussion Questions — 269
    In-Class Assignment: Objective Questions — 271

12. Approaches to Sustainable Water Management — 273
Jill Boberg

    Discussion Questions — 299
    In-Class Assignment: Objective Questions — 301

13. Product Modularity and the Design of Closed-Loop Supply Chains — 303
Harold Krikke, Ieke le Blanc, and Steef van de Velde

    Discussion Questions — 321
    In-Class Assignment: Objective Questions — 323

14. EDP Renewables North America: Tax Equity Financing and Asset Rotation — 325
Pedro Matos and Griffin Humphreys

    Discussion Questions — 355
    In-Class Assignment: Objective Questions — 357

15. Coffee Cup Woes: Starbucks' Environmental Sustainability Challenge — 359
Debapratim Pukayastha

    Discussion Questions — 377

Index — 379

An Introduction to Sustainability
and Why It Is So Important

Anoop Desai

There has been considerable interest in the general topic of sustainability of late. We are bombarded with sustainability related phrases every day. It seems as if it is unavoidable. Whether it be on the Internet or more traditional communication media such as television, radio, and newspapers, we keep running into phrases like "green energy," "green technology," "recycling," "refurbished products," etc. There has always been some amount of underlying public interest in the general notion of "environment conservation"; however, that interest has been spiking in recent years. One cannot help but wonder what has changed so dramatically. What is it about the times we live in that has prompted such a seismic shift in public attitudes toward the idea of sustainability?

In order to seek answers, it is essential to first define the term *sustainability*. What are the constituent elements of sustainability? How do these elements interact with each other? What are the economic repercussions of sustainability? It is important to understand and realize these, since any sustainability-related concept fails to make a significant impact if the economics of that concept is not thoroughly researched.

M-W Webster's dictionary (2011) defines sustainability as "of, related to, or being a method of harvesting or using the resource so that the resource is not depleted or permanently damaged." It is clear from this definition that sustainability concerns itself with conservation of resources. It doesn't specify whether those resources are natural or artificial. However, it needs to be borne in mind that most resources at their basic level tend to be natural resources. Take, for example, plastics used for a variety of industrial applications, as well as those used in consumer appliances. Plastics, however, do not occur naturally. Certain fractional distillates of crude oil are subjected to processing and chemical reactions in order to produce a variety of plastics. Celluloid, for instance, is a plastic that is used in photographic film. But where does celluloid come from? Can we just obtain it directly from nature? Does it grow on trees? Celluloid was first created by processing nitrocellulose and camphor, both of which occur naturally. Another example is that of steel, the universal building material. Where does steel come from? Can we plant steel? Can we dig the earth and obtain steel directly? In order to obtain steel, one has to have access to two basic natural substances: iron ore and carbon. Several other additives can be used to impart several specific properties to different kinds of steel. The key point to be borne in mind is that there literally is no such thing as a purely artificial substance. Everything that we use in everyday life, no matter the scale of use, can be traced back in one way or another to some natural resource. For example, steel can be traced back to iron ore and carbon, glass can be traced back to silica (silicon dioxide), paper can be traced back to wood—and therefore trees—plastics can be traced back to cellulose and crude oil. There has been a lot of discussion recently concerning rapidly diminishing global supplies of crude oil. It has been argued that if the world runs out of oil, alternate energy sources could be used to power the next industrial revolution. However, those alternate sources of energy are still in their infancy. Also, what about plastics? They are omnipresent. Almost everything, every consumable in our everyday life is somehow related to some plastic. If we do run out of crude oil, what product can be used to replace plastics?

So far, the discussion has been limited to consumables. But what about life itself? What are the resources we need to sustain ourselves, to keep us alive? We need clean air, freshwater, and food. Each of the aforementioned is a natural resource that is absolutely essential to sustain life as we know it on earth. Clean, freshwater is the life force of human beings on this planet. Yet most of our freshwater supplies, such as rivers, lakes, and streams, in addition to underground water or aquifers, are highly polluted. This pollution often takes the form of industrial waste or human and animal waste. Polluted air hampers breathing; polluted water renders water undrinkable and gives rise to a host of communicable diseases or epidemics. Lack of fresh, healthy consumable food is the cause of malnourishment and disease. Once again, please notice that each of the aforementioned resources occurs naturally. It can be appreciated that absence of natural resources would endanger our quality of life, not to mention threaten the very basis of life itself.

But what about technology? What about innovation? Can they not ease the burden on natural resources? The answer is, they can, to some extent. Since the dawn of human civilization, our quality of life has been directly tied to our ability to innovate. This innovation is formalized through various technologies that we use to help make our lives easier and richer. Beginning with the invention of the wheel and agriculture, humans have indeed come a long way when they put a man on the moon and cloned animals. The current state of technology, whether it be digital communications or smart materials, state-of-the-art cancer research, or hybrid vehicles is indeed the driving force of all civilization. Without innovation, human beings would really not be all that different from the rest of the animal kingdom. Innovation is not only the driving force of all civilization, but is the lifeblood of the economy as well. It will be noticed that throughout history, an ever advancing state of technology has been driving the global economy to ever higher levels of productivity. That productivity has translated itself into an abundance of wealth, and enhanced our standard of living. An example is apt at this juncture. The late 19th century was widely regarded as the most productive period in U.S. history. It spawned new innovations, such as the railroad, the phonograph, the electric bulb, the electric generator, and so on. The economy grew by leaps and bounds on account of this innovation, and gave rise to an extraordinarily wealthy class of people. So much so that Mark Twain referred to this period as the "gilded age." This age was followed in the 1920s by another era of innovation, which saw the widespread adoption of consumer products like the automobile, the washing machine, and the refrigerator. This brings us to the present age of digital innovation that was sparked by the microchip and the Internet. One only needs to recall the number of new technology companies spawned and brought to market during the late 20th century. The effervescence of the age found expression in the aptly coined term "dot-com mania."

However, in spite of a high level of technological sophistication, the fact still remains that our very existence continues to depend directly on natural resources. The role of technology is to enhance productivity. Productivity can be defined simply as the ratio of output to input. Natural resources comprise the input elements to all technologies conceived by man. Technology and innovation can be many things; however, so far they have not been a substitute for natural resources. Thus, it can be noted that sustainability pertains to conserving our natural resources so that future generations may be afforded a similar—or better—quality of life as the current generation.

The concept of sustainability is comprised of several individual elements. These elements act in isolation as well as interact with each other. Examples of the individual elements of sustainability include water management, food management, energy management and the greenhouse effect, and product recycling and remanufacturing. This book addresses each of the aforementioned elements in detail and discusses the economics of sustainability. Any project that deals with environmental conservation has to address the economics of implementation in order to be feasible. This is an important, often overlooked sustainability-related detail. The following discussion will introduce readers to the elements of sustainability.

If someone were to ask you the question "What makes the world go round?" what would your answer be? What is it that enables humans to go about their day-to-day activities? What is it that enables us to manufacture goods and services, sell them to consumers—indeed, what is the powering force behind innovation itself? Some might say it is imagination, but within the domain of sustainability, the answer has to be sought elsewhere. Energy apparently is the singular driving force of all human civilization. We have used energy since primordial times. The sun constitutes the earth's greatest single source of energy. Before the dawn of the industrial revolution, animal power was the dominant source of energy. This was replaced by steam power, and hence mechanical power (power derived from machines) after the invention of the steam engine, and subsequently, the internal combustion engine. Mechanical power remains our most widely used source of energy. However, what is it that powers the machines? Is there a single source of energy that powers all our industrial and transportation equipment? What is it that really drives and greases the wheels of our economy? The answer is: Gasoline and diesel drive the internal combustion engine and diesel engine, respectively (used in cars, buses, airplanes, ships, locomotives), and coal is used in electricity generation. Gasoline and diesel are obtained as a result of fractional distillation of crude oil. Together with coal, crude oil is the one most dominant source of energy used by human civilization today. Coal, crude oil, and natural gas together are referred to as fossil fuels. The reason for this is that they are the by-products of living organisms that died millions of years ago and were converted into fuel as a result of decomposition over time and under pressure. As the term implies, fossil fuels are obtained from the remains of dead decaying organisms (both plants and animals), and as such, their supply is limited. Simultaneously, the demand for fossil fuels has been rising on an ever expanding scale. For instance, of all the coal that has been mined over the last 800 years, 50% was mined during the past 40 years. This increase in demand has been fueled by the expanding energy needs of emerging economies of countries such as Brazil, Russia, India and China (together referred to as the BRIC countries), Indonesia, Mexico, South Korea, etc., not to mention the high energy reliance of the G7 countries (United States, United Kingdom, France, Germany, Italy, Japan, and Canada). This demand-supply imbalance has led to skyrocketing crude oil prices (crude oil cost about $150 per barrel in the summer of 2008), rendering it unaffordable for some to use gasoline or diesel as their primary source of energy.

The high price of crude oil constitutes only one part of the problem. The other part is just as dire. One might ask a simple question: "How do we really obtain energy from gasoline, diesel, and coal?" Energy is obtained from fossil fuels by burning them in an engine or a generator. The process of combustion releases heat, which causes air to expand. Expanding air is used to drive pistons in an engine, thus converting thermal energy into kinetic energy, and therefore mechanical power. This is the basic physical principle of how fossil fuels are used to generate energy. So far, everything looks quite beneficial. The problem we have encountered so far as obvious from the preceding discussion is the dwindling supply of fossil fuels. However, the combustion of fossil fuels results in a problem related to emissions. Fossil fuels are essentially hydrocarbons. They are compounds that are comprised of chemical bonding between carbon and hydrogen atoms. When coal, gasoline, and diesel are burnt, the combustion process consists of chemically combining oxygen from the air with hydrogen and carbon atoms (from fossil fuels), thus releasing water vapor and carbon dioxide, both of which are referred to as green house gases.

The major concern about greenhouse gases is that they entrap heat from the earth's atmosphere and prevent it from escaping into space. This trapped heat further feeds on itself, and results in a gradual but definite rise in the atmospheric temperature. Rising temperatures result in melting polar ice caps, rising sea levels, and destruction of natural habitats, creating over time an environmentally catastrophic situation. It is sometimes difficult to ascertain what is more serious, the diminishing supply of fossil fuels, or the greenhouse effect. Assuming that if one were to ignore the greenhouse effect (which would have tragic consequences in and of itself), the ever diminishing supply of fossil fuels is a problem that cannot be solved. The preceding discussion doesn't even begin to consider environmental effects and resulting health hazards of burning fossil fuels, such as atmospheric pollution, resulting in sulfur dioxide poisoning, black lung disease (resulting from emissions due to burning coal), etc. The human health hazards of such practices are a major cause of concern by themselves.

The only answer, then, is to resort to what are called alternate energy sources. Such sources of energy include—and not limited to—solar power, wind power, geothermal energy, tidal power, nuclear power, biofuels, such as ethanol (corn based, sugarcane based or peanut based), and hydroelectric power. Electric and hybrid electric cars that run solely on electricity or a combination of gasoline power and electric power could be construed as another form of alternate energy. The major advantage of using alternate sources of energy is that most of them are renewable. Unlike fossil fuels, which are limited in supply, alternate energy sources can be renewed over and over again without ever having to worry about supply running out. For instance, solar power uses solar cells to convert sunlight into direct current electricity. It is clear that solar power constitutes a renewable energy source. A similar principle is at work with wind power (which uses a windmill-based electric generator that generates electricity, based on the principle of electromagnetic induction) to convert kinetic energy of moving air (wind) into electricity. The principle of tidal power constitutes using the power of rising and falling tides to generate electricity. Ethanol is one of the most predominant forms of biofuels. Ethanol can be produced from corn or sugarcane or peanuts. Ethanol is really an alcohol that is created by the fermentation process of foodstuffs. It can be used as a substitute for gasoline in order to power cars and other forms of transportation. The major drawback of this fuel is that it draws on various edible foodstuffs, and thus could result in food shortage and food price inflation.

Water is perhaps one of the most important natural resources known to man. It has been argued in many circles that water comprises the essential life force. Throughout history, human civilizations have sprung and flourished near major sources of freshwater, such as lakes and rivers. Examples of such civilizations include the Nile valley civilization in ancient Egypt, the Indus valley civilization in ancient India, and the ancient Chinese civilization in the Yangtze basin. As a matter of fact, the city of New York was settled along the banks of the Hudson River. The prosperous Yangtze River delta generates almost 20% of the Chinese GDP.

Come to think of it, almost all forms of life as we know them on earth depend on water for their sustenance. Approximately 75% of the earth's surface is covered by water. This water is present in large water bodies such as oceans, seas, lakes, rivers and streams. And this is just surface water. Underground water sources are present in the form of water tables or aquifers. Most of the earth's surface water, however, is present in oceans, and is therefore highly saline. High salt content renders such water unfit for human consumption. Humans can only consume water with no salt content (referred to as freshwater). A miniscule percentage of all surface water found on earth is found in the form of freshwater. Freshwater can be found in rivers, lakes, streams and polar ice caps.

In addition to sustaining life, water plays a key role in almost all human activities. It is widely used in large amounts in industrial production, food production, materials processing, etc. In the context of food production, it should be noted that all food crops require freshwater to grow. In fact, irrigating land with seawater actually renders it barren over a period of time.

The total water supply available on earth is limited. The same cannot be said, however, for human population, which continues to explode. At last count, the earth had approximately 6 billion people, all of whom depend on water to sustain themselves. The explosion in human population since World War II has put severe pressure on water resources. The situation is especially dire in arid regions of the earth, such as the Middle Eastern countries. Lack of readily available freshwater has prompted many countries to resort to innovative techniques to obtain their water supply. Water desalination constitutes one such innovative practice. It consists of obtaining saline seawater and processing it to remove the salt content (through processes like reverse osmosis), thus rendering it fit for human consumption. This practice is widely followed in geographic regions of the earth facing a chronic water shortage. It should be noted in the context of sustainability that the process of water desalination is a highly energy-intensive process. The interplay between various facets of sustainability can be readily appreciated in this particular example.

Freshwater, when consumed in pure form, is the elixir of life. However, the same water, when contaminated by pollutants such as industrial, human, and animal waste, is perhaps the greatest killer known to man—and civilization's worst nightmare. Waterborne diseases come in different forms based on the type of infecting mechanism, such as protozoal, viral, parasitic, or bacterial. Examples of such diseases include typhoid, cholera, dysentery, and hepatitis, to name just a few. Quite a few of the aforementioned diseases were a major source of human fatalities until the first half of the 20th century. Typhoid fever, for instance, used to be regarded as a fatal killer until a series of remedies were discovered, which has rendered this once lethal killer comparatively ineffective and readily curable. Advances in medical technology and innovation have enabled us to obtain cures for such diseases and prolong the human life span, while simultaneously enhancing the quality of life. Given this backdrop, it is easy to realize that water filtration and purification could constitute a major business opportunity in the future, as emerging counties such as India and China put increasing pressure on global water resources.

Thus far, we have addressed the issue of sustainability in the context of natural resources, their importance to people, and their limited supply (yet seemingly unlimited demand). These questions are bound to arise: "So what are we doing about it? What are we doing to solve the problem?" One of the avenues to address the diminishing supply of natural resources is to resort to a three-pronged approach of reuse, recycle, and remanufacture. Together, they probably could constitute the holy trinity of conservation. As its name implies, reuse refers to reusing a product once it has been retired from useful life. Recycling a product differs from reusing it, in that the entire product is not used again. Instead, only its constituent material is used, this time in another product. Thus, recycling takes place primarily at the material level only. We hear of recycled parts all the time. Plastics are recycled on a regular basis; so is steel. Remanufacturing, on the other hand, is the process of taking an end-of-life (EOL) product and updating it in terms of technological relevance. A refurbished music system could be construed as an example of remanufacturing. Recycling, reusing, and remanufacturing are especially important in the context of the consumer-based economy that we live in. Each of the aforementioned processes increases the life span of a product, and thus adds value to it.

Businesses are beginning to realize the importance of conservation. They are starting to understand that the idea of sustainability is taking root in the public psyche. As a result, people are more likely to embrace (and buy) any product or service that is considered to be sustainable and shun its (unsustainable) counterpart. In terms of public companies whose shares are routinely traded on the stock exchange, the sustainability image constitutes a significant portion of the "intangible" asset value assigned to its common stock. Companies like Nike and Xerox continue to strive to make their products more sustainable. They serve as examples of increasing corporate interest in sustainability.

Sustainability is everywhere. It is everywhere, because it makes sense to conserve. Understanding the fact that natural resources are limited in supply—and irreplaceable in some cases—adds even more value to the sustainability argument. Sustainable business practices add more value to a company's balance sheet. Sustainable lifestyles make life more enjoyable. These lifestyles are responsible personal decisions that everyone should make to ensure the quality of life of future generations.

Sustainability in A Global Era

Mark Jayne

Introduction

The year 1968 is synonymous with social revolution and an emerging desire to try and reshape the world in which we live. Whether it was in relation to the peace protests surrounding the Vietnam war, student demonstrations on the streets of Paris, or urban race riots in the United States of America, 1968 was a time when people were questioning the moral and political assumptions upon which (at least Western) society was based (for a fascinating discussion of the relationship between the year 1968 and changing geographical ways of understanding the world see Watts 2001). On Christmas Eve 1968, however, images sent back from the Apollo 8 space mission would provide a new challenge to the ways in which humanity understood the world in which we live. The now famous Apollo 8 images were the first live pictures of the planet Earth transmitted from space. In addition to showing the beautiful fragility of the Earth, set against the inky darkness of space, the Apollo 8 mission also captured pictures of the Earth rising over the surface of the moon. In many ways it is difficult to overestimate the profound impact these pictures of the Earth had on the collective political consciousness of global society. In an attempt to try and capture the profound geo-historic importance of these Earthly images, rather than proving a commentary to support the pictures they were beaming back, the crew of Apollo 8 decided to read from the book of Genesis (and in particular the story of creation). Quite what people from non-Judaeo-Christian religious backgrounds made of the crew of Apollo 8's choice of reading is difficult to assess, but it is clear that the use of an evocative piece of scripture, recounting the creation of the Universe, reflected the emotive potential of these new pictures.

The pictures which reached Earth on Christmas Eve 1968 provided the iconic imagery for two socio-political forces which were set to dominate the final quarter of the twentieth century. In the first instance the image of the planet Earth has become the dominant motif associated with the complex mix of social, cultural, political and economic forces which are collectively referred to as globalization (see Cosgrove 2001). In this context,

Mark Jayne, "Sustainability in a Global Era," *Spaces of Sustainability: Geographical Perspectives on the Sustainable Society*, pp. 117-137. Copyright © 2007 by Taylor & Francis Group. Reprinted with permission.

the idea of a single planet moving in isolation through space, confirmed by the Apollo mission, served to underline the global connections which underpin human existence and life (Pickles 2004: 78–79). In a related, but very different context, the image of the Earth from space also became a crucial inspiration and symbolic icon of the environmental movement (Ingold 1993). To many in the fledgling green movement of the late 1960s, the image of the Earth from space served to indicate two things: first the idea that the Earth is a self-contained, environmentally integrated system; and second that the Earth is a finite and fragile resource which should be cared for and protected. This chapter considers the relationship between these emerging understandings of the planet as a globalizing system and debates surrounding sustainability. In relation to these aims, we will see that the emergence of notions of sustainability in an era of renewed planetary consciousness was no mere coincidence. Consequently, in this chapter we chart how a growing awareness of the global dynamics of environmental change and socio-economic interaction (see Liverman 1999) have been central factors within the formation and implementation of the principles of sustainability.

This chapter begins with a critical investigation of the concept of globalization and how latest geographical work is challenging many of the underlying assumptions associated with the analysis of global phenomena. Through the use of two case studies, the remainder of this chapter then explores the relationship between the processes of globalization and sustainability. In the first instance we consider the emergence of global political responses to sustainable development issues. Focusing in particular on the issue of climate change, this section considers the importance of developing international policies for sustainable development which cut across national boundaries, but also reveals the political difficulties associated with this process. In the second case study we consider the emergence of sub-global but supra-national strategies of sustainable development. Drawing on the example of environmental management in the Barents Sea, analysis considers the importance of developing environmental management strategies which harmonize different national policy regimes.

Geographies of Globalization: On Conceptual Hangovers and the Scale Debate

Unpacking Globalization

Globalization is one of those words which appears to be used with far more frequency than understanding. As a word, globalization entered the popular social and political lexicon in the early 1960s, but it now appears clear that the processes of globalization have been operating for many centuries (see Hirst and Thompson 1996). When trying to come to terms with what globalization actually is, it is helpful to begin by dispelling any notion that it refers to one thing or one set of processes. In this context Peter Dicken argues that globalization is actually a set of *overlapping discourses* which are used to understand

and transform the world (2002: 315). Supporting this view, David Harvey argues that it is possible to break globalization down into three interrelated categories: a process; a condition; and a political project (2000: 54). As a condition, globalization is perhaps best thought of as a mode of being within which people, economies and cultures are becoming increasingly integrated and connected. In this context, global forms of integration are based upon new modes of communication and information technology (Castells 2000); new types of social mobility and transport (Urry 2000); and the functional integration of economic systems (Harvey 1989). As a political project, globalization can be understood as a strategy of global economic and political expansion (pursued largely by Western nations and corporations) to open up new markets and sources of labour under the ideologies of free trade and neo-liberalism. In this context, the discourse of globalization is often described by politicians as a basis for transforming and improving the world, but can equally be maligned as a cause of economic decline within a given state (see Swyngedouw 1997a on the scalar discourses of globalization). Finally, when understood as a process (Harvey's preferred understanding of the term) globalization involves the *geographical reorganization of capitalism* (Harvey 2000: 57).

Understanding globalization as the *geographical reorganization of capitalism* has important implications. At one level globalization has been understood as the creation of a borderless world of infinitely mobile bodies and processes (see Ohmae 1990). In this context, globalization has gradually become synonymous with the erosion of the sovereign power and influence of nation states—as containers of political and economic power—as global processes effortlessly transgress and permeate their borders (for an excellent critical analysis of this perspective see Brenner 2004). Harvey's assertion that globalization involves the spatial reconfiguration of capitalism does, however, challenge such notions. Rather than seeing globalization as the end product in a linear movement from territorial states to non-territorial economies, Harvey interprets globalization as part of the ongoing spatial de- and re-territorialization of capitalism. In this context, rather than being an end of geography, or the power of space, many geographers are now interpreting globalization as the context within which a new array of spatial forms and modes of existence are being created which offer fresh opportunities for economic dominance and political resistance (see for example Bridge and Wood 2005; Routledge 2003; Sharp *et al.* 2000; Sparke 2004).

Recent work by geographers has consequently sought to understand how the flows and processes associated with globalization are *reshaping* not *replacing* spatial existence. In the context of this research agenda, an increasingly large number of geographers have focused upon the issues of scale and how globalization has been rescaling the geographical categories in and through which we organize the world (Brenner 1999; Jones 1998; MacLeod and Goodwin 1999; Marston 2000; Smith 1992, 1993; Swyngedouw 1997a, b; Whitehead 2003b). Historically, of course, geography has a long association with the study of scales. Neighbourhoods, regions and nations are scales around which our world has been organized and have provided the loci for a range of different geographical modes of enquiry. The contemporary *scale debate* does, however, differ from geographers' conventional approach to scale. It differs in the way in which it questions the extent to which scales can actually be interpreted as fixed geographical layers of existence. In the broad context of the spatial transformations associated with globalization, those analysing scales have started to question understandings of scales as rigid hierarchies of spatial organization, within which purportedly *natural* scales like the home, neighbourhood, city, region and nation are all neatly nested together in a kind of *Russian doll* framework (see Whitehead 2003b: 285). Instead, scalar analysts claim that globalization is seeing the disintegration of certain scales and the formation of new scalar frameworks of political, social and economic life. These new scales include urban innovation districts, city regions and supra-national coalitions (see Brenner 2004). This debate has been famously described as *the hangover after the party which was globalization*. On these terms, the scale debate is perhaps best thought of as an attempt to replace the discourse of a de-territorial, almost weightless existence associated with globalization studies, with a more careful excavation of the new geographies which are emerging as part of the contemporary transformations associated with global development.

The scale debate has recently been criticized from a range of different perspectives. Some argue, for example, that it produces a simplified vision of the world based upon a series of hierarchically organized, arbitrarily conceived levels of existence (see Collinge 2005). In this sense, it is claimed that notions of scale tend to hide the complex networks which stretch between purportedly local and global sites and connect and disrupt our established notions of scale (Murdoch 1997). Others claim that work on scale tends to reproduce a highly territorial vision of the world, which belies the aterritorial topologies (or connections) which constitute globalization (see Amin 2002). While I am sympathetic to many of these critiques, I do feel that they tend to unfairly represent the work of many scale theorists. While at one level the scale debate does find it hard to escape the seeming rigidity of the scales it describes, scale theorists have consistently emphasized the importance of understanding scales (like the neighbourhoods, regions or nations) not as absolute categories, but as relational entities. As relational entities, scale theorists argue that you can only begin to understand individual scales when you understand the social, economic and environmental relations which connect them. So for example it is only really possible to understand a regional scale by interpreting the relationship that a regional space economy has to its constituent metropolitan economies, and in turn the way the region then connects to national political arenas and directives. Additionally, while those working on scale have emphasized the abstract territorial boundaries which scales tend to construct, they have consistently argued that such scales are not eternal categories, but are flexible constructs, designed and configured as emergent ways of managing social, economic and environmental relations. The remainder of the book seeks to explore the sustainable society through its emergent scalar categories. Consequently, while globalization provides a crucial context within which I interpret the sustainability, it is not used to suggest that the sustainable society is a post-geographical world within which territorial boundaries (of all kinds) no longer matter. Instead through an exploration of the scales associated with sustainable development, this book charts the emergent geography of the sustainable society. As we will see, this emergent geography is sometimes based upon entirely new scales, and at others on the reinvention or reconfiguration of much older scales of social, economic and political life. In order to begin this scalar journey, however, the remainder of this chapter is dedicated to an exploration of the new global, or supra-national, scales which are becoming synonymous with sustainability. In particular analysis is concerned with how and why the ideal of the sustainable society is connected to visions of globalization and global political co-operation.

Globalization and Sustainability

Before moving on to consider specific examples of how the processes of globalization have intersected with the rise of policies for a more sustainable society, I want to briefly consider why globalization and sustainable development are related. A number of writers have explored the relationship between globalization and sustainable development (see in particular Held *et al.* 1999: chapter 8; Sachs 1999). In order to begin to understand the connections between globalization and sustainability, it is necessary to unpack the notion of sustainability into its component environmental, social and economic parts. This section consequently focuses on the links between globalization and sustainability through a consideration of the globalizing dynamics of environmental change, social justice and economic development (for a broader discussion of these processes see Yearley 1996).

As we have discussed at the beginning of this chapter, environmental considerations have played a crucial role in our growing consciousness of the notion of globalization. This link between globalization and the environment can be understood in relation to three key considerations: (1) a growing awareness of the global processes which make the Earth an integrated environmental system; (2) the scientific production of new evidence showing that environmental change and degradation is increasingly moving from being local and regional phenomena to a truly global issue (to the extent that environmental transformation in one place is having impacts on other more distant environments and societies); and (3) debates regarding the use and management of global environmental commons.

With respect to the first consideration, it is clear that since the late 1960s and the Apollo space missions, there has been a renewed desire to understand the Earth as an integrated environmental system. This renewed desire has, however, not only been based upon the cultural impacts of pictures of the Earth from space, but also on new forms of imaginative science. While numerous scientists have dedicated their lives to studying the processes which connect environmental systems at a global level, perhaps the most famous exponent of this brand of global environmental science is James Lovelock. In his now well-documented *Gaia hypothesis,* Lovelock claimed that life on Earth is the product of a series of intricate balances and symbiotic relationships between the biosphere, atmosphere and hydrosphere. Crucially, Lovelock argued that these interdependencies could only be appreciated fully at a global environmental level—a level at which a homeostatic environmental balance has been achieved. One of the key implications of Lovelock's hypothesis, of course, is whether, if life on Earth is the product of a careful balance of environmental systems, human induced changes to the natural environment could threaten the ability of the planet to support life. While Lovelock's methods have been routinely maligned, his notion of a global environmental system—operating in a similar way perhaps to a human body—has been a great inspiration to many in the green movement.

In terms of the second point, the precise relationship between globalization and environmental degradation is difficult to assess. In his highly original book *Planet Dialectics* (1999) Wolfgang Sachs provides a timely analysis of the relationships between globalization and sustainability. According to Sachs, it is possible to link emerging forms of global environmental problems (like global warming, ozone depletion and marine pollution), which cross national boundaries, to key economic processes which also transcend national spaces (1999: chapter 8). At one level, Sachs argues that the new, global mobility of economic activities (facilitated by telecommunications and new flexible production systems) means that an increasingly large expanse of the Earth's biosphere is now open to economic exploitation and resource extraction (Sachs 1999: 137). Beyond the physical exploitation of the environment, however, Sachs also notices how globalization has been synonymous with the spread of economic values developed predominantly within the economies of Europe, Japan and the USA. Consequently, with the deregulation of global trade and investment between nations, not only has more of the world's environment been opened up to economic use, but it has also been subjected to the *resource intensive,* fossil-based economic practices of the West (ibid.: 137–142). Based upon unfettered economic growth, competition and fossil-based growth, this is precisely the type of economic formula which many argue has caused contemporary global warming and its potentially hazardous socio-environmental consequences. Finally, Sachs argues that under contemporary globalization we have not only witnessed the spread of neoliberal economic practices, but also certain cultural norms and values. In this context, throughout the world we see the same multinational corporations selling global goods (like motor cars, hamburgers and televisions), which are being mass consumed and incorporated into an increasingly wide range of people's lifestyles (Sachs 1999: 137). The spread of such global goods, and the types of lifestyles they offer (particularly with regard to the motor car), have the potential to deepen contemporary patterns of global environmental change. The point is that contemporary global environmental problems are not just part of the processes which are making us think about globalization, global ecological threats are actually a product of globalization and the economic practices and customs it is based upon.

The third and final intersection between the environment and globalization I want to consider relates to the management of global environmental commons. In many ways the growth of the global capitalist economy has been based upon the uncosted exploitation of environmental commons like the atmosphere and the oceans. By their very nature, these commons are environmental assets which do not, and cannot, belong to any one nation state. Because of the lack of a clear ownership system, however, global commons are often exploited by different nations and corporations to serve their own narrow economic and social needs. According to Garrett Hardin's infamous *tragedy of the commons* thesis, the unregulated exploitation of common resources could lead to the disintegration of the entire resource base, making key resources eventually unusable for all groups (Hardin 1968).

James Lovelock and the Gaia Hypothesis

James Lovelock developed his Gaia hypothesis while working on the Mars Viking Programme for NASA. In many ways his understanding of the reasons why planet Earth has an abundance of life was related to his claim that the planetary environment of Mars could not support any. His Gaia hypothesis was set out in his most famous book, *Gaia: A New Look at Life on Earth* (Lovelock 1979). By studying the operations of large ecological spheres, like the atmosphere, biosphere, cryosphere and hydrosphere, Lovelock charted the interconnecting balances which these systems had forged throughout environmental history. These balances, between global levels of oxygen and carbon dioxide, and marine salinity and cryospheric transformations, he argued, are the reasons why the Earth is able to support life. On the basis of these insights, Lovelock claimed that the Earth could be thought of as a giant super-organism, which as with other organisms had achieved a balanced, life-producing state. Lovelock's choice of the term Gaia is significant in this context, originating as it does from the Greek word for the goddess of the Earth.

Key reading: Lovelock, J. (1979) *Gaia: A New Look at Life on Earth*, in Lovelock, J. (1988) *Ages of Gaia*.

While Hardin argu[ed for common ownersh]ip systems in place of privately enclosed n[atural resources, others h]ave actually resulted in a series of internatio[nal agreements on the eco]nomic management of common resources t[hrough organizations such] as the United Nations and World Wildlife [Fund. Recent decades h]as seen a proliferation of international con[ventions aimed at pro]tecting environmental commons. Many in[volve shared resp]onsibility for the management of the glob[al commons as a crit]ical component of the myriad processes ass[ociated with sustainability.]

So we can see tha[t the emerging n]otion of sustainability is connected to the [globe in impor]tant ways. In order to understand the natu[re of globalization an]d sustainability, however, we must also a[ppreciate other facto]rs. As we discussed in the previous section, many interpret globalization as the historical extension of capitalist economic relations (see Harvey 2000). But the global spread of capitalist relationships throughout the world has been a highly unequal process. Consequently, while the operation of multinational corporations and trade organizations has opened up non-Western economies to new forms of investment, this investment has come at a cost. Whether it is in terms of the overt exploitations of the colonial era, or the more subtle modes of power associated with the *new world order* (see Hardt and Negri 2000: chapter 1), globalization appears to have strategically favoured Western market economies. The uneven development of capitalism can be seen today in the great social and economic disparities which exist between Western market economies and less economically developed countries. These inequalities in social and economic welfare are sustained by uneven trading relations, market protectionism, crippling debt and the saturation of LEDCs' economies with cheap surplus goods from the West. They are also manifest in elevated levels of social poverty throughout large parts of the non-Western world. The ideologies of sustainable development have become embroiled within the debates surrounding globalization precisely because notions of sustainability recognize that just as environmental problems cross national borders, many forms of social and economic inequality originate within global socio-economic relations. It is in this vein that advocates of sustainability argue that a sustainable society can only be achieved when the combined environmental, social and economic forces which are threatening life on this planet are recognized and addressed in a holistic way.

I started this chapter by emphasizing that globalization is not one thing, but actually a set of competing discourses. This sentiment is clearly echoed in Wolfgang's Sachs analysis of globalization and sustainability when he states:

> Globalization is not a monopoly of the neo-liberals: the most varied actors, with the most varied philosophies, are caught up in the transnationalization of social relations [...] Accordingly the image of the blue planet—that symbol of globalization—conveys more than just one message.
>
> (Sachs 1999: 153)

In this chapter I want to argue that sustainability represents a different message about what globalization can be. I want to claim that certain versions of sustainability offer a different vision of globalization than those promoted within free-market liberalism. This is a vision which is based upon social justice and ecological conservation, not unregulated economic expansion.

Climate Change and the Global Response

In many ways the first example I want to consider of the relationship between sustainability and globalization is the most directly illustrative case of the complex social, economic and environmental forces which make up globalization. It is also, without question, one of the most widely discussed and debated issues within the policies and politics surrounding sustainable development. Climate change is a 'hot' political topic. It is the subject of numerous scientific studies, the object of an increasing number of TV documentaries, and has dominated the agenda of several major international conferences, from the United Nations World Summit on Sustainable Development (Johannesburg) to the G8 Summit of 2005 at Gleneagles, Scotland. Climate change is an important example of the relationships between sustainable development and globalization because at one and the same time the atmosphere is one of the most important global environmental systems—the recipient of many of the polluting outputs associated with economic development and a crucial basis for human life and health on this planet. As a meeting point (or perhaps context) for global environmental, economic and social processes and needs, it is hardly surprising that what goes on in the Earth's atmosphere has been a keenly contested area within contemporary discussions of the sustainable society.

The global atmosphere is a complex composite of gaseous elements, swirling winds, weather systems and convoluted pressure dynamics. While an important environmental system in and of itself, the global atmosphere cannot be understood as a closed system—it instead embodies a complex interface between the hydrosphere and biosphere and witnesses the continual exchange of elements and compounds between these systems everyday. The contemporary atmosphere is largely composed of nitrogen and oxygen, but it is also home to a range of important trace elements and compounds, including nitrous oxides, sulphur dioxide, ozone, methane, lead and of course carbon dioxide. The importance of the atmosphere to life on Earth is illustrated well in Bill Bryson's reflections on the planet:

> Thank goodness for the atmosphere. It keeps us warm. Without it, the Earth would be a lifeless ball of ice with an average temperature of minus 50 degrees Celsius. In addition the atmosphere absorbs or deflects incoming rays of cosmic radiation, charged particles, ultraviolet rays and the like. Altogether the gaseous padding of the atmosphere is like a 4.4-metre thickness of protective concrete and without it these visitors from space would slice through us like tiny daggers.
> (Bryson 2004: 313)

In addition to protecting us from dangerous forms of radiation, then, the atmosphere also ensures that the Earth is warm enough to sustain life. It is in this context that many people find global warming and climate change confusing.

Global warming is a naturally occurring phenomenon, driven by carbon dioxide in our atmosphere. In essence carbon dioxide acts like a giant blanket (or greenhouse) sealing in the Sun's heat, the heat which is reflected from the Earth's surface, and keeping the planet warm. To add to the confusion, not only is the greenhouse effect (at least partially) a natural phenomenon, but so too is climate change. Environmental scientists, glaciologists and climatologists constantly remind us of the great fluctuations which have occurred in the Earth's temperatures over time, as the planet has swung between ice ages and interglacial warmth. So you may well ask why all the fuss about contemporary patterns of climate change? Does it not simply represent the atmosphere doing what it is supposed to do? Should we simply accept climate change as a condition of life on Earth? Isn't trying to regulate the temperature of the planet an expensive and futile thing to do? Some people would answer yes to these last three questions (see Lomborg 2001). The problems with these assumptions start to emerge when you compare certain key atmospheric trends. Temperature records show that the twentieth century was on average a much warmer century than the nineteenth century; that the 1990s were the warmest decade so far on record; and that 1998 was the warmest year since records began. Furthermore, the

Intergovernmental Panel on Climate Change (IPCC)—the independent collection of scientists created to monitor global climate change—estimate that by 2100 the Earth will experience temperature increases of somewhere between 2 and 4.5 degrees Celsius (Lomborg 2001: 264).

The case for increasing global temperatures may appear conclusive, but you are entitled to ask is this just part of natural fluctuations in global temperatures, changes which have been occurring on this planet long before humans arrived? (For an interesting discussion of the social construction of climate change by a geographer see Demeritt 2001). We could take refuge in this question if it wasn't for the untiring work of one scientist in particular. Working from his observatory on the Hawaiian island of Mauna Loa, the climate scientist Charles David Keeling monitored levels of carbon dioxide in the Earth's atmosphere for many years. While carbon dioxide only makes up a tiny fraction of the atmosphere's gaseous content, it is one of the most important trace gases when it comes to global warming. Keeling started monitoring carbon dioxide in the late 1950s and his work showed a steady increase in the levels of the gas residing in the Earth's atmosphere (allowing for seasonal variations). Now at one level Keeling's finding were not unexpected; scientists recognized that large-scale industrialization had seen elevated levels of carbon dioxide being emitted into the atmosphere. But before Keeling's analysis, many felt that this excess carbon dioxide would simply be reabsorbed into the Earth's carbon cycle. Keeling's work indicated that not only was more carbon dioxide entering the atmosphere, but that it was also residing there over a long time period. When Keeling's graph of steadily increasing atmospheric carbon dioxide is placed alongside data showing increasing global temperatures, the cause of global warming starts to look less like a natural event and more like a human-induced phenomena (see McKibben 2003). Debate still rages about the relationship between observed increases in atmospheric carbon dioxide and global temperatures. Could the two trajectories just be coincidental, or are they intimately connected? Those scientific experts working for the IPCC appear to believe the latter scenario, arguing that 'the balance of evidence suggests that there is a discernable human influence on global climate' (quoted in Lomborg 2001: 266).

Even if we accept (and many still do not) that global warming is a product of human environmental intervention, the consequences of global warming are still far from clear. While climate change is a global phenomenon, one thing which is clear is that the projected effects of global warming will affect different geographical areas in very different ways, with often the most marginal of communities experiencing much more of the adverse consequences. The possible consequences of global warming include the loss of productive agricultural land, more frequent incidents of extreme weather events, sea-level changes and increased levels of inundation, and even the wider spread of tropical diseases (see United Nations 2003b; for a critical review see also Lomborg 2001: 287–297). Beyond the specific effects of climate change, however, it is crucial to realize that our contemporary social, economic and political systems are amazingly dependent upon climatic stability. Consequently, whether it be highly prized and valued real estate on low-lying ocean frontages, agricultural communities in Bangladesh, or even the electrical cables which power our homes (there is a fear that heat expansion caused by global warming could seriously damage such supply lines), much of the sustain-ability of our current ways of life depend on the climate not changing.

In response to these perceived threats, the last twenty-five years have seen a range of international initiatives emerging which have sought to develop more sustainable socio-economic relationships with the global atmosphere. At the centre of these initiatives has been a desire to try and instigate a global response to these global environmental problems. A global level response to climate change is seen by many as vital, because even if certain countries reduced their production of carbon dioxide, greenhouse gases from other countries could continue to drive climate change and effect other people's and nation's livelihoods throughout the world. The first major international agreement on climate change abatement was formulated at the Rio Earth Summit in 1992. The United Nations Convention on Climate Change took effect on 21 March 1994, after it had gained the signature of 165 national leaders. While this convention has still not been ratified by all of the countries that signed up to the convention, its main aim was to stabilize global carbon dioxide production so as to enable global ecosystems to adjust sustainably to climate change.

Figure 2.1. The possible consequences of global warming (taken at the Centre for Alternative Technology, Wales)

In 1997 the Kyoto protocol superseded the United Nations Convention on Climate Change. As its name suggests, the protocol was developed in the Japanese city of Kyoto. Its basic aim was to reduce the amount of greenhouse gases produced by signatory states by an average of 5.2 per cent below their respective 1990 levels. This target has to be achieved during the time period running from 2008 to 2112, or else failing states are subject to further greenhouse gas reduction penalties. Although the fact that the Kyoto protocol was signed by over one hundred states, its effectiveness as a form of international collaboration to tackle climate change has been questioned for some time. In order to come into force, the Kyoto protocol needed to be ratified by states that collectively produce 55 per cent of global greenhouse gases. For a long time, the reluctance of nations like Australia, the USA and Russia to ratify the agreement meant that the protocol was not binding on the other states that had signed it. In October 2004, however, the Russian government finally ratified the Kyoto protocol and reignited the long-stalled process of international action on climate change. The protocol has still not been ratified in the USA. Opposition to the protocol in the USA has been expressed in a number of ways. First, the George W. Bush administration continues to question the scientific link established between rising levels of industrial greenhouse gases and climate change. Second, the US government is also concerned about the economic impact which ratifying the Kyoto agreement could have on the country (Lomborg for example estimates that the Kyoto protocol could cost the global economy approximately US$150 billion a year, 2001: 322). The USA remains the world's largest emitter of greenhouse gases. While this remains the case, and the USA continues to dispute the principles of the Kyoto protocol, it appears that global attempts to tackle climate change will remain seriously compromised.

Interestingly, in the context of this book, the Kyoto protocol incorporates within its policy dictates an appreciation of the variable geographies of sustainable development. Recognizing the urgent need for economic development in many LEDCs, the Kyoto protocol only sets strict targets for carbon dioxide emissions reductions in MEDCs, while offering more flexible mechanisms for carbon reduction for other nations. The geographical sensitivities associated with the Kyoto process have, however, created further geopolitical tension, with the USA arguing that large carbon dioxide emitters such as China and India need to make firmer commitments to reductions before they can themselves consider committing to the protocol. The scientific, political and economic issues surrounding climate change are complex and it is not the purpose of this section to put forward an unqualified assault on US climate change policies (it is important to realize for example that despite US opposition to the Kyoto protocol at a federal level, many individual states have begun legislating to reduce greenhouse gas production at a state level—most famously perhaps in Governor Arnold Schwarzenegger's state of California, see Toepfer 2004). What these discussions reveal, however, are the geopolitical difficulties associated with developing global action on sustainable development issues.

The story of climate change reveals many of the tensions which exist between the processes of globalization and sustainability. At one level it appears that the economic processes associated with globalization are generating contemporary patterns of climate change and threatening global socio-environmental sustainability. At the same time, it appears that effective forms of global political action are required to tackle the socio-environmental threats associated with global warming. Such action it is argued could impinge upon economic growth in both MEDCs and LEDCs, and threaten socio-economic sustainability worldwide. It is in this context, that despite being related to global phenomena, the principles of sustain-ability, and debate concerning how to achieve the most sustainable solution to global problems, have their own geographies. Sometimes this geography is expressed at a national level, with countries like the USA refusing to ratify the Kyoto protocol. At other times, the geographical response to sustainable development issues takes a more local form, as individual neighbourhoods, cities and regions decide to address climate change in their own ways and through their own programmes (see Burgess 2005; Slocum 2004). The point is that globalization does not represent the end of geographical enquiry into sustainable development issues, but rather the need for a newly intensified study into the emerging spaces and scales of sustainability (see Eden 2005).

Supra-national Coordination and the Case of the Barents Sea Region

Emerging Socio-environmental Conflicts in the Barents Sea

As we discussed in the first section of this chapter, it is important not to think of globalization as only involving global-level events. Globalization should instead draw us in to considering the reconfigured scales and spatialities associated with contemporary socio-ecological life and existence. The remainder of this chapter considers one example of sustainable development which, while not operating at a 'global' level, is clearly occupying a space which has been opened up by the complex processes associated with globalization. The Barents Sea is a remote expanse of ocean located to the north of the Arctic Circle and which meets the shores of Norway and Russia. For a long period of time the relative remoteness and inaccessibility of the Barents Sea has meant that it has remained a relatively unspoilt ecological region (for a discussion of the geopolitics of the Barents Sea see Churchill and Robin 1992; Osherenko and Young 1989). Recent developments have, however, changed popular understandings of the Barents Sea and seen it take an increasingly prominent role in debates surrounding globalization and sustainability. Two key events have been central to this process. First, scientific surveys have indicated that beneath the Barents Sea lies the world's largest untapped oil and gas reserves. As two of the main energy sources driving the global economy, excavation of these reserves could be extremely beneficial to the economies of Norway and Russia. Second, Russia has started to explore the potential of building an oil pipeline to Murmansk (the main Barents Sea port), through which oil extracted from continental Russia could be transported via oil tankers all around the world. These twin developments have generated a great deal of anxiety among many environmental groups. Organizations such as the WWF have described the Barents Sea as one of the world's few unspoilt fragments of marine nature and argue that the extraction and transport of oil through the sea could jeopardize the delicately balanced ecosystem of the area.

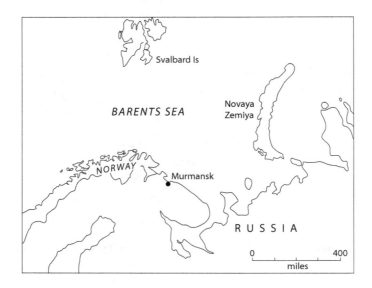

Figure 2.2. The Barents Sea

Figure 5.3. Konigfjord, Svalbard

In the context of the struggles surrounding the use and management of the Barents Sea, I want to argue that the region represents an important meeting place of globalization (in the form of trade, resource extraction and oil transportation) and sustainability (expressed in terms of the environmental integrity of the area). The story I am going to tell about the

Barents Sea, however, will reveal how in the context of the late arrival of globalization in the region, the Barents Sea has been re-forged as a supra-national zone of resource conflict and environmental management. In this way we will see that various attempts to claim the Barents Sea have involved the reconstruction and re-imagination of its geographical parameters.

Sustainable Development Planning in the Barents Sea and the WWF

In many ways the tensions over resource development in the Barents Sea illustrate the geographical nature of sustainable development issues. In one geographical region, the Barents Sea, we find the presence of untold riches of fuel and of ecological diversity. To extract oil and gas could be vital for economic development and social employment in both Russia and Norway. But to do so could also threaten the unique marine ecosystem and environments which surround the oil reserves, through the building of industrial pipelines and the dangers of large-scale oil spills. Tensions surrounding economic and environmental sustainability often arise as a result of such geographical conundrums. In response to concerns over sustainable development in the area, the WWF has been trying to think of alternative ways in which the Barents Sea can be thought about and used. Significantly in the context of this book, the WWF have developed a sustainable management strategy for addressing resource conflict and ecological conservation in the sea.

According to the WWF, environmental management in the Barents Sea has been inhibited by a very narrow conception of what the Barents Sea actually is. At one level this narrow conceptualization of the Barents Sea has been based upon understanding it primarily as the home of energy resources. Despite the continuing political obsession with fossil-based resources in the region, the WWF have worked hard to uncover an amazing wealth of ecological resources and diversity in the area. A biodiversity assessment of the Barents Sea (conducted by the WWF) revealed the wealth of biodiversity in the area (see WWF 2001). According to the WWF's report, the Barents Sea represents a highly *productive and fluctuating* environment. With its unique mix of ocean currents, ice sheets and nutrient flow, the Barents Sea is able to play host to a range of interconnected organisms and species (ibid.: 22–24). In addition to a range of fish life, this biodiversity includes ice flora and fauna, plankton, whales, walruses, seals and various varieties of sea bird (ibid.: 25–48). In addition to this rich biodiversity, the Barents Sea region is also home to two groups of indigenous peoples, who are rarely mentioned in the prevailing economic discourses of the area. The region is home to several tens of thousands of Sami and Nenets (ibid.: 25). By highlighting and recording this environmental and social data, the WWF are trying to promote a vision of the Barents Sea as a space of social, ecological and economic interest. The second reason why those working for WWF believe that the Barents Sea has been narrowly misconceived is because of its territorial division. According to the WWF, the reason why such ecological diversity exists in the Barents Sea region is because of the forms of environmental and biological interaction which occur throughout the whole region. They further claim, however, that the division of the region between capitalism and communism during the Cold War and now between the resource interests of Norway and Russia, is making it hard to think and act on the region as an integrated socio-environmental whole (WWF 2004).

In the context of this narrow and territorially divided vision of the Barents Sea, the WWF proposed a new spatial management strategy for the region (see WWF 2004). This strategy, known as *the Barents Sea Ecoregion Programme,* was launched in 2004. It is important at this point to explain what is meant by the term ecoregion as it is being deployed by the WWF. As a spatial entity, regions usually refer to sub-national territorial entities, which are used to divide up larger states into political, cultural, economic and environmental districts. This is not, however, how the WWF are currently using the term. In their Barents Sea campaign, the WWF are utilizing the idea of an eco 'region' in order to construct a supra-national space, which cuts across Norwegian and Russian territorial sovereignty in the area. In this way the ecoregion is trying to create a sense of environmental space which existed in the Barents Sea long before the emergence of modern nation states. According to the WWF, the Barents Sea ecoregion incorporates both the Norwegian and Russian sections of the Barents Sea as well as large parts of the terrestrial hinterland which contribute to the ecological balance of the area (WWF 2004).

As an international non-government organization (a type of organization which appears to be having an increasingly important role within the era of globalization, see Desforges 2004), the WWF are of course not able to enforce their policies and visions for the Barents Sea. In order for the ecoregion strategy to be enacted the WWF require the full co-operation of the Norwegian and Russian authorities. At present, while the WWF have gained support from the Norwegian government, negotiations are still ongoing with the Russian government. If delivered, however, the WWF's vision of ecoregional development would see the Barents Sea managed as an integrated social, economic and environmental space. According to this plan, any form of proposed economic development in the region would have to be carefully assessed for its potential social and ecological consequences. Research on socio-environmental relations in the Barents Sea indicates that the implementation of new modes of integrated sustainable development cannot come soon enough (see Brunstad *et al.* 2004). With extensive industrial development already occurring in the Russian Barents and plans to extend oil supply and refining facilities there, the future sustainability of the area is already under serious threat.

While obviously not representing a global response to socio-environmental problems, the WWF's ecoregion is typical of the types of supra-national spaces of sustainability that are emerging in the shadow of globalization. Such supra-national spaces of sustainable development planning can encompass environmental units like rainforests, or economic units such as the European Union. What these spaces all have in common, however, is a desire to implement sustainability in ways which cut across established territorial boundaries, in order to produce more integrated spatial systems of planning and development.

Summary

In this chapter we have explored the multiple meanings associated with the term globalization and how it is intimately connected to our discussions of sustainability. At its simplest level we have seen that the multiple processes associated with globalization force us to think beyond the narrow bounds of nation states in our attempt to understand how the world works. In relation to sustainability, we then moved on to consider how the social, economic and environmental processes which determine our relative levels of sustainability are in part constituted at a global level. From the systemic forms of environmental change associated with climate change, to the imperialistic economic relations of trade which cause so much social inequality in the world, we have seen the inextricable links between globalization and questions of sustainability.

Despite our explorations of globalization, this chapter has not suggested that we should see globalization as a form of weightless, aterritorial existence which signals the end of geography as we know it. Instead, this chapter has revealed that the processes of globalization are creating a series of new geographical spaces and scales. While many of these new spaces and scales are devoted to narrow economic goals, as our two case studies have shown, certain spaces are emerging in this era of globalization which are being designed specifically to deliver a more sustainable future for our planet. In our first case study, we saw how contemporary attempts to tackle the serious social, economic and environmental threats of globalization are operating through global networks of international agreements and protocols. In this sense, the global atmosphere is becoming a meeting point for different components of the sustainable development debate, within which the needs of oil-based industries are colliding with those concerned about the social utility and environmental integrity of the atmosphere. In the second case study, we saw how the pressures associated with global economic development are increasingly threatening biodiversity and delicate ecological systems throughout the world. In the case of the Barents Sea, we saw how such tensions are in part being fuelled and sustained by narrow nationalist claims to territorial resources. The activities of the WWF, however, serve to illustrate how certain NGOs and political groups are using the transnational visions, typically associated with globalization, to construct new supra-national spaces for sustainable development. What is perhaps most interesting about the activities of the WWF in the Barents Sea is that it is still dependent upon the active participation of territorial powers—as is still the case with climate change negotiations. In this context, it appears unwise to simply acclaim the arrival of a more integrated global space within which sustainable development can thrive, and to remember to be vigilant about the sustainable development credentials of the new spaces which are emerging in the contemporary global era.

Suggested reading

There is a bewilderingly wide range of introductory texts which can be used to familiarize the reader with the key concepts associated with the term globalization. There are, however, fewer texts which effectively link globalization with debates surrounding sustainability. Held, D., McGrew, A., Goldblatt, D. and Perraton, J. (1999) *Global Transformations – Politics, Economics and Culture* provides a good overview of a variety of themes relating to globalization and also includes one chapter on the links between globalization and environmental issues. Sachs, W. (1999) *Planet Dialectics: Explorations in Environment and Development*, chapter 8, provides one of the best critical discussions available

Suggested Websites

For an accessible discussion of the evidence, policies and possible consequences of global warming, visit the BBC's climate change page at: http://www.bbc.co.uk/climate/

For more information on key international policies designed to tackle climate change go to the United Nations Environment Programme's climate change page at: http://www.unep.org/themes/climatechange/

Information on the WWF's Barents Sea campaign is available at: http://www.panda.org/about_wwf/where_we_work/arctic/what_we_do/marine/barents/index.cfm

Discussion Questions

1. Explain in detail the relationship between globalization and sustainability.

2. How does globalization affect climate change?

3. What is the gist of the GAIA hypothesis?

In-Class Assignment: Objective Questions

1. An increasingly large expanse of the Earth's biosphere is now open to economic exploitation and resource extraction.
    a. True
    b. False
2. According to the Stern Review mid-range scenario, unless action to counteract rising temperatures and changes in climate pattern is taken, which of the following scenarios is likely to unfold in the future?
    a. The size of the economy in 2050 will be significantly larger.
    b. The size of the economy in 2050 will equal the current size.
    c. The size of the economy in 2050 will be significantly smaller.
    d. The economy does not depened on environmental issues.
3. Lomborg estimates that the Kyoto Protocol will cost the global economy approximately which of the following?
    a. $500 million annually
    b. $150 billion annnually
    c. $1 billion annually
    d. None of the above

The Invisible Green Hand
How Individual Decisions and Markets Can Reduce Greenhouse Gas Emissions

Martha Amram and Nalin Kulatilaka

Every day, we make purchase and consumption decisions. Embedded in every action is some amount of greenhouse gas emissions—and so our many small, individual actions contribute to global climate change.[1] This article outlines a strategy for mitigating the course of climate change in exactly the same way, using the strength of our market economy to trigger changes in our investment and consumption decisions. Focusing on private decisions and taking into account local conditions can bring about faster and cheaper results to the climate change problem.

At the root of the climate problem is the familiar economic concept of common goods (a common "bad" in this case). We only pay a price for the good or service consumed (e.g., clothes we wear, heating and lighting our house), but not for the GHG emissions that imposes a long-term, uncertain cost on society as a whole. As a result, our decisions are based on the desired attributes of what we consume, with scant attention paid to the undesirable attributes that come bundled with it. Although the conceptual solution to this problem is straightforward—attach a price to GHG emissions—doing so is fraught with enormous challenges.[2]

While global and national legislative agendas pursue climate change mitigation, progress at the micro level is made through local policies and initiatives that shape many private investment decisions. The abstraction that often exists in climate change actions needs to be removed. We must connect the big ideas to micro decisions and local realities. It is at this level of detail that climate change solutions can work—it constitutes the best choice for private decision makers. While local actions may be shaped or tipped by public policy or regulatory constraints, it is ultimately the many small decisions made by individuals that will drive emissions out of our purchases and consumption. By placing investment and consumption decisions at the center of our analysis, and by defining climate change solutions in response to market strengths and gaps, we demonstrate the scale, speed, and cost-effectiveness of market-based solutions. With a little prodding and the help of both national and local policies, the invisible hand can go green.

Martha Amram and Nalin Kulatilaka, "The Invisible Green Hand: How Individual Decisions and Markets Can Reduce Greenhouse Gas Emissions," *California Management Review*, vol. 51, no. 2, pp. 194-218. Copyright © 2009 by California Management Review. Reprinted with permission.

A particularly illuminating way to discuss climate change mitigation efforts uses the stabilization wedge framework proposed by Pacala and Socolow.[3] They link climate science to policy actions and suggest that stabilization efforts are best achieved through a portfolio of technologies. Seemingly cheap and immediately scalable technologies are available, so why aren't they widely adopted? The answer lies in missing and incomplete markets. What looks cheap at first glance is far more complex (and more expensive) to implement in the real world. The slow adoption of energy conservation efforts, especially in the built environment which accounts for 40% of total GHG emissions in the United States, is a case in point.[4] There are three categories of factors that contribute to the adoption gap in clean technologies: uncertainty, incomplete markets, and agency problems.

Private investment decisions are the key drivers of GHG. On a day-to-day level, emissions are largely set by behavior, which in turn is constrained by the existing capital stock. Because capital stock is so long-lived, the private investment decision is a moment of great consequence for determining further emissions levels. Public infrastructure shapes private investment decisions, but public institutions typically respond after the fact; they have a poor track record of acting *before* catastrophe strikes. This leaves us relying heavily on markets to tip private investment decisions toward clean technologies.

New and reconnected markets can tip private decisions and accelerate the adoption of clean technologies. Climate change is experienced locally and mitigation must use local opportunities. Regional weather patterns or individual catastrophes are sometimes interpreted as evidence of climate change and can galvanize communities and individuals into action. Extreme weather can destroy infrastructure, leading to higher local taxes and depleting the asset base. Local conditions—infrastructure, wealth, regulations, and more—will determine the costs of adapting to a changing climate in a given area. Because we experience climate change at the community level, local policies and market dynamics become more important, as they are ways to tailor solutions to the conditions immediately at hand.

The Efficiency Gap

The next step after assembling a list of possible technologies is to rank the available technologies by cost. Over the years, there have been a number of detailed studies that provide a clear sorting of emissions-reducing technologies.[5] Table 3.1 summarizes findings from a recent study by McKinsey & Co.

Table 3.1. Technology Solutions Ranked by Cost

	Cost per Ton of CO2 Abated		
	(1) Positive Return on Investment	(2) Below $25 per ton	(3) $25 to $50 per ton
Total Reduction (gigatons of CO2 per year)	1.3	0.8	1.0
Selected Technologies	Building Insulation	Nuclear Industrial	Motor Systems
	Lighting Systems	Livestock	Carbon Capture and Sequestration
	Air Conditioning	Forestation	Coal Plant Retrofits
	Water Heating	Enhanced Oil Recovery	Waste
	Fuel Efficiency (low-cost penetration)	Wind	Biodiesel

Technologies that reduce emissions are listed in three columns in Table 3.1. The items in the first group (column 1) have lifetime benefits that exceed their cost, providing a positive return on investment. Cumulatively, these technologies could reduce annual CO_2 emissions by 1.3 gigatons. Technologies in the middle group (column 2) cost less than $25 per ton of CO_2 abated, and the third group (column 3), in the right column, costs less than $50 per ton abated. The total savings possible from employing all of these technologies is 3.1 gigatons per year. Many of these technologies, such as building insulation or increased water heater efficiency, are available right now. Others, such as coal plant retrofits or biodiesel, are expected to reach a cost-effective scale by 2030. Some technologies such as insulation and efficient water heating continually pop up as good personal investments that happen to be good for the environment as well. So the obvious question is: if these technologies are so cheap that they actually make us money, why aren't they widely adopted, to the point of negative marginal returns? The answer to this question has been called the "efficiency gap"—the difference between the clear, favorable economics of many energy-efficiency technologies and their low rate of deployment. The efficiency gap has been noted in energy policy literature since the energy crisis of the mid-1970s.[6]

To understand the bottleneck issues, consider the market for energy efficiency—a stream of investments in heating and cooling systems, insulation, innovative engines and storage technology, and the light bulb—that reduce energy use. Experience over the past ten years has shown that investments in energy efficiency cost approximately three cents per kilowatt hour (kWh) saved. In contrast, coal, the nation's lowest-cost domestic fuel, costs seven cents to nine cents per kWh delivered to the customer. Study after study shows that the United States can provide for half its growth in energy use by investing in energy efficiency, in all regions of the country.[7] Yet we rely on utilities—not known for their sophisticated marketing—to promote energy efficiency.

Another barrier to adoption involves up-front costs. It is estimated that a one-dollar subsidy to the up-front cost of a new energy investment is eight times more effective than a one-dollar energy price subsidy.[8] Yet utilities don't benefit from the energy savings and thus provide only limited subsidies of up-front costs. Private capital can be drawn into the market for energy efficiency if there are profits in energy savings. Metering is a key technology for monetizing energy savings or on-site energy generation. Some of the most interesting clean technology startups are working in metering and information flow, technologies that are critical to creating active local energy markets.

Finally, note that the national electric grid was built to provide reliability and smooth prices. Although U.S. demand varies by time of day and season, almost all customers pay a flat rate for power. One recent study shows that utilities use their full capacity less than 5 percent of the year and that there would be sufficient off-peak capacity in the industry to recharge 84 percent of the nation's auto fleet, assuming all autos were converted to plug-in hybrids. It is difficult to think of another industry that has developed such a large physical infrastructure without using price signals to smooth the seasonal and daily shifts in demand and supply. As transportation and stationary markets for energy converge, new contracts and financial instruments could significantly reshape power demand and avoid high-cost capacity.

In sum, impressive and ever cheaper technologies are at hand, but for substantive and speedy change, we must introduce modern markets and innovative financial instruments and business models. Brown succinctly laid out the reasons for the efficiency gap, including: misalignment of incentives (the principal-agent problem); weak price signals; insufficient information; lack of capital; distortions through legislative or fiscal policies; unpriced costs; unpriced benefits; and the relative unimportance of carbon reduction compared to other investment or purchase attributes.[9]

Adam Jaffe and Robert Stavins examined the rational features of investment or consumer adoption behavior that could lead to the appearance of an energy-efficiency gap.[10] For example, they noted that technologies come in bundles of attributes, and often a prominent feature is unattractive. Many people, for example, understand that compact fluorescent bulbs save energy, but still won't buy them because they don't like the color of light the bulbs produce. In a similar vein, a recent article profiled the green choices of a ski resort. Each year the green team presented ideas that lowered both costs and energy use to the resort owner, only to see them squeezed out of the budget by less-green projects with higher returns.[11] Finally, Jaffe and Stavins note that new technologies face long adoption cycles, so a critical observer might see an energy-efficiency gap, but the reality is a normal rate of adoption—one that remains incomplete for many years. The rational adjustment response is slow because of the high transactions costs encountered by existing organizations when trying to adjust the existing capital stock using incomplete markets.

Transactions costs can be reduced by connecting new markets with new organizations and, thereby, accelerate the adoption of clean energy technologies.[12]

The residential water heater provides an example of the micro factors that produce the energy-efficiency gap. Every house has one and, on average, they consume about 15 percent of the homeowner's energy budget.[13] The EPA estimates that 60 percent of water heaters are replaced on an emergency basis.[14] The frustration of a homeowner temporarily without hot water no doubt puts energy efficiency last on the list of buying factors. In the middle of the night, or on short notice, the homeowner calls the installer to replace the water heater. However, the installer's incentives have little to do with energy efficiency. His performance is being measured on speed, the reliability of his product, and his service demeanor. Meanwhile, many parties are trying to tip the homeowner's decision: the utility offers a rebate for energy efficiency; the U.S. government offers a tax credit; and a host of nonprofit advocacy groups provide detailed information on efficient water heaters. Yet, in the panic of having no hot water, none of these factors matter greatly. Speed trumps other concerns.

The water heater decision brings up another problem. There is so little differentiation between standard gas and electric water heaters that, in 2004, the EPA declined to rate them for inclusion in their Energy Star program. Since 2004, a number of new technologies have been developed and marketed, but they have a combined market share of less than 1 percent. This further confuses the marketplace, as the installers must act as pioneering adopters for their customers—a very risky business proposition.

Now let's look at the 40 percent of homes that do not replace their water heaters on an emergency basis. Table 3.1 shows two of the factors influencing a replacement decision—the outdoor water temperature and the cost of natural gas—and the optimal replacement decision for three different locations. The results in the middle column show the number of occupants needed to warrant an early replacement. For example, in Palo Alto, California, where natural gas prices are high, early replacement is warranted in all homes. However, in Brookline, Massachusetts, where gas prices are lower, early replacement is warranted only in homes with high hot water use.

The two right-hand columns of Table 3.2 illustrate the perversity of some energy-water links. First, examine the column showing the water costs to the homeowner. The costs are the highest in Brookline, where a per gallon assessment was added to water bills to pay for a new sewage treatment plant in Boston Harbor. Palo Alto has the second-highest water rates, due to the presence of a wastewater treatment plant. Sunnyvale gets the majority of its water from a local aquifer. In all three locations, there are substantial savings to be gained by using less hot water in showers. This can be done by shortening shower time or replacing the showerhead with an ultra-low flow unit (one that uses 1.25 gallons per minute instead of the flow rate of most shower heads, which is 2.5 gallons per minute). Ironically, once hot water use is reduced by a new showerhead, there is less value in replacing a standard hot water heater with an energy-efficient one. This is shown by an increase in the number of occupants needed in each home to warrant early replacement in Brookline and Sunnyvale.

Table 3.2. The Water Heater Early Replacement Decision in Three Locations

	Cold Water Temperature (F°)	Natural Gas Price ($ per therm)	Replace water heater if # residents exceeds…	Cost of Water (¢ per 10 gallons)	Replace showerhead. Then replace water heater if # residents exceeds.
Brookline, MA	48	$1.30	5	13.7	6
Sunnyvale, CA	56	$1.47	3	2.7	5
Palo Alto, CA	62	$2.07	0	6.5	0

Note: Assumes constant natural gas prices; early replacement is warranted when a price rise of more than 3 percent per year is forecast Source: GreenNow USA

The simple water heater market is a good illustration of the energy-efficiency gap. Decisions are made for nonefficiency reasons, and there is an interaction between technologies (water heater vs. showerhead) that rationally slows diffusion of the most efficient water heater models. Further, the optimal replacement decision varies by locale and depends on a number of local factors. The rebates, tax incentives, and information from advocacy groups seem tangential to the issues driving purchase and replacement decisions in this market. The water heater market functions, but is unresponsive to energy-efficiency concerns.

As highlighted in the water heater replacement decision, there is no single market for energy services; instead, the "market" consists of hundreds of end uses, thousands of intermediaries, and millions of consumers. As a result, we do not believe the debate about market barriers or the debate about appropriate public policies to overcome market barriers can be settled by a single best policy. Instead, these issues must be addressed in a highly disaggregate fashion, considering the workings of individual markets.[15]

Reducing Emissions Now or Later? Who Will Take the Lead?

No one knows exactly how climate change will unfold and what its effects will be, and current emissions reduction is costly. Suppose breakthrough technologies that can cheaply and easily remove carbon from our atmosphere are just around the corner. Why then should we invest in emissions reduction today? The answer was provided by the *Stern Review*, a British report issued in early 2007 that translated rising temperatures and changes in climate patterns to economic losses.[16] The report examined many potential outcomes, but one mid-range scenario in particular provides a useful summary: unless action is taken today, the size of the economy in 2050 will be significantly smaller, a shrinkage that negatively affects the standard of living and worsens the poverty rate. (Predicted outcomes range from a 3 percent loss of GDP to a 20 percent reduction in capital consumption.) To avoid this catastrophe, the *Stern Review* argues that it is worthwhile to give up 1 percent of GDP today. When presented as a crisp tradeoff, the choice seems clear, and the report argues for immediate action.

However, climate change is a very uncertain process, and this uncertainty reframes the question: Do we want to sacrifice today's GDP to "insure" against future economic losses? Perhaps waiting to invest is a better choice, particularly when any reduction in GDP has a high immediate human cost. Insights into these trade-offs can be gained from considering a model on the value of waiting in the face of environmental uncertainty presented by Robert Pindyck.[17] He notes that there are two types of uncertainty: economic (the cost of various fuels and output prices) and environmental. Economic uncertainty is reversible—prices go up and down—while environmental uncertainty may not be so. Others, such as Gregg Easterbook, have argued that environmental damage (acid rain, the hole on the ozone layer, and so on) has been reversed through concerted action in the past, with faster results and a lower cost than was expected.[18] Thus waiting to solve climate change may be less costly than we surmise. Pindyck's model shows, however, that even a small amount of environmental irreversibility greatly diminishes the value of waiting. Delays in investment will accelerate climate change and

---

Low-Tech and High-Tech Solutions for Air Conditioning Load

Air conditioning is ubiquitous throughout the United States; in most regions, it is found in almost every home. Nationally, air conditioning accounts for 15 percent of total residential electricity demand.[a] In response to the heavy load air conditioning places on the power grid during peak hours and hot days, utilities are implementing smart meters that cycle air conditioners off for 15 minutes per hour[b] At the same time, a recent study shows that air conditioning loads can be reduced by 15 to 30 percent by the use of awnings that prevent sunlight from reaching into the home during the summer months.[c] Alternately homeowners can plant trees on the south and west sides of their homes, shade their air conditioning units, or install window shades.[d] Both low-tech and high-tech solutions are available to close the efficiency gap—but the connections provided by markets to stimulate change are missing.

a. EPA water heater draft 2 update.
b. "U.S. Household Electricity Report," Energy Information Administration, U.S. Department of Energy, July 14, 2005.
c. See, for example, the smart meter web page for Southern California Edison (http://www.sce.com/ PowerandEnvironment/ami/default.htm?=from=redirect).
d. "Awnings in Residential Buildings: The Impact on Energy Use and Peak Demand," University of Minnesota, Center for Sustainable Building Research, June 2007.

may exceed the threshold point that triggers irreversible change that can have disastrous effects on human civilization. The consequent costs will be borne by society at large, but in a very disproportionate manner across geographic regions, industries, and income groups. Recent scientific evidence suggests increasing concern by scientists that the point of irreversible damage is near. [19]

The Stern Review's numbers and Pindyck's argument call for immediate action. Meanwhile, newspaper headlines make the link between local weather, adverse outcomes, and climate change—suggesting that we have no time to wait.[20] But *what* immediate actions should we take? And *who* should lead the way? Any social insurance policy will have uneven and immediate costs on current citizens. How can a representative institution link these tangible costs to vague and elusive benefits? In a successful scenario, the costs are borne and disaster is averted, leaving many citizens feeling overtaxed. However, in most scenarios, the costs are borne and damage is incurred anyway—leaving citizens believing that their government failed to prepare for disaster. The issue of social insurance, as with the stabilization framework of Pacala and Socolow, raises questions about the actors. Leadership comes not only from tough macro tradeoffs, but also from micro-level policies that strengthen markets and tip private decisions to clean outcomes, making individual consumers the key actors.

Uncertainty, Private Investments, and Public Policy

A well-known result of real options theory is that investments with uncertain future benefits must only be committed when the present value of those benefits far exceed the investment cost. As uncertainty increases, so does this investment threshold. This may explain why private decision makers are reluctant to commit to clean-technology adoption. The very high levels uncertainty about future oil prices and future innovations in (and hence, costs of) new technologies create a very high investment hurdle.

What is lost in waiting to invest? In the case of green investments, the losses are faced by society at large in the form of climate change and energy insecurity. Energy markets only partly reflect these externalities. Due to incomplete markets for GHGs and security externalities, fossil fuel users do not directly face these costs. Creating markets that internalize the GHG externality will not only raise the level of benefits, but also reduce the uncertainty. This in turn, will spawn innovation in new technologies and bring down the investment costs. All three of these effects (higher level of benefits, lower volatility of prices, and lower investment costs) are reflected in these investment models.

The Central Role of the Private Investment Decisions

Our framework is shown in Figure 3.1. In order to trace through the links and flows, let's start with the physical world of climate change, shown at the bottom. In the pre-industrial world, net emissions were near zero in that carbon emissions were balanced by the capacity to absorb carbon. In our modern world, net emissions are significantly positive, as carbon emissions are far larger than carbon sinks. The large amount of net emission raises the stock of carbon in the atmosphere, which in turn raises temperature and sea levels.[21]

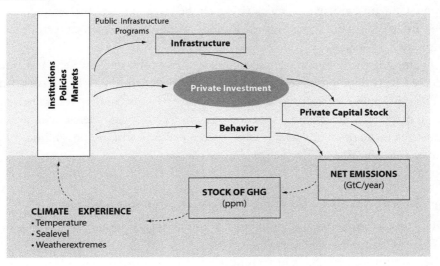

Figure 3.1. The Central Role of Private Investment Decisions

The two main drivers of human emissions are behavior and capital stock. Behavior (how much we drive, our thermostat settings, what we eat, and so on) is flexible and emissions can be reduced in the short run by simple modifications. Our capital stock (cars, buildings, and so on) is fixed in the short run, and leads to persistence in emissions. One of the clearest examples of this is an experiment run in a housing development in Lakeland, Florida.[22] Two homes were completed in 1998, with the same floor plans and orientation. Through specific energy efficient choices during the building process, one home used only 10% of the power of the other. With an expected service life of 40–60 years, there will be a persistent difference in energy consumption that even good habits cannot erase.

At the center of our framework are private investment decisions. The challenge for climate change policy is to make carbon emission concerns a significant factor in the private decision, because (as shown in the water heater example) there are many other factors that drive capital stock decisions. The status of public infrastructure also tips private investment decisions. For example, water in the Los Angeles basin comes through a large infrastructure, in which water is transported from north to south through the California Aqueduct and from the Colorado River on the eastern border of the state to the coastal city. These water supplies—vulnerable to earthquakes and energy intense—have allowed the rapid population growth of Los Angeles. Without the infrastructure, Los Angeles would not be nearly as large; the presence of the public infrastructure tipped the private housing decisions.

The left column closes the feedback loops. Markets, regulatory policies, and public institutions set the stage for the economic considerations of private and public capital stock decisions. Climate outcomes, such as rising temperatures or extreme weather, influence policies. Effective organizations and institutions at the top level can tip private decisions and ultimately emissions.

In Figure 3.1, the lines connecting the markets and private investment and consumption decisions are darker than those connecting institutions and public infrastructure. For several reasons, public institutions and infrastructure play a secondary role to private decisions and markets in the dynamic selection of clean technologies. Here are some examples that cause us to question the match between the decision-making possibilities of public entities and the uncertain trajectory of climate change.

The problem is clearly seen, but government action is not forthcoming:

- U.S. sewers need $500 billion in repairs, but local governments are reluctant to raise taxes for a problem voters can't see.[23]
- The southeastern U.S. is just months away from running out of water and local officials are taking actions to drastically cut water use. A problem that was entirely foreseeable: A recent headline reads, "Recipe for a Water Crisis: Plan. Fail. Repeat." Georgia's water shortage took decades to develop, with action thwarted by official's short attention spans, feuds between states, and the false assurance of rainy years."[24]
- Similarly, the lack of water in the western United States led to this headline: "The Perfect Drought—Will population growth and climate change leave the West without water? The West is the fastest-growing part of the country. It's also the driest. The Future is Drying Up."[25]
- Scientists predict the collapse of 75% of the world's fisheries by 2050, which will have devastating consequences for the 1 billion people who rely on fish for their major source of protein.[26] Yet, coordinated global action this international problem—one much more simple than climate change—is entirely lacking.

The problem is not seen by constituents:

- Nobel Prize winning economist Daniel Kahneman has observed: "The potential consequences of climate change are completely abstract, as far as the population is concerned. Nobody is feeling a thing, yet. To mobilize public opinion to get anything done is extremely difficult. One has to question whether the social arrangements by which we live are adequate to cope with the threat."[27]

Problems with maintaining the public policy/private markets balance after extreme events:

- Private insurers exited the California earthquake insurance market after the 1994 Northridge earthquake. The state established its own program, but a high deductible ($50,000) and a large premium ($1,700 to $2,800 per year) have kept 89% of Californians from buying in. Meanwhile, Californians keep making the private decision to build and buy homes in an area with high earthquake risk.[28]
- In Louisiana, high insurance premiums and greater restrictions have made insurance unattractive to homeowners after Katrina. Insurers are not willing to cover parties new to the state, and this has hurt recovery efforts.[29]
- Florida's governor pleased voters by mandating lower insurance premiums. Insurance companies responded by dropping customers. The one company that is picking up customers is under-capitalized, leaving insured and uninsured homeowners in the state vulnerable to financial ruin at the next large hurricane.[30]
- The policies are "captured" by private interests:
- The ethanol program in the U.S. is a conspicuous example of a well-intentioned program gone massively wrong; it is a lose-lose solution (higher costs for motorists and taxpayers and increased GHG).

Of course, the list could go on, but our basic point is clear: climate change is a tough issue for public institutions. Its uncertain trajectory, diffuse costs, and complex sources make climate change ill-suited for public decision making in a timely or adaptive manner. Decaying public infrastructure is vulnerable to extreme weather events, and the path to recovery is not clear. Meanwhile, citizens continue to make private decisions, and markets help them rapidly adapt to changing conditions. For these reasons, we emphasize the market and private decision based feedback loop in Figure 3.1.

New and Reconnected Markets to Support Private Decisions

One of the most compelling features of the Pacala and Socolow's stabilization triangle analysis is the menu of technologies that are ready to scale to achieve substantial emissions reduction. Also, there is an immediate and compelling set of market opportunities to reduce emissions, including companies and new markets already at work. Our survey of opportunities covers five areas: service-based business models; the creation of new markets through legislative and regulatory fiat; the use of standards and transparent metrics to accelerate market growth; modernization of the utility business model; and restricted choices to integrate markets.

Table 3.3. New Markets for Emissions Reductions

Market	Rationale	Scale of Adoption
Renewable Energy Credits (RECs), Regional Greenhouse Gas Initiative (RGGI)	State mandates renewable power generation and allows utilities to buy and sell RECs to meet goals	29 states, 10–20% of local energy markets
White tags	State mandates energy efficiency and allows utilities to buy and sell tags to meet goals	Connecticut mandate; under consideration in other states. Needs verification protocol.
Carbon credits (U.S.)	Voluntary trading: More than 300 companies have joined the Chicago Carbon Exchange (CCX).	34 million tons of CO_2 eliminated since 2003; $2–$5 per ton.
Carbon credits (Europe), CDM and ERU	Mandatory trading necessitated by complying with climate change agreements	100 million tons of CO_2 traded per month; $24–$26 per ton.
Green pricing.	Utility customers voluntarily agree to increased rates to secure power from renewable energy	Retail option in 44 states. Incremental cost of 1.5–2 cents per kWh. Average adoption rate of 1.5% to date.

BEACONPLACE.COM

...ing Carbon Markets

...emissions have been created at the state and regional

...luction can be used in several ways. First, they smooth ...in renewable technologies, as the renewable energy ...ovide flexibility over geography and time. Second, the ...loption when used in project financing. One example ...voluntary carbon credits from the Chicago Carbon ...e company has been able to monetize 5 to 7 percent ...lizes. Third, the markets are used to spur innovation. ...kets are changing energy from a single-dimensional commodity to a multi-dimensional product. Imagine a host of subscripts on a unit of energy:

^{E}BTU, RE, EE, PCO2, PK, ST

BTU = price of energy content

RE = price premium for renewable energy (REC)

46 | Global Sustainability and Innovation

EE = price premium for energy efficiency (white tag)
PCO$_2$ = price premium for carbon reduction (carbon credit on CCX)
PK = price premium for peak demand reduction
ST = state where power is generated (so as to comply with local mandates)

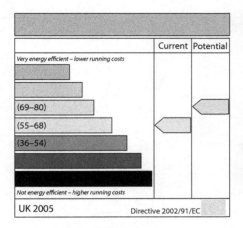 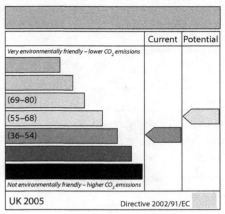

Figure 3.2. EU Home Performance Ratings Make the Carbon Footprint Visible

Each sub-script represents an attribute of energy that is traded in a market. The challenge is to find business models that connect investment and consumption decisions to this rich market infrastructure.

Standards

Information gaps often lie at the heart barriers to adopting energy efficiency, but standards can be used to insert information into the flow of commerce. For example, in early January 2008, the Massachusetts State Senate passed a bill that would require home sellers to obtain an energy audit, so as to inform potential buyers of the home's energy costs.[31] The European Union has had a similar measure in place since January 2006.[32] Figure 3.2 shows a certificate required for each home in the EU at the time of sale or rent. In the box on the left, the home is given an energy-efficiency score, and a score for its potential. Each is accompanied by a letter grade. In the sample certificate, the home is rated D but could improve to a C. For this home, the A grade remains elusive, most likely due to retrofit challenges. The box on the right shows the home's CO$_2$ emissions rating.

The EU standard does not require homes to meet a minimum level of performance, but merely having access to this information can change the decisions of buyers and renters. Homes that are more energy efficient should command higher prices in the long run, as the energy savings are capitalized into the value of the house. With more rapid turnover, the rental markets will more quickly differentiate, and rental prices will fall for homes and apartments with expected high energy costs. Table 3.4 shows the multiple efforts to create standards in the U.S. market.

Table 3.4. Selected U.S. Standards Efforts

Standard	Sponsor	Adopters
LEED	U.S. Green Building Council	Commercial buildings nationwide
GreenPoints	BuildItGreen	New construction in California
Home Energy Rating System (HERS)	Resnet	Nationwide
Home Performance with Energy Star™	U.S. Department of Energy	Local organizations, 19 states
International Performance Measurement and Verification Protocol (IPMVP)	Coordinated with U.S. government agencies	Worldwide protocol for commercial sector

There is a host of supporting technology to implement these standards, including test equipment, data loggers, and software models. This in turn has led to the creation of several start-up companies that specialize in these technologies, as well as market entry by established companies. Clean tech does not always mean power generation, and as markets and standards develop, it will require the presence of supporting infrastructure technologies.

BEACONPLACE.COM

ity Business Model

carbon emissions. Forty percent of our na- ectric and gas utilities, under their pricing most of the United States the only way a re power. Their flat or even declining tiered wever, Oregon and California have passed n profits, making it possible for utilities in vation of their customers.[34] Ten states have ings.[35] Decoupling allows utilities to avoid urchase energy conservation and energy ef-

gy conservation is through upward tiered frequently suggested as way for customers to ds. Strong opposition to time-of-use pricing

48 | Global Sustainability and Innovation

can be expected, however, from those industries that have heavily subsidized electricity rates under our current flat pricing scheme.[36] Tiered pricing may be a more palatable political alternative because it provides low-cost energy for a base level of usage. As Table 3.5 shows, upward tiered pricing encourages energy conservation. In addition, research shows that upward tiered pricing can be made revenue-neutral for the average customer, even as it encourages high-consuming customers to be the early adopters of on-site renewable technology.[37]

Table 3.5. Tiered Pricing Changes Consumption

Consumption Tier kWh/day	Pricing to Encourage Consumption Otter Trail Power Co. (Northern Minnesota) cents/kWh	Consumption Tier kWh/day	Pricing to Discourage Consumption PG&E (Northern California) cents/kWh
Less than 16	8.2	Less than 13	11.4
16–66	5.4	13–16	13.0
More than 66	5.1	16–26	22.7
		26–38	31.7
		More than 38	36.4

Restricted Choice Sets

Landlords buy appliances, but renters actually pay the electricity bills. The housing developer saves $2,000 in construction costs, but it is the home-buyer who will ultimately pay an extra $600 per year in heating bills due to inadequate insulation. The uninformed consumer goes to the appliance store and saves $300 by buying the low-cost unit, only to pay $75 per year in extra energy costs. An effective way to close these principal-agent and information gaps is through appliance and building standards. Appliance standards are estimated to save 1 to 3 percent of the nation's energy use, but have an outsized impact, as sometimes a demand decrease of this size can remove congestion in the natural gas markets and drop prices by 10 to 20 percent.[38]

During the energy crises of the 1970s, California adopted the nation's tightest building codes. Since then, California's per-capita energy consumption has remained flat, while the U.S. average per-capita consumption has increased by 50 percent.[39] About half of California's savings are estimated to come from the state's building standards and efficiency standards for appliances. The other half of the savings came from regulator incentives to encourage customer energy conservation. California's experience shows that local building codes—while not nearly as glamorous as solar and wind power—have played an important role in decreasing energy use.

Markets differ from technologies in how they aggregate. While the emissions-reduction capacity of technology is simply additive, market changes are synergistic. The many new and reconnected markets listed above work together, and the presence of one market innovation can cause others to grow faster and on a larger scale. Other market areas, not mentioned here, could do the same.

Broken Markets and Policy Stumbles

It is useful to briefly consider two policy stumbles, attempts to close market gaps that failed: solar hot water subsidies in California and energy-efficient mortgages. In response to the energy crises of the 1970s, California heavily subsidized solar hot water[a] This technology is much more energy-efficient than photovoltaics, and a good match for this sunny state. However; the program's implementation stumbled at contractor certification and training and at inspection of installations. This led to a great deal of fraud (only half of installations still worked five years later) and tainted the solar hot water market. Only in 2007, thirty years later did the state legislature reconsider solar hot water subsidies.

Energy-efficient mortgages also failed to catch on; a pilot initiative closed with little notice in June 2007. The goal of the program was to expand the borrowing capacity of homebuyers who wanted to make home improvements to increase energy efficiency The federally guaranteed program added up to $15,000 to the amount borrowed for a home purchase. However it simply proved to be too difficult to get the paperwork done during home-buying transactions. In a situation with multiple bids on a home, program users were at a disadvantage because their paperwork was processed more slowly than that of other bidders. Coordination between the federal agency the banks, and mortgage brokers was difficult, so even eco-friendly mortgage brokers did not advertise the program.[b]

These two policy failures highlight how the natural stream of commerce dominates the transaction flow. To be effective, policies must slot in to an existing stream of commerce and work at its speed. Expecting to change the stream of commerce for a new attribute—energy savings or energy efficiency—won't work.

a. Source: Private communication, Tomek Randio, CEO of MortgageGreen, October 2007.
b. Margaret Taylor et al., "The Role of Technological Innovation in Meeting California's Greenhouse Gas Emission Targets," Managing Greenhouse Gas Emissions in California, California Climate Change Center at UC Berkeley, March 2006.

ESCOs

In other new or fragmented markets, the market players and issues have been connected through a service provider. Systems integrators played this role in past eras of the computer industry, providing expert knowledge to companies struggling with their hardware and software purchases. As the water heater example demonstrated, there is a similar business opportunity in the energy markets.

In energy markets serving industrial and large commercial customers, this service model already exists in the form of ESCOs (energy service companies). An ESCO will perform an on-site audit, recommend energy and water-reduction strategies, and then, if the client approves, implement the changes. Water heating is a large end-use for this market segment, and there is a strong interaction between energy and water savings. ESCOs are paid as consultants and/or for savings performance. (Table 3.6 shows the current innovative market variations on the ESCO business model).

Table 3.6. Innovative ESCO Business Models

Service Model	Sample Companies	Target Customers
ESCO acts as a system integrator in designing and implementing energy efficiency projects. In some cases, financing is provided and linked to cost savings.	Amaresco, Noresco	Large business customers, Schools, municipalities, commercial and large residential groups.
Implement and finance HVAC system improvements through cost savings.	H_2O Technologies, Trane, Siemens, Honeywell, Johnson Controls	Similar to above but primary focus is on HVAC systems.
ESCO owns solar installation on customer's roof, sells solar power back to customer.	Sun Run, SunEdison	Home owners, small and medium-size businesses
Nonprofit entity (funded by utilities) acts as full-service ESCO, including financing.	Oregon Energy Trust, Efficiency Vermont	Homes and businesses
Peak demand aggregators	Energy Curtailment Services, EnerNOC, Consumer Powerline, Converge, Energy Connect	Businesses in targeted geographic power markets
Nonprofit ESCO as role model of better practices.	Cambridge Energy Alliance (see box for details), Clinton Foundation	Cities

The market for ESCO endeavors is growing rapidly. For example, peak-demand aggregators currently manage 2,600 MW of peak capacity, the energy equivalent of four and a half average-sized coal plants. There are also some recent and important innovations in the ESCO business model. The Clinton Foundation, for example, pools together projects in a single city for a smoother construction schedule and increased purchasing power. The contract passes this savings on to the building owners, significantly reducing the cost of energy efficiency. Another innovative variant on the ESCO service model is the reworking of a fixed cost to a variable cost. Sun Run and SunEdison, for example, don't sell solar systems; they sell solar power. Instead of incurring a large up-front cost, the customer buys power as it is generated from these companies; the companies themselves own and maintain the rooftop systems. A similar fixed-to-variable cost transformation in the insurance markets (pay-as-you-drive insurance) has decreased miles and emissions by 9 percent.[40] Pay-as-you- drive insurance is available in Europe, and a pilot project is underway in

---

Cambridge Energy Alliance

The City of Cambridge, Massachusetts, has embarked on a path to reduce electricity demand by 50 megawatts and to reduce fossil-fuel consumption by 5 percent in five years. The Cambridge Energy Alliance (CEA), a non-profit organization was created to design and implement program that would use a new financing model to fund the energy-efficiency retrofits needed to achieve the energy reduction goal.

CEA's financing model relies heavily on the New England Independent System Operator (ISO-NE).[a] Following the path forged by PJM (the system operator in New Jersey, Pennsylvania, and Maryland), last year ISO-NE began allowing energy efficiency measures to compete with supply-side resources in the forward capacity market (FCM). Since demand reduction can be achieved at a lower cost than new generating capacity, electricity costs to consumers in the ISO-NE region could decline as the cost of the procurement per megawatt decreases—these reductions are called "negawatts". The CEA will aggregate the demand reductions from residential customers, businesses, and other organizations throughout the city and sell the negawatts in the FCM.

CEA plans to begin with the "low-hanging fruit" like switching from incandescent to CFL lighting, and work with ESCOs to implement and finance energy efficiency projects with seven- to twenty-year payback periods. The range of energy savings initiatives will range from high efficiency central and window air-conditioning units, to solar hot water heaters, and retrofitting all buildings in the city with real-time energy consumption information to allow consumers to make their energy use decisions based on real-time information. (http://ceic.cambridgeenergyalliance.org/).

a. As with other regional system operators, ISO-NE monitors the purchase and sale of electricity on the wholesale electricity market and ensures sufficient electricity is generated to meet the system load.

Washington State. Further proposals include pay-as-you-pump insurance, so that drivers must frequently confront the full cost of driving. For both solar panels and auto insurance, the innovative service model uses a variable charge to elicit the desired behavior.

Finding Local Solutions to a Global Problem

Living through climate change in a particular locale shapes private decisions. The local environment has implications beyond simply the wea in my back yard") sentiment prevents local situatin; leading to lawsuits.[41] Without these power sources, with the water heater/showerhead example, local infra private decisions.

As citizens and residents, we are most concerned w we typically have only a passing interest in the big snov also make large inferences about events taking place Katrina and the California wildfires to climate chang certainty, it is easy to confuse local weather variations of the local experience is a large amount of uncertai change, and uncertainty tends to delay investment tailored to conditions on the ground can reconnect m

Table 3.7. Varied Local Consequences of Climate Change

West	Great Plains	Mid-Atlantic
Smaller snowpack in mountains; less stored water for agriculture and urban areas.	Higher temperatures increase water needed for agriculture by 50%; water conflicts decrease farm income.	More frequent hurricanes in major metro areas; expected damages $50–$66M per event.
Lack of water causes farmland values to fall by 30% or more.	Drought-stressed eco-systems susceptible to invasive species; crops destroyed.	Sea level rise of 20 inches; damages of $8B–$58B.
Rising seawater; $1.5B cost to protect San Francisco Bay Area.	Decreased soybean productivity (70% decline); decreased wheat productivity (10–50% decline).	Rising ocean temperatures and poor water quality; $63B regional ocean-based economy at risk.

While it might seem that the enormous, fully integrated climate change models are the sum of many local outcomes, they are in fact high-level models that rely on smoothly growing trajectories. These trajectories differ from each other by relatively small amounts. Population growth might be 1.8 percent per year in one scenario and 2.1

percent in another. However, with 50- and 100-year horizons, these small differences are compounded, becoming a large change. By contrast, a local recession—such as the 20 percent drop in employment experienced in Silicon Valley during the dot-com bust—has no effect in the broader models. However, in real life, it did impact responses. A local recession can halt new construction for a few years, and when building resumes, a different technology set is available for capital stock decisions. Because of the long service lives of buildings and machinery, this blip in investment patterns has local energy-use consequences for years to come. So, unlike the smooth climate change models, the local experience is path-dependent; the particular history of a locale matters for future investment decisions.

Local Outcomes for Climate Change, Local Concerns

The University of Maryland recently completed a study on the variation of climate change by U.S. region. Table 3.7 shows a few illustrative findings for three diverse regions.[42] It highlights the large regional variations and the many sources of local stresses from a climate shift. The examples also illustrate the tight connection between water and climate, with water issues hitting local economies hard. This is not surprising, as nearly 40 percent of water use is for power generation and 80 percent of the cost of water goes toward energy.[43]

Similar studies have been completed by a number of state and regional agencies, and a common finding is that climate change presents a daunting challenge for local resources and infrastructure.

Local Resources Color Local Experience

Two facts collide at the local and state levels. Scientists are calling for climate change policies that reduce emissions immediately. However, state budgets are quickly falling into deficits due to the economic crisis. Unlike the federal government, states can't run large persistent deficits, so there is a conspicuous lack of resources for new climate change policies just as the call to action has become acute.

The lack of local resources will change outcomes. One small but clear example can be seen in the resort town of Sisters, Oregon. Surrounded by the Cascade Mountains, Sisters sits directly below a large mountain lake that is nestled in a rock bed. Currently the rocks are held together with frozen ice. However, as climate change progresses, experts worry that the ice will warm and the lake waters will race down the mountain and flood the town.[44] The population of Sisters (2,400 people) is too small to galvanize authorities into building a dam to slow the floodwaters. Perhaps the town will put a sensor on the mountain lake to gain a twenty-minute warning of the impending flood. Local residents and institutions can see but not prevent this potential disaster. If Sisters or another town in the area experienced a significant flood, property values for vacation homes would fall sharply, further eroding the local tax base. The mountainside would be scraped clean of trees and foliage, setting the town up for a stream of mudslides in years to come. Extreme weather and local resources are significantly mismatched.

Public Infrastructure Failures, Private Investment Solutions

As with the extreme drought in the southeastern U.S., a similar prolonged drought in Australia illustrates the role of private investment decisions in adapting to new conditions.[45] During the Australian drought, a water desalinization plant was constructed in Perth, but another one in Sydney was rejected by voters. The government has responded to the water shortage by instituting rationing rather than significant increases in water prices. As a result of stalled infrastructure development and water rationing, Australian companies have developed a number of interesting technologies to make the best use of scarce water. These include small-scale wastewater recycling systems for agricultural use (purchased by farmers); in-home gray-water treatment systems (purchased by home owners); and office water coolers that create water through condensation (purchased by companies). These technologies illustrate the strong climate/ markets/private-decision feedback loop presented in Figure 3.1.

In sum, local climate change is the version we actually experience, and the constraints and challenges encountered at the local level are more acute than aggregate climate change models can capture. The mismatch between local resources and the magnitude of the challenges presented by extreme weather is worrisome. The technology innovations from Australia illustrate the shift toward market and private investment decisions in these cases, but they do not completely address the large public infrastructure gap that is becoming more apparent as the climate shifts.

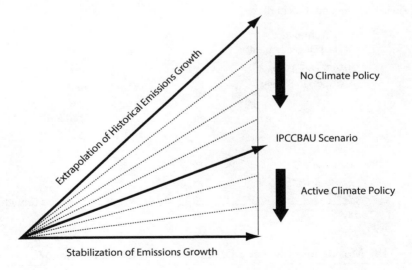

Figure 3.3. Expected Decarbonization Reduces the Role of Active Policy

A New Geometry for Climate Change Policy

Bringing together the insights about markets and private investment decisions creates a new stabilization framework. A convenient starting point is Jeffrey Greenblatt's reconciliation of the stabilization-triangle framework with the baseline scenarios of the IPCC.[46] The IPCC baselines are referred to as "business as usual" and represent a climate future if no policy action is taken. Figure 3.3 provides a summary.

The IPCC's BAU scenario does not depict an economy that is simply running on autopilot. It represents expectations that: across the economy, units of commodity input per unit of output continue to decline at historical rates; and there is significant adoption of low-emission technologies, at "rates that exceed historical precedent." In contrast to the top line (labeled the current path), the BAU scenario actually *includes* the effects of global modernization and substitution of renewables. Yet the very dismal prognosis from the IPCC is that the BAU outcome will bring significant increases in temperature, ice melt, and rising seawater.[47] So the BAU scenario represents a flurry of activity—but not enough of it.

Figure 3.3 is divided by the BAU scenario. Four of the seven wedges prescribed by Pacala and Socolow are above the IPCC's BAU scenario. Three additional wedges are needed for emissions stabilization, and these will require policy action. However, both above and below the BAU line, our economy must undergo significant change to deliver emissions reductions.

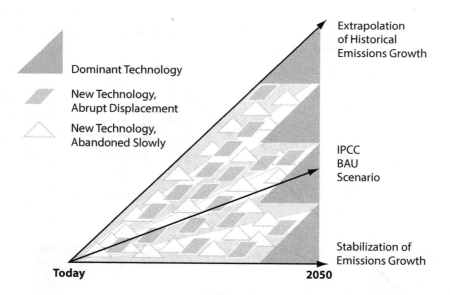

Figure 3.4. The Mosaic of a Market-s Based Policy for Climate Change

Figure 3.4 is a depiction of how markets and private investment decisions will fill the stabilization triangle. Instead of the smoothly growing wedges of Pacala and Socolow,

Figure 3.4 shows an almost chaotic mosaic of shapes and colors: the result of rapid technology change in a dynamic market. Two geometric shapes in the mosaic represent the demise of new technologies. The "wedgelet," or small triangle, represents a small growth in adoption followed by sudden displacement. (An example of such rapid demise is the fax machine, once considered a breakthrough in communications; its sales plummeted once the Internet took hold.) The diamond represents a small growth in adoption followed by a slow demise. (Think of phones with corded handsets; these are slowly replaced as wireless handsets continue to take over the home market.) Over time the stabilization triangle becomes a collage of new and then abandoned technologies.

On the right side of the figure are large right triangles. These are the fruits of intensive R&D programs for clean technologies. The stabilization triangle is only the first stage in emissions reduction. Under this two-stage plan, aggressive emissions reductions are avoided until 2055, when—it is fervently hoped—new technologies will reduce emissions more cheaply. The right triangles represent technologies poised for rapid and large deployment at mid-century.

The market dynamics create a collage of shapes above and below the IPCC's BAU scenario. To those experiencing this economic transformation, it will appear that some emissions-reduction technologies are cheap (these are in the area above the BAU line) and some are more expensive, requiring policy action to tip private decisions (these are shown in the area below the BAU line). Regardless, the marketplace becomes a rapid arbiter of winning and losing technologies.

Table 3.8. A Broad Spectrum of Innovations Will

Market Taxonomy	Company Examples	
Services	**Recycle Bank**	Pay cu
Transportation	**Tesla Motors**	Electric
Efficiency Infrastructure	**EnerNOC**	Aggregate
Power Generation	**Nanosolar**	Laser-print
Energy Storage	**United Technologies**	
Building Materials	**Serious Materials**	Low
Recycling and Waste	**Green Citizen**	Corpora
Water	**Air2Water**	Ocean water

While the Pacala and Socolow wedge approach focuses on energy sources and energy efficiency, the opportunities to reduce emissions span a much wider set of decisions and technologies. Table 3.8 is an abstraction of the full range of technologies that will be part of the market dynamics in the next half-century.

The Invisible Green Hand | 57

The left column shows the many market segments of clean, emissions-reducing technology. The two row marked with stars are the types of clean technology in the wedge framework. The six new rows represent significant opportunities for emissions reduction. Examples of companies currently offering these services or products are to the right. While some of the carbon-reducing features of the innovations are obvious, others are less so. This is a salient feature of the market-based approach: with broad-based standards, innovations will emerge from all quarters to supply reductions. For example, Recycle Bank pays residential customers to recycle and splits the proceeds of selling the waste stream with local municipalities. Serious Materials is a company that reduces the energy input and thus the carbon emissions of drywall (used in construction) by 90 percent.[48]

While we currently have a carbon-based infrastructure, innovations are underway at many levels to make the shift to a non-carbon future. The challenge is to drive capital to investments build around a clean infrastructure—one where individual consumers make behavior and consumption choices that emit less greenhouse gases. Our framework shows how policies that are triggered by local concerns and that leverage opportunities can help direct private investments to green technologies. Emerging markets for GHGs and new business models provide information and mechanisms for such capital flows. These micro, private actions will add up to a new landscape for climate change mitigation policy.

Notes

1. For a vivid description of the impact of our daily actions, see Alex Shourmatoff, "An Eco-System of One's Own," *Vanity Fair* (May 2007).
2. Carbon prices can be internalized, either through the imposition of a carbon tax or by setting emissions limits and use markets to allocate across users (cap and trade). The uncertainties in the mechanisms through which GHG effluents affect climate, as well as the uneven distribution of impacts, makes designing and implementing a global climate mitigation scheme extremely difficult. For a critical analysis of climate change policy architectures, see Joseph Aldy, Scott Barret, and Robert Stavins, "Thirteen Plus One: A Comparison of Global Climate Policy Architectures," KSG Working Paper Series No. RWP03–012, 2003.
3. Stephen Pacala and Robert Socolow, "Stabilization Wedges: Solving the Climate Problem for the Next 50 Years with Current Technologies," *Science,* 305/5686 (December 2004): 968–972. This approach is also consistent with that of the IPCC.
4. Pew Center report.
5. An early study is Alan Manne and Richard Richels, *Buying Greenhouse Insurance* (Cambridge, MA: MIT Press, 1992). See also Vattenfall, Climate Map 2030, <www.vattenfall.com/cli-matemap/>, January 2007; McKinsey & Co., "Reducing U.S. Greenhouse Gas Emissions: How Much at What Cost?" <http://mckinsey.com/clientservice/ccsi/>, November 2007.
6. Alan Manne and Richard Richels, *Buying Greenhouse Insurance* (Cambridge, MA: MIT Press, 1992).

7. See Alliance to Save Energy, <www.ase.org/> for references. The European Council for an Energy Efficient Economy (<http://eceee.org/>) addresses similar issues in a broader international context.
8. See <www.ase.org/> and <www.acore.org> for details.
9. Marilyn Brown, "Market Failures and Barriers as a Basis for Clean Energy Policies," *Energy Policy,* 29/14 (November 2001): 1197–1207.
10. Adam Jaffe and Robert Stavins, "The Energy Efficiency Gap: What Does It Mean?" *Energy Policy,* 22/10 (October 1994): 804–810. In an extension of this work, Jaffe, Newell, and Stavins provide evidence on the effectiveness of energy-efficiency investments. A. Jaffe, R. Newell, and R. Stavins, "Energy-Efficient Technologies and Climate Change Policies: Issues and Evidence," *Resources for the Future,* 1999.
11. "Little Green Lies," *Business Week,* October 20, 2007, pp. 45–52.
12. See Carl C. Koopmans and Dirk Willem te Velde, "Bridging the Energy Efficiency Gap: Using Bottom-Up Information in a Top-Down Energy Demand Model," *Energy Economics,* 23/1 (January 2001): 57–75. Koopmans and te Velde recognize the role of market failures and information gaps in the estimation of the energy-efficiency gap.
13. <www.energystar.gov/index.cfm?c=new_specs.water_heaters>.
14. Donald R. Wulfinghoff, *Energy Efficiency Manual* (Wheaton, MD: Energy Institute Press, March 2000).
15. One of the early papers to recognize this issue was William H. Golove and Joseph H. Eto, "Market Barriers to Energy Efficiency: A Critical Reappraisal of the Rationale for Public Policies to Promote Energy Efficiency," Lawrence Berkeley National Laboratory working paper LBL-38059, 1996.
16. The Stern Review on the Economics of Climate Change, Office of the Treasury, British government, <www.hmtreasury.gov.uk/independent_reviews/stern_review_economics_climate_change/sternreview_index.cfm>.
17. Robert S. Pindyck, "Uncertainty in Environmental Economics," NBER Working Paper No. W12752, December 2006, available at SSRN: <http://ssrn.com/abstract=949761>.
18. Gregg Easterbrook, "Some Convenient Truths," *Atlantic Monthly* (September 2006).
19. "U.N. Chief Seeks More Leadership on Climate Change," *New York Times,* November 18, 2007; "In Greenland, Ice and Instability," *New York Times,* January 8, 2008. See also "Summary for Policymakers of the Synthesis Report of the IPCC Fourth Assessment Report," IPCC, <www.ipcc.ch/pdf/assessment-report/ar4/syr/ar4_syr_spm.pdf>.
20. See James Hansen, "Tipping Point: Perspective of a Climatologist," Chapter 1, *State of the Wild,* Wildlife Conservation Society (2008). The scientific evidence is widely reported in the popular press. A random sampling from the *New York Times:* "Inch by Inch, Great Lakes Shrink, and Cargo Carriers Face Losses," *New York Times,* October 22, 2007; "Precipitation Across U.S. Intensifies Over 50 Years," *New York Times,* December 5, 2007; "World Food Supply Is Shrinking, U.N. Agency Warns," *New York Times,* December 18, 2007; "U.N. Chief Seeks More Leadership on Climate Change," *New York Times,* November 18, 2007; "In Greenland, Ice and Instability," *New York Times,* January 8, 2008. See also the U.N. report cited earlier.

21. Paul Falkowski et al., "The Global Carbon Cycle: A Test of Our Knowledge of Earth as a System," *Science,* 290/5490 (October 2000): 291–206; Nicholas Gruber, Charles D. Keeling, and Nicholas R. Bates, "Interannual Variability in the North Atlantic Ocean Carbon Sink," *Science,* 298/5602 (December 2002): 2374–2378.
22. See "Building America Research is Leading the Way to Zero Energy Homes," <www.eere.energy.gov/buildings/building_america/pdfs/37547_zeh.pdf>.
23. "Deteriorating Sewer Systems: A Dirty and Costly Secret," *The Wall Street Journal,* November 20, 3007.
24. "Recipe for a Water Crisis: Plan. Fail. Repeat," *Atlanta Journal-Constitution,* December 16, 2007.
25. *New York Times Magazine,* October 21, 2007, pp. 68–72, 104, and 155–157.
26. For a good summary of the issues, see "Collapse: End of the Global Fish Stock by 2050?" Globalization 101, The Carnegie Endowment, December 2006.
27. "From Global Warming to Weight Watching," *The Milken Institute Quarterly,* 8/3 (3rd Quarter 2006): 29.
28. "After the Quake, Who Will be the Chump?" *San Jose Mercury News,* September 10, 2007.
29. "Risky Business: Insurers See the Big Easy as the Big Loser," *Fortune,* August 20, 2007, pp 77–85.
30. "Wishing the Wind Not To Blow," *The Economist,* August 11, 2007, p. 28.
31. "Bill Seeks Home-Energy Use at Sale," *Boston Globe,* January 11, 2008. Many other communities are discussing similar measures.
32. <www.ecorating.co.uk/eco-house/energy-performance-certificate/default.aspx>.
33. "Energy Basics 101," EIA, <www.eia.doe.gov/basics/energybasics101.html>.
34. Martin Kushler et al., "Aligning Utility Interests with Energy Efficiency Objectives: A Review of Recent Efforts at Decoupling and Performance Incentives," American Council for an Energy-Efficient Economy, October 2006, report U061.
35. American Gas Association.
36. Severin Borenstein, "Wealth Transfers from Implementing Real-Time Retail Electricity Pricing," NBER Working Paper No. W11594, September 2005, available at Social Science Research Network: <http://ssrn.com/abstract=800449>.
37. Thomas Huff and Christy Herig, "Electricity Rate Structures Can Be Used to Promote Solar PV: Lessons from California," June 19, 2002, <clean-power.com>. For example, in northern California the early adopters of photovoltaics are frequently at the top price levels shown in Figure 12. For these homes, photovoltaics (at an average 22 cents per kWh) are immediately cost-effective.
38. "Leading the Way: Continued Opportunities for New State Appliance and Equipment Efficiency Standards," American Council for an Energy-Efficient Economy, March 2006. See also Jaffe, Newell, and Stavins, op. cit.
39. California Energy Commission presentation on March 4, 2007, to the California Assembly's Utility and Commerce Committee, <www.energy.ca.gov/2007publications/CEC-999-2007-010/CEC-999-2007-010.ppt#10>.

40. Ian Parry, "Pay as You Drive Insurance: A Better Way to Reduce Gasoline than Gas Taxes?" *Resources for the Future,* April 2005.
41. "Lawsuits Aim to Slow New Coal Boom," Associated Press, January 14, 2008; "Strangers as Allies: Fight Against Coal Plants Is Creating Diverse Partnerships," *New York Times,* October 20, 2007.
42. "The U.S. Economic Impacts of Climate Change and the Costs of Inaction," Center for Integrative Environmental Research, University of Maryland, October 2007.
43. "Energy-Water Nexus Overview," Sandia National Laboratories, 2006.
44. "Lake Threat Looms Above Sisters," *The Nugget,* August 1, 2007.
45. Kathryn Marks, "Australia's Epic Drought: The Situation Is Grim," *The Independent,* April 20, 2007, <www.independent.co.uk/news/world/australasia/australias-epic-drought-the-situa-tion-is-grim-445450.html>.
46. Jeffery Greenblatt et al., "Wedge Analysis of IPCC SRES and Post-SRES Scenarios," Fourth Annual Conference on Carbon Capture and Sequestration, U.S. Department of Energy, May 2005.

Discussion Questions

1. Discuss the role of different entities (private and public) in reducing emissions.

2. Discuss the role of markets in reducing emissions.

3. What is the Cambridge Energy Alliance?

4. What are the different consequences of climate change in the U.S. based on geographical region?

5. Discuss the role of the private sector in solving problems that the public sector has failed to solve.

In-Class Assignment: Objective Questions

1. The Home Energy Rating System (HERS) is sponsored by the U.S. Green Building Council.
    a. True
    b. False
2. Greenpoints is sponsored by
    a. Resnet
    b. U.S. Department of Energy
    c. Builditgreen
3. RECs and RGGI have been adopted by 29 states.
    a. True
    b. False
4. Tesla Motors is an example of a company in the buildings materials industry.
    a. True
    b. False

Note on Energy

Sean Harrington and Dennis M. Rohan

Introduction

On December 18, 2007, U.S. President George W. Bush signed the Energy Independence and Security Act. This new law dramatically increased the Corporate Average Fuel Economy (CAFE) standards, set stricter efficiency regulations for lighting equipment and electrical appliances, and laid the groundwork for increased investment in biofuel development. The bill was emblematic of an increased focus on energy policy around the world. Risks associated with climate change and a desire for energy independence were the driving forces behind this trend.

Such policy changes and increased consumer appetite for environmentally sustainable products and services led to an enormous growth in capital flow to "greening" the energy sector. A study by McKinsey & Company projected U.S. investments in clean technology[1] would grow from $55 billion in 2006 to $227 billion in 2016. Clearly, the conditions were ideal for innovation and entrepreneurship.

This note is designed to give a broad overview of the energy sector, highlighting trends and market dynamics in 2008. It is intended to be used as a basic primer for entrepreneurs and investors interested in the energy sector, not an exhaustive or definitive text for scientists and engineers. As such, especially given the breadth of the topic, only the most salient topics and facts are included.

---

1 Clean technology, or "cleantech," refers to any technology that helps create electricity or fuels in a manner that limits the negative impact on the environment, or any technology that increases the energy efficiency of appliances or processes. The term does not have a strict universal definition, however, and safe drinking water technologies are sometimes included within cleantech.

Sean Harrington & Dennis M. Rohan, "Note on Energy," pp. 1–50. Copyright © 2008 by Stanford Graduate School of Business. Reprinted with permission.

Background: Physics and Units of Measurement

Energy refers to the capacity to do work and it cannot be created or destroyed; rather, it can only be transferred from one form to another. For example, there is potential energy stored in a ball that is held above the ground, which is then converted to kinetic energy when the ball is released and allowed to fall. Energy is primarily measured in calories, British thermal units (BTU), kilowatt-hours (kWh), or joules (J). Since the scientific community predominantly uses the "Système International" (SI) of units, we will use joules most frequently throughout this note. However, since nearly all electricity is quoted in terms of kWh, we will do the same here. Also, when comparing various fuels, millions of BTU are often used as the preferred units. (Please see Exhibit 1 for common conversions and SI prefixes.)

The rate at which work is performed, referred to as power, is typically measured in watts (W). One watt is equivalent to one joule of energy per second. A typical household incandescent light bulb uses 60 W of electricity and someone jogging is working at 200 to 400 W, depending on their mass and speed. A million watts, or a megawatt (MW), is enough power to supply about 750 typical homes in the U.S., and a nuclear power plant operates at roughly 1,000 MW.

In 2008, the most common way to produce useful work was through the combustion of fossil fuels. This process can be described by the following chemical equation:

$$C_xH_y + (x + \tfrac{y}{4})O_2 \rightarrow xCO_2 + (\tfrac{y}{2})H_2O$$

In words, this can be translated to: fuel + oxygen → heat + carbon dioxide + water. It is important to note that only a certain portion of the heat generated from the combustion process is converted into useful work. For example, in conventional coal-fired power plants, heat from burning pulverized coal is used to boil water, which creates steam to drive the electricity-generating turbines. However, throughout the process, much of the heat escapes into the air (raising the ambient temperature) and is not converted to useful work. This phenomenon is described by the second law of thermodynamics and cannot be avoided. In fact, only about one third of the heat content of the coal at coal-fired plants is converted into electricity. Efficiency is a measure of the amount of heat that is converted into useful work in an energy system. Internal combustion engines (ICE)—standard in most cars—are only characterized by efficiencies of 20–30 percent. From the second law of thermodynamics, we know that the efficiency of any thermal energy system is always less than 100 percent.

When analyzing and comparing various energy storage systems (e.g., batteries), an important consideration is energy density, or the amount of energy per unit of volume or mass. Energy density is an important metric when determining economic or logistical feasibility—for example, when choosing a battery technology for electric vehicles, energy density has a significant impact on car design and performance, fuel efficiency, and cost. Energy density by volume can be quoted in terms of cubic meters (m³) or liters (L), which are the SI units, or more often the gallon or barrel is the preferred unit when referring to fuels. Mass is measured in kilograms (kg) or pounds (lb).[2]

Assessment of the energy efficiency or environmental impact of a technology must include an examination of the entire lifecycle of the system, and when comparing various options it is essential to be consistent with units. This may seem obvious, but in fact, misconceptions and misnomers arise easily when the requisite analytical rigor is not exercised. For example, some U.S. cities operating public buses that use hydrogen fuel cells advertise them as having "zero emissions." Although it is true these buses create no emissions directly, in many cases carbon emissions were released in the process of generating the electricity required to produce the hydrogen used to fuel the buses.

The Incumbent Sources of Energy

Entrepreneurs considering opportunities in the energy sector (even those focusing entirely on cleantech) must first have a solid grounding in the economics and science of the existing primary sources of energy used in the world. In 2008, fossil fuels—oil, coal, and natural gas—still dominated the energy sector, supplying approximately 85 percent of the world's power needs (see Exhibit 2). As such, these fuels heavily influenced the price of energy in the global economy. Nuclear and hydroelectric plants contributed 6 percent and 3 percent, respectively, to the global supply of energy, and had long been important sources of electricity in many countries. However, growth had been slow in these sectors (about 2 percent per year) and was not expected to increase dramatically.[3] on the other hand, although they supplied only a small fraction of global needs in 2008, renewable energy sources, such as wind, solar, and ethanol, were growing at rates of 20–40 percent per year.[4]

Petroleum (Oil)

Given its abundance, high energy density, and the ease with which it can be transported, petroleum, or crude oil, has been the dominant source of energy in the world since the 1950s, supplying 37 percent of the world's energy needs in 2006. It is usually extracted from reservoirs located beneath the surface of the earth or ocean floors, using a drill and oil rig. Geologists use various sophisticated sensing techniques, including seismic surveys, to

---

2 Strictly speaking, pounds are not a unit of mass, but rather of weight. People often confound weight and mass, but they are not equivalent. Weight is a measure of the gravitational forces acting on an object, while mass is independent of gravity. On Earth, weight and mass are roughly proportional (1 kg = 2.2 lb) because the gravitational force acting upon objects on the surface of the earth is relatively constant, but this relationship does not hold more broadly in physics. For the purposes of this paper, the distinction is not critical.
3 *The Outlook for Energy: A View to 2030,* ExxonMobil, 2007.
4 *Statistical Review of World Energy,* BP, 2007.

locate potential pockets of petroleum, but these methods are not guaranteed to be accurate and there is always risk associated with oil exploration.

Once an oil reservoir has been accessed by a bore hole, the thick black liquid can be pumped to the surface. Various technologies exist to increase the percentage of oil extracted from a reservoir, otherwise known as the rate of recovery. The volume of oil is usually measured in barrels, where one barrel contains 42 U.S. gallons and approximately 5.8 million BTUs of potential energy. Petroleum is then shipped to a refinery where it is processed into more useful products, including gasoline, diesel, jet fuel, kerosene, and asphalt. Since crude oil is not useful in its original form for most purposes, it must be purified and separated into its component parts through the refining process. Refined products are then distributed for retail sale. Only 10 percent of petroleum is used directly as a fuel in its unprocessed form.[5] Given the economies of scale inherent in oil exploration, extraction, processing, and distribution, the industry is dominated by large multinational corporations that generally participate in all levels of the value chain. In some cases, these organizations are owned and operated by national governments, and in other cases, they are private entities. Nine of the top ten oil companies in the world—in terms of proved reserves—were state-owned in 2008.

The price of oil is the single most influential factor in the economics of the energy sector. Since petroleum can be shipped virtually anywhere around the globe using oil tankers, trucks, and pipelines, its price is set in the global market.[6] Fluctuations in oil pricing affect not only the energy sector, but also impact nearly every facet of the economy, since virtually all goods and services have some energy input requirements. The most obvious example of this effect experienced by consumers is the price of gasoline, which is heavily correlated with the price of oil. Also, since some power generation facilities or heating mechanisms (e.g., the boiler in a commercial building) can use more than one fuel type, including oil, there are inter-fuel competition dynamics at play, and fluctuations in the price of oil will trickle down and impact the price of other fuels, including natural gas. There are also a myriad of second-order effects, which are not always so obvious. For instance, as oil becomes more costly, all manufactured goods made primarily from plastic will rise in price—owing to the fact that these products' primary input is petroleum. Further, any liquid alternative fuels, such as ethanol, butanol, or biodiesel, need to be price competitive with oil (gasoline), or otherwise rely on government subsidies, rebates, or tax incentives.

There is an important distinction between the physical market for oil and the futures market. Futures contracts are legally binding agreements between two parties who commit to trade a certain quantity of crude oil at a specified future date. These contracts are bought and sold on the so-called paper market, and many corporations' trade oil futures as a

---

5   Ibid.

6   Reference is generally made to only one global price of oil, even though, in fact, there are multiple markets trading oil, each with a slightly different clearing price at any given time. For the purposes of most analyses, this simplifying assumption is safe to make because the price differences are not significant.

means of hedging against risk of fluctuations in fuel prices. Others trade in the futures market in an effort to earn profits through speculation—betting the price of oil will rise or fall in the future. The majority of oil futures contracts are traded on the New York Mercantile Exchange (NYMEX) or the International Petroleum Exchange in London. The physical market, or spot market, is where the product itself, as opposed to a financial instrument, is sold by oil companies. The futures and physical markets have typically been in equilibrium, with clearing prices tracking fairly closely together. However, in 2008, some argued that this balance was being undermined by short-term speculators, who were driving up the price of oil in a way that was inconsistent with the fundamentals of supply and demand.[7]

Between 2005 and 2008, the price of oil was near historic highs. The rise began in September 2003, when the price climbed steadily up from $25 per barrel to over $60 per barrel in August 2005. It remained between $60 and $90 per barrel until late 2007 and then reached an all-time high on January 3, 2008, topping out at just over $100 per barrel. When adjusted for inflation, the price of oil had only reached such levels once since the start of the twentieth century, and this occurred at the height of the Iranian revolution between 1979 and 1981[8] (see Exhibit 3).

Although innumerable factors influence the price of oil, at the highest level it is governed by the fundamental laws of supply and demand. In the long run, supply is limited and there was much debate about the total amount of oil reserves remaining in the world. Some analysts believed that "peak oil"—the time at which global production reaches its maximum and begins a terminal decline—was imminent or had occurred already by 2008. The concept was first introduced by M. King Hubbert in 1956, thereafter known as the "Hubbert peak theory," and was used to accurately predict peak production in the U.S. (excluding Alaska and Hawaii) in the late 1960s (see Exhibit 4). Optimistic calculations indicated the global oil peak would occur between 2020 and 2030,[9] and some questioned the relevance of Hubbert's theory to global supply. In the short term, supply was controlled by the Organization of Petroleum Exporting Countries (OPEC),[10] a cartel of mostly Arab countries that controlled approximately two-thirds of the world's oil reserves, and a handful of non-OPEC oil-rich countries, including Russia, Norway, Mexico, Canada, and the U.S. In 2008, proved oil reserves would last approximately 40 more years at then-current production rates (see Exhibits 5 and 6).

---

7 Robert Winnett, "Soaring Prices: Speculators hijack the oil market," *The Peninsula*, 2005.

8 This period is referred to as the "1979 Energy Crisis." Since Iran and Iraq produced 10 percent of the world's oil supply at the time, and much of it was severely impacted by the war between them, this caused prices to increase dramatically. Fear also drove prices higher than was warranted by the reduction in supply alone.

9 Cambridge Energy Research Associates (CERA), *http://www.energybulletin.net/22381.html*, archived on 14 Nov 2006.

10 The member countries of OPEC in 2008 were Algeria, Angola, Indonesia, Iran, Iraq, Kuwait, Libya, Nigeria, Qatar, Saudi Arabia, the United Arab Emirates, Venezuela, and Ecuador.

World demand for oil had grown fairly consistently since the 1960s at a compound annual growth rate (CAGR) of 2.4 percent (see Exhibit 7). This growth rate in consumption was predicted to continue into the foreseeable future, with the transportation sector contributing most significantly to this trend, given the lack of viable, large-scale fuel alternatives (see Exhibit 8). Geographically, Asian countries were projected to contribute most to the global rise in oil consumption in the coming decades, with China and India leading the way (see Exhibit 9). In 1993, China went from being a net oil exporter to importer, and from 1995 to 2006 its usage of petroleum grew at a CAGR of 8.2 percent, tracking closely with its GDP growth.[11] Although India used only one-third the amount of oil as China in 2006, its consumption had increased by an average of 5.0 percent per year since 1995, and, as a greater proportion of Indians became car-owners, this effect would only be amplified.

In sum, a finite supply of oil and growth in demand was leading to long-term pricing pressure. Unlike the 1979 energy crisis, which was caused by an enormous exogenous shock (Iran-Iraq war), the price increases of 2005 to 2008 were not expected to reverse. The U.S. government's Energy Information Administration (EIA) predicted oil prices would remain above $50 per barrel and reach as high as $100 per barrel indefinitely (see Exhibit 10). Further, the trading of oil futures offered a good insight into the market's beliefs of upcoming prices, and in January 2008, contracts to deliver oil in 2016 cost $86 per barrel (see Exhibit 11).

Non-Conventional Oil

With the petroleum price increases of 2003 to 2008, alternative and more expensive methods for producing crude oil became economically viable. The two most important forms of such non-conventional oil are oil sands[12] and shale oil, which were both financially unattractive when crude oil was $25 per barrel. Extracting the oil (or bitumen, which is a type of hydrocarbon similar to oil), which is mixed in with the sand or shale, is energy intensive and requires large quantities of water. The sand or shale is strip-mined from the surface of the earth and subsequently refined into gasoline, diesel, and other products, or it is processed into synthetic crude. With the latest technology in 2008, a barrel of synthetic crude produced from oil sands consumed about 1.0 GJ of energy (usually from natural gas), which implied that for every unit of energy consumed, 5 to 6 units of energy were produced. About 2 to 3 units of water were used for every unit of synthetic crude produced.[13]

The two largest non-conventional oil reserves in the world were the Athabasca oil sands in Canada and the Orinoco extra-heavy oil deposits in Venezuela. Proved reserves for the

---

11   *Statistical Review of World Energy*, BP, 2007.
12   Oil sands are sometimes referred to as tar sands or extra-heavy oil (in Venezuela). The more proper name is bituminous sands, although the term is not frequently used.
13   *2006 Fact Sheet on Oil Sands*, Alberta Energy, 2006.

Athabasca site were estimated at 174 billion ba[rrels, an extremely conservative]
figure since it represented only 20 percent of the [total bitumen in place. Newer]
technology was achieving recovery rates as high [as 50 percent, and other sites]
likely contained more bitumen than the Athaba[sca region, though the quality]
was lower because the deposits were located d[eeper underground. Assuming]
the price of oil were to stay above $100 per [barrel and technology continued]
to increase in efficiency, some estimated that t[he Canadian oil sands could]
represent two-thirds of the world's total petrole[um reserves.]

Coal

Coal is a carbon-rich, black or brown rock th[at is used to generate electricity.]
It is removed from the earth using surface or [underground mining methods and]
transported to power plants, where it is pulve[rized and burned, with the resulting heat]
used to drive steam turbines. Given its relatively low energy density—approximately 24 MJ per kg compared with 42 MJ per kg for crude oil—coal is quite expensive to move long distances, and for international shipping, transportation costs can represent up to 70 percent of the price.[14] As such, coal is more frequently used in relatively close proximity to the site of extraction. It is typically transported by rail cars, often to regional markets where coal is traded, each with a different clearing price at any given time. Although average coal prices nearly doubled from 2003 to 2006 due to a run-up of demand, the EIA projected a gradual decrease in price levels until 2020 (see Exhibit 12 for historic coal prices).

There are various classifications of coal, based on carbon (and thus energy) content and level of moisture. The so-called low-rank coals are generally softer, earthier-colored and have lower carbon content, while high-rank coals have a darker, more lustrous appearance and have higher energy densities (see Exhibit 13).

Coal is the most abundant and widely distributed fossil fuel, with estimated proved reserves of about 900 gigatonnes by the end of 2006 according to the EIA and BP, which would last about 147 years at the rate of consumption in 2008. Coal was mined in over 100 countries, with the largest reserves located in the U.S., Russia, China, and India (see Exhibit 14). In the U.S., over 90 percent of its production was used in electricity generation, and in China, over 80 percent of all electricity was generated from coal. Worldwide, about 40 percent of electricity came from coal.[15]

Although it had long been the cheapest fuel for electricity generation, coal was also the most detrimental in terms of contributing to climate change. Compared to oil and natural gas, coal has a higher ratio of carbon to hydrogen atoms. Therefore, for every unit of energy generated from the combustion of coal, there is a greater amount of carbon dioxide emitted—relative to other fossil fuels (see Exhibit 15). Since carbon dioxide is a greenhouse gas and was considered the greatest factor in climate change, conventional

---

14 World Coal Institute, http://www.worldcoal.org, accessed January 15, 2008.
15 *Statistical Review of World Energy*, BP, 2007.

coal-fired power plants came under scrutiny. This led to investment and innovation in new technologies and processes—known as "clean coal"—which aimed to reduce the environmental impact of coal-based electricity generation. The leading clean coal technology was called Integrated Gasification Combined Cycle (IGCC), which converted solid coal into synthesis gas (known as syngas) using a high-temperature chamber. IGCC plants produced about 20 percent less carbon dioxide than conventional plants and were more efficient, but also cost 15–20 percent more to build.[16] However, since these plants were much more easily retrofitted for carbon dioxide capture and sequestration technology,[17] some argued this cost would be easily recouped if a carbon tax or cap and trade program were introduced.[18] By the end of 2007, only two IGCC plants were in operation in the U.S., but several more were expected to be built in the subsequent decade.

Natural Gas

Natural gas is a gaseous fossil fuel primarily composed of methane (CH_4), which provided 23 percent of the world's energy needs in 2006. It also contains ethane (5–15 percent), propane and butane (less than 5 percent), along with other impurities. Before it is combusted or transported through pipelines, natural gas is processed to remove virtually all other materials besides methane.

One of the primary uses of natural gas was electricity generation, which was done using gas turbines and steam turbines. Gas was also used extensively in industrial processes that have large thermal requirements or used the fossil fuel as a primary input of production, such as the petro-chemical sector. Since methane is the shortest and lightest hydrocarbon, it produces less carbon dioxide than other fossil fuels. For every unit of heat produced, burning natural gas emits 30 percent less carbon dioxide than burning petroleum and 45 percent less than burning coal.[19] Natural gas was also sold directly to commercial and residential consumers for heating, cooking, and powering other appliances, such as washing machines and clothes dryers. Utilities delivered natural gas to customers using pipelines.

There is an abundance of proved natural gas reserves—enough to last about 63 years at 2006 levels of production—but the majority of the supply was concentrated in a small number of countries (see Exhibit 16). Over half of all proved reserves were in Qatar, Iran, and Russia, while less than 5 percent were located in North America. The main challenge with natural gas was getting it to market. Although it was economical to transport natural gas using pipelines, this distribution model did not extend across oceans, which created regional markets. The most extensive natural gas pipeline system in the world was in North America (see Exhibit 17), although a similar type of network was being developed in

---

16   *American Energy Review*, US Coal, 2005.

17   Technology was being developed, but not yet in use in 2008, that would capture carbon dioxide at power plants and store it in deep holes dug into the earth.

18   Carbon tax and cap and trade programs are discussed in more detail on pages 13-14 of this note.

19   *Natural Gas Issues and Trends*, U.S. Energy Information Administration, 1998.

Europe. Only 20 percent of natural gas was consumed outside the country where it was produced, with Russia leading the way in terms of exports, supplying the majority of European nations with the fuel.[20] The price of natural gas was often tied, either directly or indirectly, to the price of oil. Given the ability of certain power generation facilities and industrial processes to use either oil or natural gas, an increase or decrease in the price of oil could trickle down into the price of natural gas. Some producers of natural gas, including Russia, actually tied the price explicitly to the price of oil.

However, there were increasing efforts to ship natural gas across oceans in its liquefied form (LNG). When cooled to approximately -163 °C (-261 °F) at standard atmospheric pressure, natural gas liquefies, which greatly facilitates its storage and transportation. Specially designed cryogenic compartments on large seagoing vessels are used to keep the LNG adequately cooled while it is shipped. At the destination, a re-gasification terminal reheats the natural gas, and then it is transported from there using the existing pipeline infrastructure. Due to the expensive equipment required for the process (a plant costs $1–$3 billion, a receiving terminal costs $0.5-$1 billion, and vessels cost $200-$300 million), LNG had historically been a relatively small market. In 2006, about 7 percent of world gas consumption came from LNG, as only financially-rich, energy-poor countries, such as Japan, South Korea, and Spain, could afford to import it (see Exhibit 18). However, LNG was one of the fastest-growing energy markets, especially as the value chain benefited from increased efficiencies, other fossil fuels became more expensive, and carbon taxes or cap and trade systems were considered by nations around the world. From 1988 to 2004, liquefaction capacity grew 5.6 percent annually and this growth was expected to reach as high as 13 percent in the following decade.[21]

Nuclear Power

Nuclear power is obtained by harnessing the heat energy released in controlled nuclear fission. The process can be represented mathematically using Einstein's famous formula: $E = mc^2$, where 'E' is the energy released, 'm' is the mass of the material being converted to energy through fission, and 'c' is the speed of light. The energy that is unlocked can be converted to electricity using steam turbines, which is the most common type of nuclear power and accounted for about 16 percent of the world's supply of electricity in 2006, but can also be used for direct heating or propulsion, as is done by some U.S. navy submarines.

Uranium is found naturally in the crust of the earth and is mined and then enriched before being used as fuel in nuclear power reactors.[22] One of the main advantages of

---

20 *Statistical Review of World Energy*, BP, 2007.
21 James T. Jensen, *Global LNG Markets—The Challenge in Meeting Forecast Growth*, February 25, 2005.
22 Enrichment is the process of increasing the concentration of a certain type of isotope within the mined material. For example, naturally occurring uranium contains only about 1 percent of the isotope U-235 and 99 percent U-238. Since U-235 is the only uranium isotope fissionable through thermal neutrons, its concentration is enriched to 3-5 percent for most nuclear power plants.

nuclear power generation was its low fuel requirements. A 1,000 MW nuclear facility used only about 1 ton of uranium each year, while a coal-fired power plant with similar power rating required about 33,000 tons of fuel annually. Under 2006 consumption rates, about 70 years' worth of uranium remained in the world that could be extracted at the historic price level of approximately $130 per kg. If the price were to double, estimated economically recoverable reserves would increase to 300 years' worth. Since uranium represented such a small portion of the overall cost to produce nuclear power, a doubling of its price would increase the cost of electricity by only about 7 percent. Canada and Australia were among 17 countries producing uranium for nuclear reactors in 2005, leading the way with 28 percent and 23 percent of world supply, respectively.[23]

The USSR was the first nation to build a nuclear power plant connected to an electricity grid in 1954, but England and the U.S. were quick to follow, opening inaugural facilities in 1956 and 1957, respectively. Nuclear power production grew rapidly through the late 1980s, with a CAGR of 21 percent from 1965 to 1987. However, growing concerns about nuclear proliferation, long-term storage of nuclear waste, and fears of catastrophic accidents—heightened by the 1979 meltdown at Three Mile Island and the Chernobyl disaster of 1986—caused a worldwide movement to stop all new plant production. This effort was quite successful and growth in nuclear electricity consumption slowed to a CAGR of 2 percent from 1987 to 2006.[24] The top producers in 2006 were the U.S. (29.5 percent of world production), France (16.1 percent), which produced nearly 80 percent of its electricity from nuclear plants, and Japan (10.8 percent).[25] (See Exhibit 19 for world nuclear power statistics.)

However, with climate change and potential carbon-constraining policies looming, nuclear power appeared to be regaining increased favor in 2008, since it provided essentially carbon-free electricity generation. Approximately 18 companies seeking approval to build 30 new nuclear power plants in the U.S. had filed license applications with the Nuclear Regulatory Commission. In January 2008, the United Kingdom announced an aggressive plan to build a new generation of nuclear facilities to help with growing energy needs and expected the first plants to be operational by 2020. Although many experts argued the safety of nuclear plants had improved greatly since Chernobyl, the controversy surrounding the safe storage of nuclear waste was unabated. The so-called "not in my backyard" problem was emotionally charged, as residents fearful of long-term risks and pernicious side effects vehemently opposed any proposed storage sites in their proximity.[26] This was a particularly difficult problem given that the radioactive levels of most nuclear

---

23   *Statistical Review of World Energy*, BP, 2007.
24   *Statistical Review of World Energy*, BP, 2007.
25   Ibid.
26   In the U.S., Yucca Mountain served as the best example of this problem. Located in a remote and unpopulated region of central Nevada, Yucca Mountain was the proposed site for permanent storage of all U.S. nuclear waste, but opposition to the plan caused numerous delays and it was unclear whether the site would ever be opened.

waste decay extremely slowly, with half-lives of 5,000 to 20,000 years. There was additional ongoing concern over nuclear proliferation, as waste from nuclear power plants (plutonium and highly enriched uranium) could potentially be used to make bombs. One alternative proposed was onsite reprocessing of the spent fuel rods, whereby uranium and plutonium could be reused as many as 60 times by employing a technology called PUREX (Plutonium and Uranium Recovery by Extraction) (see Exhibit 20).

Another option under development was the Integral Fast Reactor (IFR), which also utilized onsite fuel reprocessing. Research on this new reactor design first began at Argonne National Laboratory in Chicago in 1984, but funding for the program was discontinued by President Clinton in 1994, only three years before its completion. There were considerable advantages that the proposed new technology offered, including a 99 percent fuel utilization rate. As such, the resultant waste material from the reactor had a half-life of only 200 years—similar to the natural uranium used as fuel—and contained no plutonium. However, since the research was never completed, commercial feasibility was not proven and there were concerns that high capital costs would price the electricity out of the market. With the renewed enthusiasm for nuclear power in the U.S. and the U.K. in 2008, there was the potential to revive development of IFR technology.

Hydropower

Hydropower is energy captured from the force of moving water to perform useful work. This power has been exploited for thousands of years, initially in the form of water wheels or watermills, which used the energy from rivers to do mechanical work, such as grinding grain into flour. In modern times, hydroelectricity is the most common type of hydropower, and accounted for 16 percent of all electricity generated in 2006 (see Exhibit 21).

In most cases, hydropower plants utilize large dams to control the flow of a river's water, allowing them to adjust electricity generation based on demand. Hydroelectricity has no fuel cost and emits zero carbon, which are its major advantages relative to fossil fuel-based alternatives. On the other hand, many experts argue the damming of rivers and subsequent flooding has lasting negative impacts on the local ecosystem. Further, the flooding often requires some type of forced population relocation. For example, the $25 billion Three Gorges Dam project in China—which, at its projected 18,200 MW power capacity, would become the largest hydroelectricity plant in the world when fully operational in 2011—was expected to displace 1.5 million people. Finally, there was limited projected growth in hydropower, as many of the world's largest rivers were already dammed.

Electricity Generation and Distribution

Unlike almost all other commodities, electricity has essentially no shelf life. As soon as it is produced, it must be delivered to the end consumer and used. Although there are some systems in place to help store excess electricity that is generated and not consumed, such as pumped water storage, these techniques are limited in scope because of cost and feasibility constraints. This storage limitation impacts every aspect of the economics of electricity generation and distribution. Approximately 17,500 TWh of electricity is generated annually (see Exhibit 22).

Electricity Value Chain

The electricity generation and distribution value chain is immensely complex, and its structure varies greatly from country to country and from state to state. In general, however, it can be broken into four main parts: generation, transmission, distribution, and retail (see Exhibit 23). In some countries, an electric utility may own the entire infrastructure, from generation to retail, which implies a natural monopoly. Usually in such scenarios, the electrical power industry is highly regulated or nationally owned so that monopoly pricing does not occur. In other countries, the industry is deregulated and there is competition between multiple firms, especially within the generation and retail processes. Thus, consumers choose the company from which they would like to purchase their power, and generators sell to distributors and retailers through a wholesale electricity market, where the spot and forward prices are determined by supply and demand.

Generation

Power plants generate electricity using a variety of primary energy sources, including coal (40 percent of world electricity in 2004), natural gas (20 percent), oil (7 percent), nuclear (16 percent), hydropower (16 percent), and other renewable sources (2 percent), such as geothermal, wind, and solar. Generation assets are typically located away from major population centers since they are unsightly and produce emissions (those burning fossil fuels). The electricity generated is delivered from the plants to the transmission network.

Transmission

Electrical power transmission is the process of moving electricity long distances using power stations, cables (transmission lines), and substations. The transmission lines are usually raised overhead using towers, are not insulated, and conduct AC electricity at very high voltages (usually 110 kV or above). The network of transmission lines is often referred to as "the grid" and supplies local distribution networks at certain nodes. Substations located at these distribution nodes are used to step down the voltage such that it can be delivered to local residential and commercial users. Although the cost of transmission is very small relative to wholesale electricity prices, there is a meaningful loss in energy that occurs when the power is delivered a great distance. For example, in the U.S., average transmission and distribution losses are about 7.2 percent.[27] Transmission is still a heavily regulated monopoly in virtually all cases.

Distribution

Electricity distribution networks are comprised of medium-voltage power lines (less than 50 kV), substations, transformers, and low-voltage distribution wiring (less than 1000 V). They can be configured in various ways and differ significantly from country to country. Usually, the energy leaves the substation in a line known as the primary, which can be mounted on ubiquitous utility poles or buried underground. Transformers are used throughout the network to step up or step down voltage, as needed, and the distribution line typically terminates at a meter located at the end-user's home or building. Like transmission, distribution is a regulated monopoly.

Retail

Retailers exist only in countries or regions that have deregulated electricity markets. These companies purchase power from various sources of generation on wholesale markets and then sell it directly to the consumer, paying fees for the transmission and distribution of that electricity along the way. It is up to the retailer to determine how they will structure the pricing and promote its offering to the consumer.

Deregulation of the North American Electricity Market

The Energy Policy Act of 1992 and the Federal Energy Regulatory Council (FERC) orders 888 and 999 of 1996 set the stage for deregulation of the generation and retailing of electricity in North America. This move was symbolic of a common trend toward electricity deregulation around the world, especially in the U.K. and throughout Europe. However, since individual states (and similar jurisdictions in Canada and Mexico) had their own energy regulating bodies, new deregulation policies varied from state to state and evolved at different rates.

Before 1992, vertically integrated utilities controlled all aspects of power supply and the retail function barely existed, consisting only of ensuring that loads were being met and customers were being properly billed at regulated prices. Initially, utilities met all their load requirements with their own generation assets, but with time, they developed relationships with neighboring utilities and would buy and sell electricity among themselves, based on varying supply and demand requirements. These relationships were informal—each utility would estimate load requirements for the following day and with a few phone calls, the trades were made.

---

27   *Technology Options for the Near and Long Term,* U.S. Climate Change Technology Program, 2003.

After 1992, supported by private investors, construction of new independently owned generation assets began. Similarly, financial institutions and existing energy companies that did not own any generation assets formed electricity marketing groups aimed at capturing value in the newly formed retail sector. However, exuberance for electricity deregulation was significantly subdued by the California Energy Crisis of 2000 and 2001. California was the first state to deregulate, but for many reasons, including gaming and manipulation of wholesale electricity markets by new owners of generation assets (such as Enron), the policy resulted in a debacle. Consumer prices varied wildly, there were rolling blackouts, and the major incumbent utilities were left in financial shambles before the state stepped in to resolve the issue. By 2008, most jurisdictions in North America had adopted some form of partially deregulated electricity markets, but with controls to ensure the problems experienced in California were not repeated. Wholesale markets were generally less regulated, allowing for competition in generation efficiencies, while retail prices had stricter controls.

Electricity Pricing and Peak Demand Constraints

Demand for electricity varies on a minute-by-minute basis, but follows fairly regular daily and seasonal cycles. Demand usually peaks in the late afternoon (especially in warmer climates where air conditioning represents a large portion of power consumption) and reaches a minimum in the middle of the night (when most people are asleep and many commercial operations are inactive). The curve showing variable demand over the course of a day is called a load profile and its shape and magnitude vary between jurisdictions, depending on climate, the breakdown of customer types, and many other factors (see Exhibit 24 for an example load profile). In some areas, there are also significant variations in demand across seasons, especially in places that experience large swings in temperature throughout the year. For example, in New York City electricity demand is significantly higher in the summer due to air conditioning loads.

Peak demand is the single most important factor in electricity pricing and generation efficiency. This is due to the fact that electricity cannot be easily stored in meaningful quantities and it is very difficult to dynamically vary generating capacity in the short term. For example, it is not economically feasible to turn nuclear power plants off and on or adjust the power output; once put into operation, a nuclear facility will generate near constant energy supply indefinitely. Coal-fired plants are not easily brought offline either, but for seasonal variations in demand, this does occur. These types of plants are called base-load. On the other hand, hydropower can be varied dynamically by allowing more or less water to flow through the dam. Adjusting power output is called shaping, and each primary energy source has a different capacity for this. Unfortunately, the cheapest methods of electricity generation also tend to be less easily shaped. As such, in order to meet peak load, expensive generating assets must be used. Additionally, if peak demand increases with time, even if the average total consumption decreases, energy providers must build expensive new plants. Peak-load plants are generally fueled with oil or natural gas.

What this implies is that, from a cost and emissions perspective, not all electricity can be treated equally. Reducing demand at peak periods in the day and year has a much greater impact than similar reductions made during off-peak times. Further, incremental electricity used in the middle of the night—for charging electric vehicles, for example—is nearly carbon and cost neutral since the base-load plants are generating unused excess power at those times.

The feasibility of shaping is also dictated by the relative cost of fuel (variable costs) compared to the upfront capital investment (fixed costs). Power plants that have lower fuel costs and higher capital costs tend to supply base-load energy. For example, nuclear plants have nearly negligible variable expenses in comparison to the upfront construction costs and are thus more suited for base-load. In contrast, fuel costs are much greater for oil-fired facilities, which are used only to meet peak demand. Although hydropower has essentially no variable cost, it does not generally provide base-load power because there is a limited supply of water that can be used over a given time period. Sun and wind electricity generators have no variable costs, but since power levels are uncertain, these sources are deemed intermittent, and are not considered base-load or peak-load—rather, they supply energy whenever possible.

Since the majority of electricity was still supplied by burning fossil fuels in 2008, wholesale and retail prices were impacted by primary fuel prices. If the price of coal increased, it would eventually lead to heightened retail electricity prices. However, there was a latency effect (time delay), as there was first an increase in wholesale prices, and then utilities would eventually raise retail tariffs. Wholesale and retail prices depended on the mix of primary energy sources used to generate electricity and varied geographically based on what was available. Retail prices were generally segmented by customer type: residential, commercial, and industrial. Residential users tended to pay the highest tariffs, with commercial rates slightly lower and industrial prices nearly half those paid by residential customers. Residential prices were often tiered, such that consumers paid more per kWh as they used more.

Time-of-Use and Net Metering

Given the issues associated with meeting peak demand, there was increasing use of electricity meters that not only measured the quantity of power supplied to a customer, but also when it was consumed. Time-of-use (TOU) metering enabled utilities to use price signals to help reduce peak demand. They could charge higher "on-peak" tariffs, which would encourage consumers to reduce consumption during those times (through efficiency or advance planning). Many utilities began installing "smart meters" that would supply TOU data dynamically using wireless communication channels. Traditional "dumb" meters were read on-site (usually monthly or annually) by utility employees or contractors. Although studies suggested a continued rapid growth in smart meter installations (23 percent CAGR into 2013[28]), TOU pricing was still rarely used and its efficacy unproven on a large scale.

Net metering was another way in which utilities aimed to smooth load profiles, by enabling customers with small generating capacity—usually from solar or wind facilities—to upload excess power onto the grid. These customers generally received credit for any power they supplied, which was applied to their electricity bill. Solar power was the most common source of energy in net metering, and it was especially helpful because it supplied extra power to the grid during daylight hours (peaking in the afternoon), when demand was often highest.

Demand Response

Demand response refers to systems designed to help induce reductions in electricity usage by consumers at critical times. TOU pricing is a specific example of demand response, but the broader category was also receiving much attention. Many technologies were emerging to facilitate, automate, and augment demand response. For example, some energy-consumptive commercial customers, such as large retail chains, installed a series of control devices that could communicate with power suppliers and automatically adjust lighting, heating, cooling, and power supply to appliances in response to periods of peak demand. In exchange, these ratepayers would receive cheaper electricity or rebates based on levels of peak power reduction.

Profit Decoupling

Where retail electricity pricing is unregulated or is linked to total volume of sales, utilities are incented to drive increased consumption—to the detriment of the environment and ratepayers. Profit decoupling addresses this by regulating retail rates in such a way that utilities' bottom lines are no longer linked to total revenue. Instead, utilities are guaranteed a fixed return on assets, irrespective of sales. In conjunction with profit decoupling, states often mandate that utilities spend a specified amount on energy efficiency programs and even provide financial incentives for reducing total consumption. California was the first to adopt a profit decoupling policy in 1982, which has served as a success story. From 1982 to 2005, per capita electricity consumption remained nearly constant in California (around 7,000 kWh per person per year), while the overall U.S. rate increased by over 30 percent during the same time period (from about 9,000 kWh to 12,000 kWh per person per year). Despite the success of decoupling in California, only four other states had adopted similar policy by 2007 (Oregon, Idaho, Maryland, and Minnesota). However, many others were considering it.

---

28 "Smart Metering to Experience a 24% Annual Growth Rate Through 2013," ABI Research, December 10, 2007.

Carbon Cost: Cap and Trade or Taxes

In 2008, the U.S. congress was considering several pieces of legislation that would effectively put a price on carbon dioxide and other greenhouse gas (GHG) emissions. In Europe, there were already limitations in place, based on standards put forth in the Kyoto Protocol. The two main ways policymakers approached carbon pricing were taxes and cap and trade programs. Governing bodies could impose a straight tax on GHG emitters, but many economists favored the cap and trade model, which would set a limit on emissions and then allow companies or other groups the opportunity to trade carbon credits. Using the latter approach, those polluting more than the specified cap would need to purchase allowances earned by those operating under the emissions limit. As such, the market would determine the price of emissions and incentives would be in place to drive innovation and competition to be more efficient. The cap and trade model also guaranteed specific levels of reduction in emissions, while it was difficult to predict how much carbon would be abated through the use of a tax.

If carbon emissions—traditionally treated as an externality—were internalized as a cost item in power generation, this would have a significant impact on electricity pricing and preferred primary energy sources. For example, if carbon emissions were priced at $50 per tonne, which was not an unreasonable estimate based on projections, electricity generated from conventional coal-fired plants—the cheapest option in 2007—would become more expensive than electricity from wind, natural gas, and nuclear.[29] (See Exhibit 25 for projected wholesale electricity pricing by energy source as a function of carbon emissions pricing.)

Electricity Storage

In a number of areas within the energy sector, improved electricity storage had become a high priority. Given the constraints imposed by peak power, there were enormous incentives to develop electricity storage technology that could be used at the utility scale. This need was heightened in regions that were developing greater levels of renewable electricity generation, especially those focused on solar and wind power, because of the intermittence intrinsic to these sources. Further, in the automobile industry, with ever-increasing prices in petroleum-derived fuels, there was a growing interest and investment in electric vehicles and plug-in electric hybrid vehicles, which were heavily dependant on advancement in battery technology.

In 2007, according to the Cleantech Group, electricity storage attracted the second highest level of cleantech venture funding in the U.S. and Europe behind power generation. There were dozens of start-ups developing various battery designs, but the leading technology appeared to be lithium ion based. These batteries used various forms of lithium ions in the cathode (generally graphite in the anode) and had originally been developed for laptop computers, cameras, and other hand-held devices. Many companies, including an

---

29   Electric Power Research Institute.

investor favorite, A123Systems, which used nanotechnology in conjunction with the basic lithium ion design, were starting to produce batteries at a commercial scale for utilities and automakers.

Energy Consumption by End-Use

Energy consumption varied between different sectors of the economy and end-use analysis was usually divided into four main categories: transportation (28 percent of U.S. consumption), industrial (32 percent), commercial (18 percent), and residential (21 percent). The world breakdown of consumption by sector was similar to the U.S. figures above (see Exhibits 26 and 27).

Transportation Sector

The transportation sector used petroleum almost exclusively as its energy source, consuming about 60 percent of total crude oil supply in the world in 2005.[30] Most of the fuel used was in the form of gasoline or diesel, while negligible quantities of natural gas and electricity were also consumed by vehicles on the road (see Exhibit 28). The main advantage of diesel engines was increased fuel efficiency, since they operated at higher temperatures.

Transportation fuel prices around the world were heavily influenced by taxes. For example, in the U.S., average state and federal fuel taxes totaled 46.9 cents per gallon of gasoline in 2007, representing about 15 percent of the retail price.[31] Europe had the highest fuel taxes, which could represent as much as 70 percent of the amount charged at the pump. These taxes were seen as an energy policy tool to help promote alternative technologies or the development of biofuels. Advocates also argued that such policy put pressure on manufacturers to produce more efficient vehicles.

Given that petroleum was the fossil fuel with the least proved reserves (in terms of years of consumption remaining), and the transportation sector contributed about 33 percent of total GHG emissions, there was tremendous development and innovation in road vehicle technologies in 2008. These advancements generally fell into two different categories. One group of researchers and entrepreneurs were focused on developing alternative fuels, such as biofuels, while others worked to create vehicles that would rely on electricity as the primary source of energy. Biofuels, such as ethanol and butanol, could be used in the existing fleet of vehicles on the road operating on internal combustion engines (ICE), which was the primary advantage of this approach. However, there were many concerns and risks associated with biofuels, which are discussed in greater detail later in this note.

The world's largest automakers were developing new lines of electric vehicles (EV), hybrid electric vehicles (HEV), and plug-in hybrids (PHEV) in 2008. These vehicles were

---

30 Key World Statistics, International Energy Agency, 2007.
31 State Motor Fuel Excise Tax Rates, American Petroleum Institute, 25 July 2007.

expected to produce less carbon emissions (on a grams of carbon dioxide per mile basis) and cost less to operate (at 2008 gasoline prices), which were the major benefits. On the other hand, the major downside was that existing vehicles could not easily be retrofitted to use electricity. Further, some argued the technology would take decades to be adopted because it required significant changes in driving behavior. One entrepreneur who believed the shortcomings of EVs could be effectively eliminated was Shai Agassi, the former COO of software-giant SAP. In January 2008, he announced a partnership between his newly formed company, the government of Israel, and Nissan-Renault, the fourth largest car group in the world. With $200 million of funding, the trio laid out a plan to convert Israel's entire vehicle fleet to EVs by 2020. Another fuel option for vehicles that was gaining favor was compressed natural gas (CNG), which was derived by compressing methane to about 1 percent of its normal volume. This fuel produced less carbon per unit of energy compared to gasoline or diesel and was generally cheaper, but required special storage tanks that took up more space within the vehicle (usually in the trunk). (See GSB Case OIT-74—*Environmental Enhancements in Road Vehicle Technologies* for more.)

Industrial Sector

The industrial sector could be loosely divided into heavy and light manufacturing. In heavy manufacturing, there were large quantities of energy used to produce new materials and products. The energy often came in the form of burning fossil fuels directly. Natural gas was the most common fuel for such purposes, since its combustion could most easily be adjusted or stopped and restarted. Process heating and machine drive were the end-uses in heavy manufacturing that consumed the most energy. Typical examples of energy intensive heavy manufacturing were steel and concrete production. In these industries some companies were pursuing opportunities to increase energy efficiency through the use of combined heat and power (CHP), which was a method of utilizing the excess heat generated from the burning of fossil fuels in a useful way. It could either be used to heat the building or piped in the form of steam to local urban areas to reduce alternative heating requirements. In regions of the world where manufacturing was a dominant component of the economy, such as China, energy consumption tended to grow more quickly than GDP.

In light manufacturing the energy intensive raw materials, such as steel and iron, were used to produce final goods. In this type of production, electricity was the main energy source. As the manufacturing processes relied more on automation and less on labor, they became more energy-intensive, although not nearly to the same degree as heavy manufacturing.

It is also important to note that about 10 percent of all primary energy consumed in the world was used for non-energy related purposes, mainly by the petrochemical sector, which produced plastics, other durable materials, and chemicals from fossil fuels. For example, ethylene was the most produced organic compound in the world, with supply exceeding 75 million tonnes in 2005. Ethylene was extracted from natural gas and crude oil using fractional distillation followed by steam cracking, which was an energy-intensive process and also used fossil fuels as raw inputs.

Commercial Sector

The commercial sector primarily used electricity and natural gas for lighting, heating, and cooling. Growth in energy needs for the commercial sector closely tracked employee growth. Since many commercial buildings were not originally designed or built to be energy efficient, much effort was going into retrofitting these buildings to help reduce energy consumption. Consultants or auditors could be hired to inspect offices, schools and other commercial sites to assess and improve energy efficiency. The two most impactful changes that could be made were to lighting and insulation, where advances in technology offered significant reductions in electricity usage as well as heating and cooling requirements. Such changes also helped lower operating expenses, which provided an added benefit and incentive.

For new buildings, a whole new category of "green construction" had emerged. The U.S. Green Building Council developed a set of standards called LEED (Leadership in Energy and Environmental Design) that were used to rate new construction projects. Buildings earned points for various design features—such as waterless urinals and skylights to allow natural sunlight—that helped lessen its resource requirements, and an award system was linked to certain point thresholds. The top certification level was LEED platinum, and the popularity of these standards was growing rapidly. For example, the Stanford GSB was set to begin construction of its new campus in 2008 and had publicly stated its goal of achieving LEED platinum status.

Residential Sector

The residential sector mainly used electricity and natural gas for lighting, heating, cooling, and operating appliances. The source energy and system for heating varied across geographies. For example, in the western and southern U.S., most homes and apartments were heated using centralized natural gas-fired furnaces that forced warm air throughout the living space. However, in other parts of the U.S., localized electrical heaters were used, and in the northeast of the country, heating oil (similar to diesel) was the main fuel combusted in residential furnaces (see Exhibit 29).

Like commercial buildings, homes and apartment buildings were generally not very energy efficient and a whole industry of products and services was evolving to address this problem. Home auditors could offer a plethora of energy-saving devices and improvements, relating to insulation, windows, lighting and appliances. So-called "smart appliances" were available and through programmed settings, they used electricity only when absolutely needed, which was not the case for most "dumb appliances." One of the simplest and most impactful efficiency gains was in lighting, with the introduction of compact florescent lighting (CFL). CFL bulbs produced about the same amount of light but used one-fifth the amount of energy compared to typical incandescent bulbs. Since lighting accounted for 16 percent of residential electricity consumption, the energy and cost savings were significant (see Exhibit 30). Additionally, in terms of heating and cooling, heat pumps were also gaining in popularity as a more energy efficient alternative, especially in Europe.

A heat pump uses a differential in temperature between the outside environment (usually the soil or nearby water source) and the area being heated or cooled. Effective heat pumps are able to achieve coefficients of performance between 3 and 4, which implies that for every unit of electrical energy used for pumping about three to four times as much thermal energy is transferred.

Renewable Energy

Broad acceptance of anthropogenic climate change and a desire for energy independence was driving exceptional growth in the renewable energy sector. When examining various renewable energy sources, it was critical to examine the cost of the renewable energy relative to the incumbent alternatives. In order to achieve deep penetration, the renewable power needed to be price competitive. Although some consumers were willing to pay a certain premium to mitigate GHG emissions, this segment of the market was still relatively small in 2008. Some renewable energy technologies were supported by local or federal governments through subsidies, rebates, or preferential tax treatment, which helped them compete with nonrenewable options. In such cases, non-market dynamics could highly influence future viability of certain technologies.

Solar

Solar radiation is the most abundant source of renewable energy in the world. Energy provided by the sun hits the surface of the earth at a rate of about 89,000 TW, which is equivalent to nearly 6,000 times the total world energy consumption in 2007. However, the difficulty of converting that energy into useful work in a cost-effective manner has hindered its widespread use. Nevertheless, enormous capital investment and research effort has been put into improving solar technologies, and progress is continually being made.

There were two main types of solar technologies: thermal and photovoltaic (PV). With solar thermal systems, energy from the sun is used to heat water. This type of set-up can be used as a direct substitute for gas-fired residential hot water tanks. Alternatively, steam from the heated water can be used to drive a turbine and generate electricity. Utility-scale solar power plants used various techniques—such as parabolic troughs—to concentrate the warmth of the sun and increase power generation efficiency. A number of high-profile start-ups emerged in this arena from 2005 to 2008, promising wholesale electricity prices from solar thermal systems that approached those associated with nonrenewable energy sources.

PV systems convert sunlight directly into electricity. In simple terms, photons from the sun strike the PV arrays prompting electrons to jump to a higher energy level, which induces an electrical current. There are different materials that can be used in the modules, but the vast majority of panels installed in 2007 were made from crystalline silicon (c-Si), a proven technology with over 30 years of commercial production. However, over a dozen venture-backed start-ups had emerged with PV technologies that used a variety of different materials to produce thinner and lighter panels. This new category was dubbed "thin film" solar and generally offered cheaper cost per unit of power generation, but with lower efficiency (power per unit area.) Efficiency was not as critical as price per kW for potential utility-scale solar facilities because sufficient land area was available in uninhabited regions, where these generating arrays would be located.

Buoyed by heavy government subsidies and tax incentives in certain countries, especially Germany, Japan, and the U.S., global installed PV capacity was growing dramatically. In 2006, the average annual growth rate was 31 percent for the previous decade, and from 2004 to 2006, annual growth rates exceeded 40 percent[32] (see Exhibit 31). With this growth came a whole suite of new businesses offering services and products to help support the solar industry, including installation, monitoring, and metering. To help defray the initial capital costs associated with buying solar facilities, some companies offered commercial and residential customers the opportunity to purchase PV-generated electricity as a service. In this model, the service provider owned the roof-mounted panels installed on the customers' home or building and charged a given rate for the electricity produced, as laid out in a power purchase agreement.

Biofuels

Biofuels are liquid or gas fuels derived from recently living organisms, including plants, animals, and their byproducts. They were considered promising as potential long-term alternatives to petroleum-based fuels, such as gasoline, diesel, and jet fuel, heavily used in the transportation sector. The most commonly produced biofuel in the world, by far, was ethanol (pure alcohol). The U.S. and Brazil were the top two producers in the world, each supplying nearly 5 billion gallons per year, which accounted for 90 percent of global production in 2007. Ethanol could be mixed with gasoline (up to a certain concentration) and used in ICE engines, and in 2008, most cars on the road in the U.S. used gasoline containing 5–10 percent ethanol already. To accommodate high concentrations of ethanol (above 50 percent), minor engine modifications were required. Some new vehicles were being sold with this added "flex-fuel" capability.

The first generation of ethanol was made by fermenting the starches from plants—primarily corn (in the U.S.) and sugarcane (in Brazil). Although a renewable energy source, this type of ethanol (especially corn-based) had many critics who highlighted three main criticisms. First, production of this type of ethanol was energy intensive. For every unit of fossil-fuel energy that went into corn-based ethanol production, only about 1.3 units were available in the resulting fuel output.[33] Second, corn and sugarcane ethanol production used important world food supplies as feedstock, which had driven up prices. Third, there was not enough arable land in the world to scale production of this type of ethanol to levels that would make a reasonable impact on gasoline consumption. Experts estimated

---

32   *Statistical Review of World Energy*, BP, 2007.
33   "Green Dreams," J.K. Bourne Jr. and R. Clark, *National Geographic*, October 2007, p. 41.

that if all the corn in the U.S. were used exclusively to produce ethanol, it would still only displace 10–12 percent of gasoline consumption in the U.S.[34]

Investors and entrepreneurs were betting that cellulosic ethanol would address the shortcomings of its first generation predecessor. Cellulosic ethanol used any biomass or other carbonaceous material—not just starches—as feedstock for its production. Since wood chips, corn stover, and other agricultural byproducts were already available in vast quantities and were going unused, these materials were initially targeted as inputs for production. Proponents of cellulosic ethanol believed that "energy crops", such as switchgrass, would eventually be grown commercially and used as inputs because of the increased energy efficiency they offered over corn and sugarcane. One study showed that every unit of fossil-fuel energy used in ethanol production with switchgrass yielded 5.4 units of new energy. As of 2008, a handful of companies claimed to have processes that produced cellulosic ethanol at prices that were competitive with corn-based ethanol and gasoline. However, it was expected to take two to five years for any of these startups to validate their claims with plants operating at full scale. (See GSB case OIT-52—*A Technical Note on Ethanol as a Motor Fuel* for more.)

Butanol was also being pursued as a potential alternative, but was earlier in the development lifecycle. Pure butanol could be used as a direct substitute for gasoline in ICE vehicles and was more easily distributed using existing pipelines and trucks compared to ethanol. However, cost and feasibility of large-scale production were still constraints. As was the case with nearly all biofuels, development of butanol was deeply rooted in the biosciences, since microorganisms and enzymes were central to technological breakthroughs.

Wind

Modern, large-blade wind turbines were first developed in the 1980s and were often clustered together on agricultural land, forming wind farms. The electricity generated was put onto the grid or used to supply power needed by nearby farming operations or residents. Installed wind power capacity tripled from 2001 to 2006 and was expected to continue growing at a rate of about 40 percent per year over the course of the subsequent five years. Germany, Spain, and the U.S. produced the most wind power and in 2006 the total world capacity was 74 GW—nearly 20 times that of solar PV.[35] Demand was so high that turbine manufacturers were backlogged with orders.

Electricity generated from wind was nearly cost-competitive with nuclear power, and unlike solar power, it did not require government subsidies. Operating costs were negligible, and, other than the energy used to construct the turbines, wind power generated no GHG emissions. However, only certain locations had wind patterns suitable for economically feasible electricity production. Also, wind farms needed access to transmission lines,

---

34 "Ethanol Fuel Presents a Corn-undrum," Deane Morrison, University of Minnesota, September 18, 2006.

35 *Statistical Review of World Energy*, BP, 2007.

which further constrained site selection. Since wind power was intermittent, utilities could not rely on it, and so it rarely constituted more than 10 percent of any utility's total supply. Despite these limitations, some calculations indicated a potential total capacity of 72 TW of commercially harvestable wind in the world, which represented nearly seven times the global electricity consumption in 2006.[36]

The wind power industry had not seen major technological breakthroughs since the development of the modern turbine. Some incremental improvements were continually being made to improve efficiency, although turbine prices had been increasing due to growing material costs and high demand.

Geothermal

Geothermal power is the natural heat of the earth and was harnessed to generate electricity using steam turbines or used for direct heating in industrial processes or residential districts. For electricity generation, holes were drilled to access pockets of hot water deep in the earth, allowing the steam from the boiling water to drive turbines located at the surface. Installed geothermal capacity was 9.6 GW in 2006 and had grown only modestly in the preceding decade, averaging 3 percent increases per year. The U.S. and the Philippines were the top producers of geothermal electricity, although Iceland was notable in that 90 percent of its inhabitants used inexpensive geothermal power for direct central heating. New technology called enhanced geothermal systems (EGS), which pumped water beneath the surface of the earth instead of tapping into preexisting reservoirs, was expected to boost growth modestly. Estimates of total global potential varied, but one comprehensive study by MIT estimated total capacity could reach 100 GW of electricity in the U.S. alone by 2050.[37]

Tidal, Wave, and Ocean Thermal

Tidal power is a specific type of hydropower and can be used to produce electricity by harnessing the energy of water rising and falling due to tides. Barrages and stream turbines are most commonly used and the power is far more predictable than solar or wind. Two major barrages existed in the world in 2008—one at the Bay of Fundy in Canada (18 MW) and the other where the Rance River meets the English Channel in France (68 MW).

Wave power can be used to generate electricity by capturing the energy from a series of buoys rising up and down due to the consistent motion of ocean waves. The first commercial plant (2 MW) was scheduled to be completed in northern California by 2012.

Ocean thermal energy conversion (OTEC) is a technology that takes advantage of the temperature differential between deep and shallow water to generate electricity with a heat

---

36   Cristina L. Archer and Mark Z. Jacobson, "Evaluation of Global Wind Power," *Journal of Geophysical Research—Atmospheres*, 2005.

37   "The Future of Geothermal Energy: Impact of Enhanced Geothermal Systems (EGS) on the United States in the 21st Century," MIT, 2006.

engine. The largest OTEC research institute was located on the Kona coast of Hawaii, and tiny pilot facilities existed in Japan and India.

Entrepreneurial Opportunities in The Energy Sector

In 2008, the energy sector was undergoing a massive transition. Billions of dollars were being invested in clean technology. Renewable energy sectors were growing at annual rates of 20–40 percent. In developments that might have seemed unthinkable a decade earlier, even major oil companies and automakers, like BP and GM, were putting resources toward green energy solutions. With such dynamism and turbulence in the marketplace, entrepreneurial opportunities were abounding.

New ventures in the energy sector fell loosely into four categories. First, there was a plethora of companies working on technologies harnessing renewable energy sources, including solar, wind, and biofuels. Second, a new portion of the energy sector was focused on building technological solutions to help make incumbent, nonrenewable power sources cleaner—primarily through the reduction of GHG emissions. Science and engineering were at the nexus of organizations in these first two categories, but business-savvy employees were also needed to help commercialize the products. Third, ancillary products and services supporting the value chain for clean technology were critical, and many did not require deep scientific advancement or defensible intellectual property. For example, solar panel installation was a much-discussed, high-growth industry, yet companies in this field were essentially construction services providers, without any proprietary technology.

Finally, while companies in the first three categories were focused on the supply side of the equation, there was a fourth grouping of entrepreneurs narrowing their sites on the demand side—building companies that aimed to reduce consumption and shape end-use behavior. Trends in policy were catalyzing growth in this area, but plain old free-market economics were also helping. Since most consumers and corporations faced large and increasing energy costs, financial incentives for conservation were significant. In fact, a 2007 research project by McKinsey & Company concluded that nearly half of all GHG abatement options (in the mid-range case) had a negative cost. In other words, over the lifecycle of these technologies, the savings gained through energy efficiency outweighed the cost of implementation.[38] However, the challenge to tapping into these negative-cost abatements was aligning incentives, since the entity responsible for the initial capital expense was frequently not reaping the subsequent savings. In such cases, the business model was the critical piece of IP.

---

38   "Reducing U.S. Greenhouse Gas Emissions: How Much at What Cost?" U.S. Greenhouse Gas Abatement Mapping Initiative, McKinsey & Company, December 2007.

Exhibit 4.1. Conversion Table and SI Unit Prefixes

Name of Unit	Symbol	Definition	Relation to SI Unit
Volume			
cubic metre (Ξ I unit)	m³	≡ 1 m × 1 m × 1 m	
litre	L	≡ 1 dm³ [3]	= 0.001 m³
gallon (U.S. fluid; Wine)	gal (US)	≡ 231 cu in	= 3.785 411 784 L
barrel (petroleum)	bl; bbl	≡ 42 gal (US)	= 158.987 294 928 L
Mass			
kilogram	kg		
tonne	t		≡ 1000 kg
pound	lb		= 0.453 592 37 kg
Energy			
joule (Ξ I unit)	J	≡ N·m = W·s = V·A·s	= kg·m²/s²
calorie	cal		≡ 4.1868 j
kilocalorie; large calorie	kcal; Cal	≡ 1000 cal	= 4.1868 kJ
British thermal unit	BTU	≡ 1 lb/g × 1 cal × 1 °F/°C	= 1.055 055 852 62 kJ
kilowatt-hour	kWh	≡ 1 kW × 1 h	= 3.6 MJ
barrel of oil equivalent	bboe	≈ 5.8 MBTU	≈ 6.12 GJ
ton of oil equivalent	TOE		= 41.868 GJ
quad		≡ 1015 BTU	= 1.055 055 852 62×1018 J
Power			
watt (S I unit)	W	≡ J/s = N·m/s	= kg·m²/s³
BTU per hour	BTU/h	≡ 1 BTU/h	≈ 0.293 071 W
horsepower (Imperial)	hp	≡ 550 ft Ibf/s	= 745.699 871 582 270 22 W

Name	y otta -	Zetta-	exa-	peta-	tera-	giga-	mega-	kilo-	hecto-	deca-
Symbol	Y	ς	E	P	T	G	M	k	h	da
Factor	10^{24}	10^{21}	10^{18}	10^{15}	10^{12}	10^9	10^6	10^3	10^2	10^1
Name	deci-	centi-	milli-	micro-	nano-	pico-	femto-	atto-	zepto-	yocto-
Symbol	d	c	m	µ	n	P	f	a	z	y
Factor	10^{-1}	10^{-2}	10^{-3}	10^{-6}	10^{-9}	10^{-12}	10^{-15}	10^{-18}	10^{-21}	10^{-24}

Source: Wikipedia

Exhibit 4.2. World Energy Supply—2006

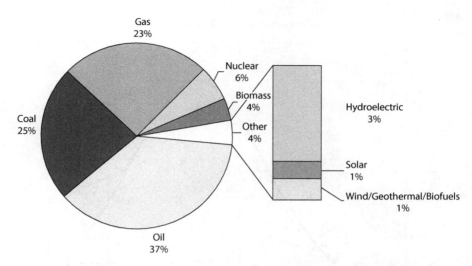

Sources: *Global Status Report on Renewables*, REN21, 2006. *Statistical Review of World Energy*, BP, 2007.

Exhibit 4.3. Historic Price of Oil

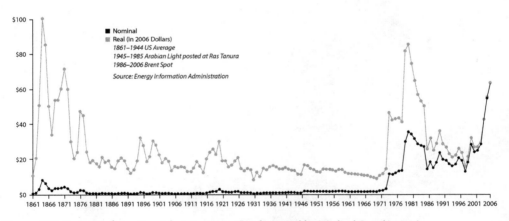

Source: U.S. Energy Information Administration. Graph created by Michael Ströck, 2006

Exhibit 4.4. Hubbert's Peak Theory for U.S. Oil Production

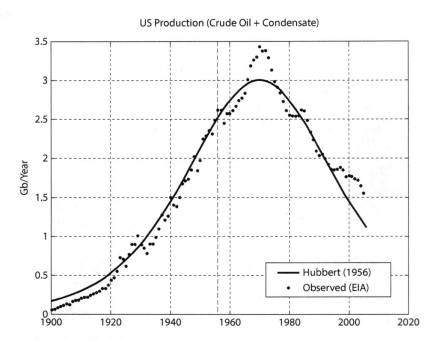

Source: Samuel Foucher, March 6, 2007

Exhibit 4.5. Proved Oil Reserves—2006

	Reserves - Thousand million barrels	Share of total	R/P ratio
USA	29.9	2.5%	11.9
Canada	17.1	1.4%	14.9
Mexico	12.9	1.1%	9.6
Total North America	**59.9**	**5.0%**	**12.0**
Venezuela	80.0	6.6%	77.6
Other S. & Cent. America	23.5	1.9%	15.9
Total S. & Cent. America	**103.5**	**8.6%**	**41.2**
Kazakhstan	39.8	3.3%	76.5
Russian Federation	79.5	6.6%	22.3
Other Europe & Eurasia	25.0	2.1%	10.8
Total Europe & Eurasia	**144.4**	**12.0%**	**22.5**
Iran	137.5	11.4%	86.7
Iraq	115.0	9.5%	157.6
Kuwait	101.5	8.4%	102.8
Saudi Arabia	264.3	21.9%	66.7
United Arab Emirates	97.8	8.1%	90.2
Other Middle East	26.7	2.2%	26.9
Total Middle East	**742.7**	**61.5%**	**79.5**
Libya	41.5	3.4%	61.9
Nigeria	36.2	3.0%	40.3
Other Africa	39.5	3.3%	19.0
Total Africa	**117.2**	**9.7%**	**32.1**
China	16.3	1.3%	12.1
Other Asia Pacific	24.2	2.0%	15.6
Total Asia Pacific	**40.5**	**3.4%**	**14.0**
TOTAL WORLD	**1208.2**	**100.0%**	**40.5**

Note: Reserves/Production (R/P) ratio—If the reserves remaining at the end of the year are divided by the production in that year, the result is the length of time that those remaining reserves would last if production were to continue at that rate.
Source: *Statistical Review of World Energy*, BP, 2007

Exhibit 4.6. Top World Oil Producers and Consumers

\multicolumn{3}{c}{Top World Oil Producers, 2006}		
Rank	Country	Production
1	Saudi Arabia	10,719
2	Russia	9,668
3	United States	8,367
4	Iran	4,146
5	China	3,836
6	Mexico	3,706
7	Canada	3,289
8	United Arab Emirates	2,938
9	Venezuela	2,802
10	Norway	2,785
11	Kuwait	2,674
12	Nigeria	2,443
13	Brazil	2,163
14	Algeria	2,122
15	Iraq	2,008

Top World Oil Net Exporters, 2006		
Rank	Country	Net Exports
1	Saudi Arabia	8,651
2	Russia	6,565
3	Norway	2,542
4	Iran	2,519
5	United Arab Emirates	2,515
6	Venezuela	2,203
7	Kuwait	2,150
8	Nigeria	2,146
9	Algeria	1,847
10	Mexico	1,676
11	Libya	1,525
12	Iraq	1,438
13	Angola	1,363
14	Kazakhstan	1,114
15	Canada	1,071

Exhibit 4.6. (continued)

Top World Oil Consumers, 2006		
Rank	Country	Consumption
1	United States	20,588
2	China	7,274
3	Japan	5,222
4	Russia	3,103
5	Germany	2,630
6	India	2,534
7	Canada	2,218
8	Brazil	2,183
9	Korea, South	2,157
10	Saudi Arabia	2,068
11	Mexico	2,030
12	France	1,972
13	United Kingdom	1,816
14	Italy	1,709
15	Iran	1,627

Top World Oil Net Importers, 2006		
Rank	Country	Net Imports
1	United States	12,220
2	Japan	5,097
3	China	3,438
4	Germany	2,483
5	Korea, South	2,150
6	France	1,893
7	India	1,687
8	Italy	1,558
9	Spain	1,555
10	Taiwan	942
11	Netherlands	936
12	Singapore	787
13	Thailand	606
14	Turkey	576
15	Belgium	546

Note: All values in thousands of barrels per day
Source: U.S. Energy Information Administration, 2006

Exhibit 4.7. Historic World Oil Consumption

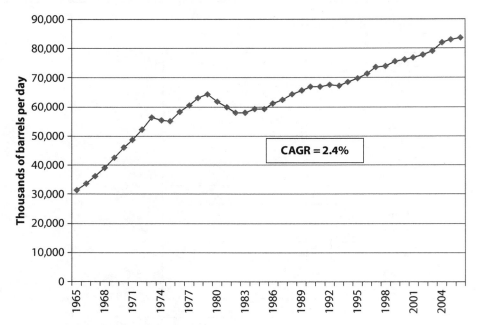

Source: *Statistical Review of World Energy,* BP, 2007

Exhibit 4.8. World Oil Consumption by Sector

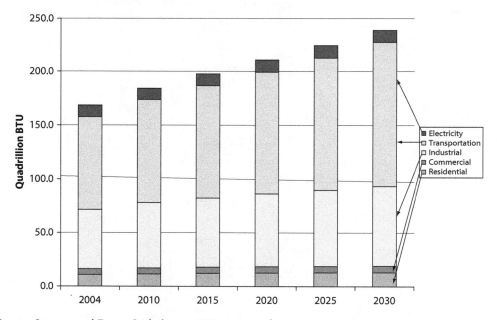

Source: *International Energy Outlook 2007,* U.S. Energy Information Administration, 2007.

98 | Global Sustainability and Innovation

Exhibit 4.9. World Oil Consumption by Region

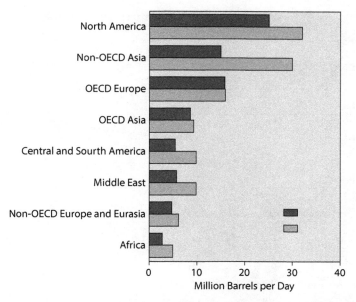

Sources: *International Energy Annual 2004* and *System for the Analysis of Global Energy Markets (2007),* U.S. Energy Information Administration.

Exhibit 4.10. Oil Price Projections

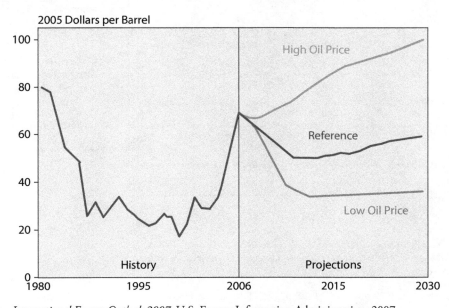

Source: *International Energy Outlook 2007,* U.S. Energy Information Administration, 2007.

Exhibit 4.11. Oil Futures Pricing—Light Sweet Crude as of January 23, 2008

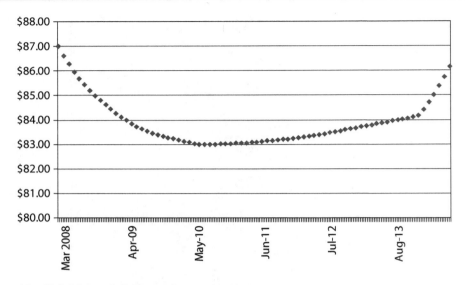

Source: New York Mercantile Exchange, January 23, 2008.

Exhibit 4.12. Historic Coal Prices

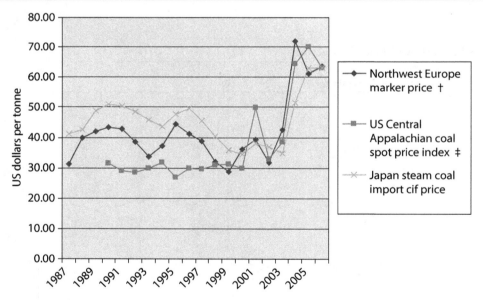

‡ Price is for CAPP 12,500 BTU, 1.2 SO2 coal, fob. Source: Platts.
Note: cif = cost+insurance+freight (average prices); fob = free on board.

‡ Price is for CAPP 12,500 BTU, 1.2 SO2 coal, fob. Source: Platts.

Note: cif = cost+insurance+freight (average prices); fob = free on board.

Source: McCloskey Coal Information Service

Exhibit 4.13. Types of Coal

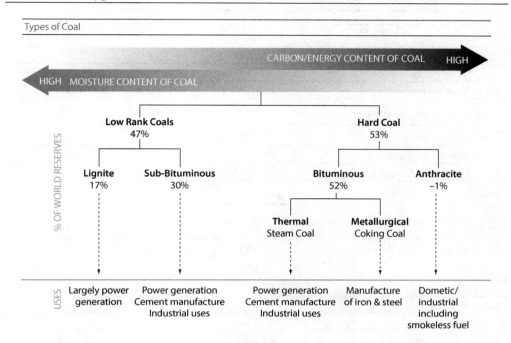

Source: World Coal Institute

Exhibit 4.14. Proved Coal Reserves by End of 2006

(Million tonnes)	Total	Share of Total	R/P ratio
USA	246,643	27.1%	
Canada / Mexico	7,789	0.9%	
Total North America	**254,432**	**28.0%**	**226**
Total S. & Cent. America	**19,893**	**2.2%**	**246**
Russian Federation	157,010	17.3%	
Other Europe & Eurasia	130,085	14.3%	
Total Europe & Eurasia	**287,095**	**31.6%**	**237**
South Africa	48,750	5.4%	
Other Africa	1,586	0.2%	
Middle East	419	0.0%	
Total Africa & Middle East	**50,755**	**5.6%**	**194**
Australia	78,500	8.6%	
China	114,500	12.6%	
India	92,445	10.2%	
Other Asia Pacific	11,444	1.2%	
Total Asia Pacific	**296,889**	**32.7%**	**85**
TOTAL WORLD	**909,064**	**100.0%**	**147**

Notes: Proved reserves of coal—Generally taken to be those quantities that geological and engineering information indicates with reasonable certainty can be recovered in the future from known deposits under existing economic and operating conditions.

Reserves/Production (R/P) ratio—If the reserves remaining at the end of the year are divided by the production in that year, the result is the length of time that those remaining reserves would last if production were to continue at that rate.

Source: *Statistical Review of World Energy,* BP, 2007

Exhibit 4.15. Emission Levels for Fossil Fuels

	(pounds per billion BTU)		
Pollutant	Nat. Gas	Oil	Coal
Carbon Dioxide	117,000	164,000	208,000
Carbon Monoxide	40	33	208
Nitrogen Oxides	92	448	457
Sulfur Dioxide	1	1,122	2,591
Particulates	7	84	2,744
Mercury	0	0.007	0.016

Source: *Natural Gas Issues and Trends,* U.S. Energy Information Administration, 1998.

Exhibit 4.16. Proved Natural Gas Reserves—2006

	Trillion cubic metres	Share of total	R/P ratio
USA	5.93	3.3%	11.3
Canada	1.67	0.9%	8.9
Mexico	0.39	0.2%	8.9
Total North America	**7.98**	**4.4%**	**10.6**
Total S. & Cent. America	**6.88**	**3.8%**	**47.6**
Russian Federation	47.65	26.3%	77.8
Other Europe & Eurasia	16.48	9.0%	35.8
Total Europe & Eurasia	**64.13**	**35.3%**	**59.8**
Iran	28.13	15.5%	*
Qatar	25.36	14.0%	*
Saudi Arabia	7.07	3.9%	96.0
United Arab Emirates	6.06	3.3%	*
Other Middle East	6.85	3.7%	*
Total Middle East	**73.47**	**40.5%**	*
Algeria	4.50	2.5%	53.3
Nigeria	5.21	2.9%	*
Other Africa	4.47	2.5%	*
Total Africa	**14.18**	**7.8%**	**78.6**
Australia	2.61	1.4%	67.0
China	2.45	1.3%	41.8
Indonesia	2.63	1.5%	35.6
Malaysia	2.48	1.4%	41.2
Other Asia Pacific	4.65	2.6%	32.0
Total Asia Pacific	**14.82**	**8.2%**	**39.3**
TOTAL WORLD	**181.46**	**100.0%**	**63.3**

Notes:

* denotes more than 100 years

Reserves/Production (R/P) ratio—If the reserves remaining at the end of the year are divided by the production in that year, the result is the length of time that those remaining reserves would last if production were to continue at that rate.

Source: *Statistical Review of World Energy,* BP, 2007

Exhibit 4.17. U.S. Natural Gas Pipeline Network

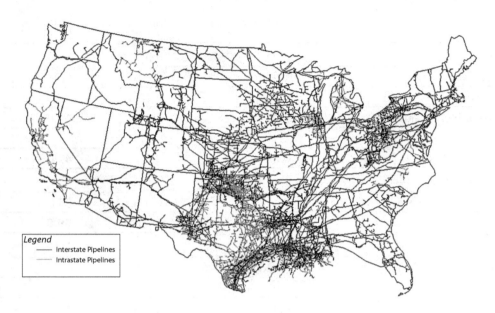

Source: Energy Information Administration, Office of Oil & Gas, Natural Gas Division, Gas Transportation Information System

Exhibit 4.18. LNG Imports in 2006

Billion cubic metres

To	Total imports
North America	
USA	16.56
S & Cent. America	
Dominican Republic	0.25
Puerto Rico	0.72
Mexico	0.94
Europe	
Belgium	4.28
France	13.88
Greece	0.49
Italy	3.10
Portugal	1.97
Spain	24.42
Turkey	5.72
UK	3.56
Asia Pacific	
China	1.00
India	7.99
Japan	81.86
South Korea	34.14
Taiwan	10.20
TOTAL EXPORTS	211.08

Source: *Statistical Review of World Energy*, BP, 2007

Exhibit 4.19. Nuclear Power Producers—2004 Statistics

Producers	TWh	% of World total
United States	813	29.6
France	448	16.4
Japan	282	10.3
Germany	167	6.1
Russia	145	5.3
Korea	131	4.8
Canada	90	3.3
Ukraine	87	3.2
United Kingdom	80	2.9
Sweden	77	2.8
Rest of the World	418	15.3
World	**2730**	**100.0**

Installed Capacity	GW
United States	99
France	63
Japan	45
Russia	22
Germany	21
Korea	16
Ukraine	13
Canada	12
United Kingdom	12
Sweden	9
Rest of the World	45
World	**357**

Exhibit 4.19. (continued)

Country [based on first 10 producers)	% of nuclear in total domestic electricity generation
France	78
Sweden	50
Ukraine	48
Korea	37
Germany	28
Japan	26
United Kingdom	20
United States	20
Russia	16
Canada	15
Rest of the world*	8
World	16

Source: *Key World Energy Statistics 2006,* International Energy Agency, 2006.

Exhibit 4.20. Nuclear Fuel Cycles

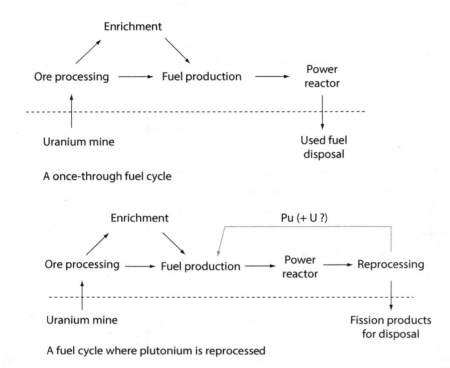

A once-through fuel cycle

A fuel cycle where plutonium is reprocessed

Exhibit 4.21. Hydroelectricity Production

Country	Energy Production (TWh)	Installed Capacity (GW)
China (2007)	486.7	145.3
Canada (2006)	350.3	89.0
Brazil (2006)	349.9	69.1
USA (2006)	291.2	79.5
Russia (2006)	157.1	45.0
Norway (2006)	119.8	27.5
India (2006)	112.4	33.6
Japan (2006)	95.0	27.2
Sweden (2006)	61.8	-
France (2006)	61.5	25.3

Sources: China Electricity Council, 2007 and *Statistical Review of World Energy*, BP, 2007.

Exhibit 4.22. Electricity Generation by Energy Source

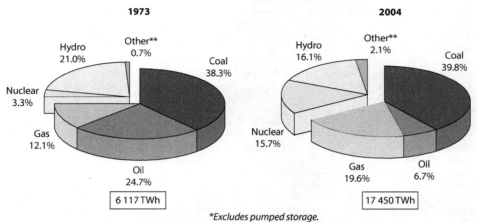

*Excludes pumped storage.
**Other includes geothermal, solar, wind, combustible renewables & waste.

Source: *Key World Energy Statistics 2006,* International Energy Agency, 2006.

Exhibit 4.23. Electricity Delivery Diagram

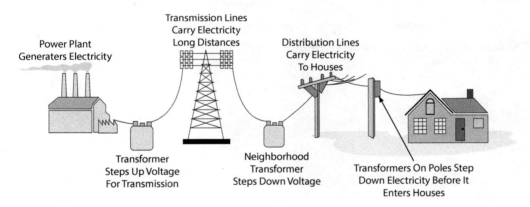

Exhibit 4.24. Example Electricity Load Profile—California

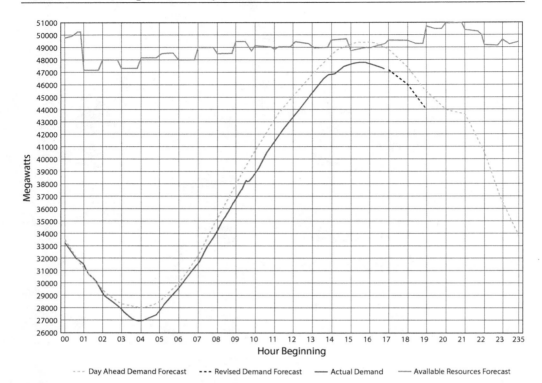

Source: California Independent System Operator.

Exhibit 4.25. Electricity Cost by Energy Source with Carbon Pricing

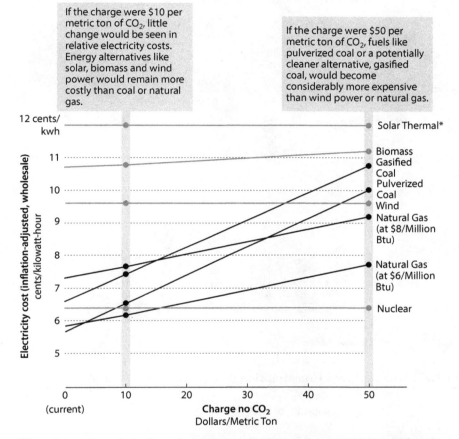

*The anticipated cost of solar thermal power is uncertain. Estimates average 19 cents per kilowatt-hour, but can range from 12 cents (best-case scenario, shown) to 26 cents.

Source: *The New York Times,* November 7, 2007.

Exhibit 4.26. U.S. Energy Consumption by End-Use Sector—2006

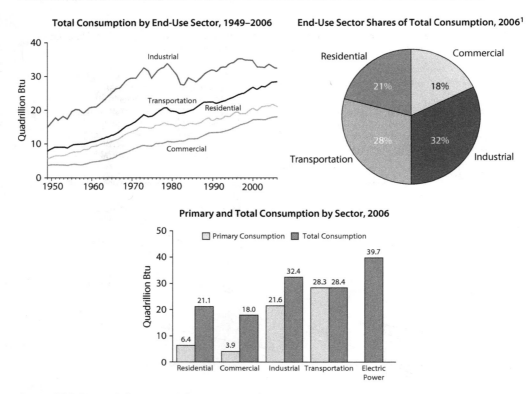

Source: U.S. Energy Information Administration.

Exhibit 4.27. World Energy Consumption by End-Use Sector—2005

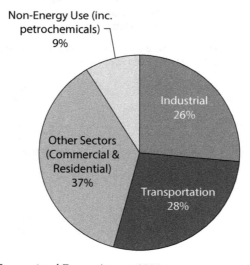

Source: *Key Statistics 2007*, International Energy Agency, 2007.

Exhibit 4.28. End-Use Sectors by Major Source of Energy

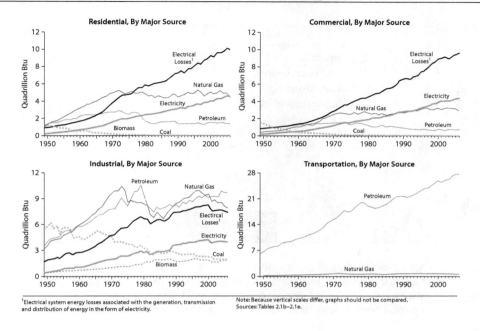

Source: *Annual Energy Review*, U.S. Energy Information Administration, 2006.

Exhibit 4.29. U.S. Household Energy Consumption and Expenditures

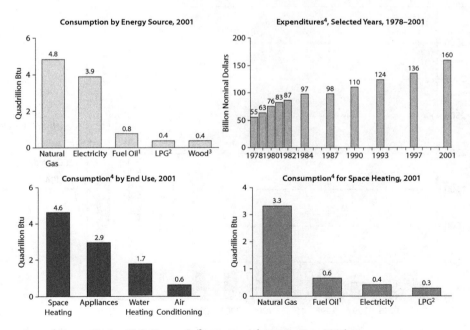

Source: *Annual Energy Review*, U.S. Energy Information Administration, 2006.

Exhibit 4.30. U.S. Residential Electricity Consumption—2005

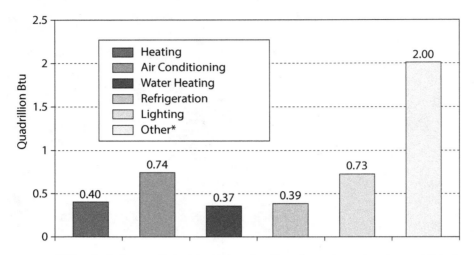

*Other includes small electric devices, heating elements, and motors not listed. It also includes color televisions, cooking stoves, clothes dryers, freezers, clothes washers, dishwashers, personal computers and furnace fans.
Source: Energy Information Administration, *Annual Energy Outlook 2007*, Table A4.

Exhibit 4.31. Installed PV Capacity

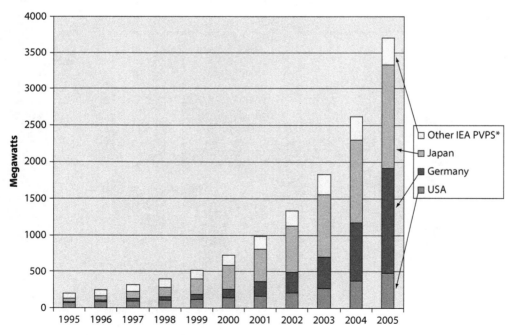

Note: IEA PVPS countries accounted for 90% of installed PV capacity in the world in 2005. Source: *Statistical Review of World Energy*, BP, 2007.

Discussion Questions

1. Explain the role played by different fossil fuels as a source of energy.

2. Discuss the different components of electricity generation and distribution and their importance.

3. Discuss the advantages and disadvantages of cap-and-trade policy vis a vis taxes with respect to carbon cost.

4. Discuss in detail various sources of renewable energy and their importance.

In-Class Assignment: Objective Questions

1. Most internal combustion engines on cars based on 2008 standards operate at efficiency equal to which of the following?
    a. 100%
    b. 50%-80%
    c. 20%-30%
    d. 5%-7%
2. Energy density is a limiting factor when determining economic or logistical feasibility of a system.
    a. True
    b. False
3. The energy density of coal is approximately which of the following?
    a. 10MJ/kg
    b. 50MJ/kg
    c. 75MJ/kg
    d. 24MJ/kg

5 Fuel Cell Technology and Market Opportunities

Andrea Larson and Stephen Keach

If methods of generating electricity threaten the well-being of people and the environment in industrialized countries, developing countries face an even greater risk. Growth in population, consumer demand, industry, and infrastructure all contribute to energy use that is increasing more rapidly than in industrialized countries. Developing countries are less able to deal with the economic cost and environmental and health problems related to burning fossil fuels[1] to produce energy.

One of the most promising technologies for reducing the problems associated with energy generation is the fuel cell. Fuel cells are an electrochemical energy source with the potential to wrest a significant market share from traditional electrical power generation technology. They are cleaner, more reliable, and in theory more efficient than fossil fuel power plants[2] and internal combustion engines. Stationary fuel cells have been put to work as commercial, industrial, and residential electricity sources, and portable fuel cells have been demonstrated in a broad range of applications including automobiles and cellular phones. Many analysts believe that fuel cells will be at the forefront of a dramatic change in the way power is produced and in the fuels we use. Some foresee large-scale use of fuel cells by 2020. Others take a more cautionary stance, pointing out that although the necessary technological improvements are readily achievable, commercial success will require mass production based on consumer demand. As costs are driven down by innovation and mass production, fuel cells have the potential, over the next 50 years, to supplant many traditional power generation sources, providing more reliable energy and cleaner air while decreasing dependence on fossil fuels such as coal and oil.

---

1 *Fossil fuel* is another term for hydrocarbons such as coal and petroleum.

2 *Fossil fuel power plants* refers to electricity generating facilities that burn hydrocarbon fuel to power steam or gas turbines.

Andrea Larson and Stephen Keach, "Fuel Cell Technology and Market Opportunities," pp. 1-23. Copyright © 2001 by Darden Business Publishing. Reprinted with permission.

This note provides an introduction to fuel cell technology, including a brief history on how the various kinds of fuel cells work, their current and potential applications, environmental issues, and advantages and drawbacks. It will explore technological and manufacturing challenges, including fuel production and infrastructure. Commercialization issues covered include time to market, potential market share, cost, and other driving forces.

How Fuel Cells Work

A fuel cell is an electrochemical device that produces electricity through the reaction of two chemical components: For example, oxygen and hydrogen react in a fuel cell to produce electricity and water. In a hydrogen fuel cell, the hydrogen flows over a negatively charged electrode (anode), and—in the presence of a catalyst such as platinum—it is separated into electrons and protons (H+ ions) (Figure 5.1).

Figure 5.1. A hydrogen fuel cell.

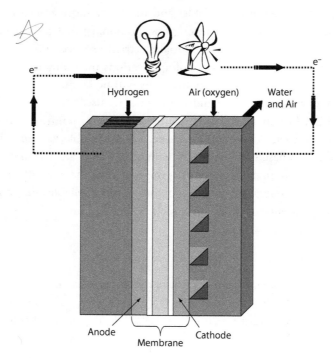

Source: Created by case writer.

The electrons become available as direct current for use in an electrical circuit. The protons pass through a membrane to a positively charged electrode (cathode). Here, again in the presence of a catalyst, they combine with oxygen and returning electrons to produce

120 | Global Sustainability and Innovation

water. The oxygen in the process typically comes from the air. The hydrogen may be produced by the electrolysis of water, or it may be generated (re-formed) from hydrocarbon fuels such as natural gas, propane, or gasoline. Water is the only emission when pure hydrogen is used as a fuel. Direct methanol fuel cells use methanol, typically derived from natural gas, directly without re-forming.

Fuel cells are more efficient than many traditional methods of electricity generation. Gas turbines, steam turbines,[3] internal combustion engines, batteries, and fuel cells all convert energy from one form to another. Power generation involving combustion converts a fuel's chemical energy into thermal energy, then converts it again into mechanical energy and—for electrical use—must convert it yet again into electrical energy. In each stage of conversion, some energy is lost. Traditional coal power generation and the most efficient gasoline-powered internal combustion engines register about 37% and 30% efficiency, respectively. Newer, combined-cycle natural gas turbines can reach 55% to 60% efficiency. Fuel cells and batteries are more efficient because they convert chemical energy directly into usable electricity. For a given power output, fuel cells are lighter and have a longer range than batteries, and refueling is much quicker than recharging.[4] Furthermore, fuel cells are up to 70% efficient in transforming the potential energy of fuel into electricity with only water as an emission. When heat from fuel cells is used in cogeneration, efficiency levels can reach 85% or higher.

Fuel Cell History

William Grove of Great Britain first discovered the principle of the fuel cell in 1839. His model produced electricity by combining oxygen and hydrogen in four cells filled with sulfuric acid and using platinum electrodes. He then used the electricity to demonstrate electrolysis, splitting water into hydrogen and oxygen in another cell. The technology grew to include phosphoric acid fuel cells, alkaline fuel cells, molten carbonate fuel cells, solid oxide fuel cells, and, more recently, proton exchange membrane and direct methanol fuel cells. Commercially, fuel cells languished in the shadow of fossil fuel-powered electricity generation, internal combustion engines, and other power sources for more than a century before their first modern application. In the 1960s, NASA began using alkaline fuel cells on space missions. Since then, NASA, the U.S. Department of Energy, government agencies in Canada, Germany, Japan, and other countries, and hundreds of other organizations and commercial enterprises have advanced the technology to a point where fuel cells have been demonstrated successfully in a broad array of applications and may soon become commercially viable on a large scale.

---

3   Steam turbines are used to generate electricity in typical fossil fuel power plants.
4   Sharon Thomas and Michael Zalbowitz, "Fuel Cells—Green Power," Los Alamos National Laboratory, 1999.

Figure 5.2. A solar (photovoltaic) cell providing energy through a fuel cell system.[5]

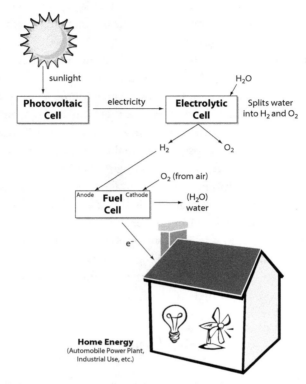

Source: Created by case writer.

Fuel Cell Markets

Power interruptions due to the vulnerability of central power plants and transmission lines cost the United States as much as $80 billion annually [6]

---

5   Photovoltaic cells, wind turbines, and other renewable energy sources produce electricity that can be used in a process called electrolysis to make hydrogen and oxygen from water. The hydrogen then powers a fuel cell that provides electricity whenever it is needed—not solely when the wind is blowing or the sun is shining.

6   From a Worldwatch Institute news release regarding Seth Dunn's book, *20th Century Power System Incompatible With Digital Economy*, July 15, 2000, http://www.worldwatch.org/alerts/000715.html (accessed October 5, 2009).

The potential market for fuel cells is immense. A small share of the transportation and power generation markets could mean several billion dollars within the current decade.[7] *Winslow Environmental News* estimated that the market for fuel cells would grow from $190 million in 2000 to $7.9 billion in 2005.[8]

With a few exceptions, fuel cell manufacturers were in a research, development, and demonstration phase in 2003 and did not yet sell any products on a commercial scale. Manufacturing costs were the largest barrier to nonautomotive fuel cell applications. For automobiles, manufacturing costs were a significant obstacle, but development of an infrastructure to produce and deliver suitable fuel might have been a more important issue. The cost of fuel cells is decreasing rapidly as components are refined, designs are improved, and manufacturing plants are built for mass production.

These are among the issues influencing the commercial acceptance of fuel cells:

- Environmental awareness, particularly concern over global climate change
- Increasing demand for electricity
- Demand for high-quality, reliable electricity for high technology applications
- Demand for distributed power[9]
- Fuel cost and availability
- Deregulation of the electric power industry
- Continued government subsidies to early fuel cell purchasers[10]

Rocky Mountain Institute (RMI) analysts argue that the initial focus for fuel cell commercialization should be buildings. RMI calculates that fuel cells can turn at least 50% of hydrogen's potential energy into:

> …highly reliable, premium-quality electricity and the remainder into 70°C pure water, ideal for heating, cooling, and dehumidifying buildings, using a modular balance of the system black box [that] several capable firms are already developing. In a typical building, such services would help pay for the natural gas and a fuel processor. With the fuel expenses thus largely covered, electricity from early-production fuel cells should be cheap enough to undercut even the operating cost of existing coal and nuclear power stations, let alone the extra cost to deliver their power.

---

7 Fuel Cells 2000's Benefits of Fuel Cells, http://216.51.18.233/fcbenefi.html (accessed June 15, 2003).

8 *Winslow Environmental News,* January 2001: 2.

9 Distributed power is power produced at or near the place where it is used, rather than transmitted over long distances from a central generating site.

10 The U.S. Department of Energy has offered as much as $1,000 per kilowatt to assist fuel cell purchasers as a part of its commitment to fuel cell development and demonstration. The U.S. government allocated more than $100 million for fuel cell-related programs in 2001.

For the future, RMI posits that fuel cells are likely to become very cost-competitive:

> When fuel cells are manufactured in very large volumes, using such innovative designs as (for example) molded roll-to-roll polymer parts glued together, they could become extremely cheap, probably less than $50 per kilowatt, which is about a fifth to a tenth the cost of today's cheapest combined gas-fired power stations.[11]

But other fuel cell experts point out the present state of the market presents a very different picture. Natural gas turbines can be built for about $400 per kilowatt and coal-fired plants for about $1,200 per kilowatt, while the only large fuel cells currently available are sold for about $3,500 per kilowatt, after a government subsidy of $1,000 per kilowatt.

As technological improvements and mass production reduce the cost of fuel cells, they may prove economically and environmentally attractive to developing countries even before they are widely accepted in industrialized countries. Many developing countries do not have an adequate distribution system for electricity. They may find that delivering fuel for fuel cells at homes, offices, and factories is more economical than providing miles of wires to transmit electricity from central power plants. But fuel cells will compete in these markets with other forms of distributed power generation, such as microturbines and diesel generators. The winning technology will probably be the least capital-intensive. If fuel cells eventually become competitive on a cost basis, positive environmental attributes could tip the scales in their favor.

Size of the Industry and Industry Segments

The *Fuel Cells 2000* directory listed more than 700 companies and organizations. Of these, more than 100 were developing fuel cells for commercial purposes. The market for fuel cells may be divided into three segments: small portable cells, midsize stationary and vehicular cells, and larger stationary cells. Companies such as Motorola have developed very small fuel cells and Manhattan Scientifics for applications like cell phones and laptop computers. Ballard Power Company and many other companies have demonstrated fuel cells for vehicles. Stationary fuel cells to provide electricity for general distribution, residences, buildings, and factories are available from United Technologies (UTC) and have been demonstrated by several other companies. An important market niche for stationary fuel cells includes electronics manufacturing, computer facilities, hospitals, and other businesses that require high-quality, high-reliability electricity. These types of enterprises already pay a premium for uninterruptible power supplies. By the end of 2000, UTC had

---

11   Amory B. Lovins and Brett D. Williams, "A Strategy for the Hydrogen Transition," paper given at the 10th Annual U.S. Hydrogen Meeting of the National Hydrogen Association, April 1999: 4–5.

already shipped more than 200 of its 200 kilowatt phosphoric acid fuel cells for use at such sites.

Allied Business Intelligence of Oyster Bay, New York, predicts that stationary fuel cells worldwide will generate over 15,000 megawatts by 2011.

Applications and Remaining Issues

As shown in Appendix A, fuel cells may be used in a wide spectrum of applications. They have been demonstrated in cell phones, laptop computers, portable generators, scooters, bicycles, automobiles, locomotives, computer centers, hospitals, financial institutions, office buildings, an unmanned solar airplane, and a broad range of military applications.

Very small fuel cells, less than one kilowatt, will compete with advanced battery technologies, such as lithium ion batteries, in portable applications such as cellular phones, power hand tools, auxiliary power units, and laptop computers. Motorola and Manhattan Scientifics have demonstrated fuel cell-powered cell phones that run far longer without refueling (inserting a new hydrogen or methanol cartridge) than battery-powered units run before needing recharging.

Midsize fuel cells from 1 to 100 kilowatts may compete with internal combustion engines (gasoline and diesel), batteries, microturbines, and battery/internal combustion engine hybrids for applications such as airport vehicles, forklifts, golf carts, buses, automobiles, portable generators, and residential power.

Fuel cells of greater than 100 kilowatts will compete with gas turbines and advanced internal combustion engines that may boost efficiency by taking advantage of combined heat and power (CHP) opportunities, as well as grid-supplied power generated from coal, natural gas, oil-fired boilers, and and nuclear plants. Uses for large fuel cells include computer centers, office buildings, manufacturing facilities, and small utility or supplemental distributed power generation plants. Future applications could include locomotives, submarines, and ships.

While they cannot yet compete on a cost basis for most applications, some fuel cells offer technologically viable competition for existing energy systems, including stationary generation where power quality and reliability are at a premium. For most other fuel cell applications, technological hurdles are still important but are eclipsed by questions about cost of production and marketing: how soon and how inexpensively can fuel cells be produced in quantity and how long will it take to gain market acceptance? Significant progress toward cost reduction is common news every week in the fuel cell industry,[12] and several companies are in the process of building manufacturing facilities.

---

12   The Hydrogen and Fuel Cell Investor Web Site (http://www.H2FC.com (accessed (accessed June 15, 2003) and Fuel Cells 2000, www.fuelcells.org (accessed October 5, 2009) are good sources for daily news of fuel cell industry progress.

One recurring technical question is, "What fuel will fuel cells run on and how will the infrastructure be developed to store and deliver it?" Opinions vary about which fuel is best. Some experts believe that, at least in the short run, hydrogen from hydrocarbons such as gasoline, methanol, natural gas or propane will dominate the fuel cell industry; however, hydrogen storage and production technology is growing apace with the fuel cell industry itself. Many fuel cell developers and their partners in the car manufacturing and energy industries feel that hydrogen is the energy source of the future. RMI has suggested that, since hydrogen production and delivery is possible using available technology, building an interim fuel infrastructure (for example, providing methanol for automobiles) may be a waste of money. Other scientists argue that, while existing hydrogen technology is workable, by no means are all the problems solved: Making the fuel available on a large scale at a competitive price involves complex issues. In 2003, it was still very difficult to predict which system will prove best and most acceptable in the market place.

When Will We Begin to See More Fuel Cells in the Market?

UTC's International Fuel Cells Division (IFC) sells 200 kilowatt PAFCs for premium stationary power, co-generation, waste gas recovery, and off-grid applications. Other companies make and sell demonstration, beta test units, small portable generators, and educational fuel cells.

Most major automobile makers have prototype fuel cell vehicles. In 2003, Honda, Toyota, and Nissan expected to have commercial models by 2003; GM expected to have a production-ready vehicle by 2004. Daimler-Chrysler, Ford, and Mazda are likely early competitors as well. XCELLSIS Fuel Cell Engines (a joint venture between DaimlerChrysler, Ballard, and Ford) and Nova BUS have demonstration buses in service.[13]

Researchers at Los Alamos National Laboratory (LANL) expect to see significant progress in commercialization of stationary fuel cells before 2005. They assume that near-term market penetration will be limited to niche applications like information technology and electronics manufacturing, which require high power quality and reliability. **Appendix C** lists companies that had fuel cells and fuel cell devices available in 2003 and shows planned commercialization years for a few fuel cell companies.

Environmental Issues: Advantages/Drawbacks

Environmental issues vary depending on the fuel used and the way it is produced. Concern over greenhouse gasses and other forms of air pollution is a driving force behind the development of fuel cells. A fuel cell powered by hydrogen produces only water as an

---

13 Fuel Cells 2000; "Fuel Cells in Passenger Cars," Office of Transportation Technology, U.S. Department of Energy, July 2000.

emission. But the process of producing hydrogen may pollute if hydrocarbons are used as the feedstock. For example, some emissions result if gasoline or natural gas[14] is re-formed to extract hydrogen for use in a fuel cell. If grid[15] electricity is used in an electrolytic cell to produce hydrogen from water, then the emissions from the power plant(s) that feed the grid must be accounted for. A gasoline-powered fuel cell is cleaner and more efficient than an internal combustion engine. **Appendix C** describes production method, safety, availability, and distribution, and environmental issues for several types of fuel that are employed in fuel cells.

Summary of Advantages

- Applicability for distributed power generation: power is generated at the site where it is used rather than central generation and distribution through a grid. On site generation avoids transmission line losses of 10% to 15%. Other small-scale power generation technologies have this advantage as well.
- Low maintenance and reduced down time. Fuel cell systems have few moving parts and can achieve 99% availability.
- Off-grid independence: Brownouts and blackouts associated with distribution grids are eliminated by onsite generation. Fuel cells and other small-scale generation systems provide a useful online backup where uninterruptible power is needed.
- Quality of electricity: Constant flow of current with few fluctuations—important for many scientific and technological applications.
- Environmentally benign: Hydrogen fuel cells have only water as an emission.[16]
- Energy independence: In the long run, broader fuel and energy choices could reduce dependence on oil imports for many countries. Hydrogen to power fuel cells may be generated by electrolysis, using electricity and water. The electricity may come from any source. If the aim is the reduction of pollutants, then the best sources are renewables such as solar photovoltaic cells, wind generators, or hydroelectric power plants. If the aim is efficiency, other sources of electricity may be put to optimal

---

14 Emissions involved in the extraction, transport, and refining of gasoline from oil should be accounted for. See footnote below on steam re-forming of natural gas and reinjection of CO_2.

15 The *grid* is a commonly used term for the distribution system (transmission wires, transfer stations, etc.) that delivers electricity from a central power plant. Hydrogen can be generated using relatively low-cost electricity during periods when other demands for electricity are low (off-peak). If electricity is supplied to the grid through hydroelectric, wind, or other renewable sources, then the hydrogen and subsequent fuel cell electricity are emissions-free.

16 It is likely that the manufacture of fuel cells generates some emissions. Emissions may result from the production of hydrogen as well: Whereas producing hydrogen by using electricity from solar or other renewable sources to electrolyze water emits only oxygen, hydrogen production from hydrocarbons or uses grid electricity is likely to involve emissions.

use by using off-peak power[17] for electrolysis. A variety of other hydrocarbons such as oil, coal, methane, and even landfill gas may be used directly in fuel cells or as feedstock for producing hydrogen.
- Lack of noise: Fuel cells are quiet.
- Efficiency: Fuel cells are more efficient than most fossil fuel power plants and internal combustion engines.[18]
- Weight and range: For portable applications, fuel cells are lighter and have a longer run time than batteries, and when they do run out of fuel they don't require a long, inefficient recharging process, only refueling.

Obstacles to Widespread Use of Fuel Cells

Cost and Technological Requirements

Those fuel cells that were available in 2003 were quite expensive. Less expensive components, such as membranes and electrodes, and mass production would greatly reduce overall costs. Several companies were in the process of building manufacturing plants. Specific cost-reduction issues, particularly reduced use of platinum as a catalyst, and development of new membrane materials were the subject of intense research. Several new patents were issued for techniques to eliminate or reduce platinum use in fuel cell components. Other technological research priorities included water management, high-purity fuel requirements in proton exchange membrane cells and alkaline cells, methanol crossover in direct methanol cells, and component resistance to extreme heat and corrosion in high-temperature fuel cells.

Infrastructure

The need for a large-scale change in fuel delivery infrastructure is often cited as the most significant barrier to the use of fuel cells in vehicles. But small hydrogen generators have already been developed for installation in individual garages.[19] The Rocky Mountain Institute suggests that hydrogen could be re-formed from natural gas at office buildings and other work places. This hydrogen would be used to run fuel cells that produce electricity for the building and heat and cool it with "waste" heat. Excess hydrogen production

---

17   Off-peak refers to electricity generated when demand is low, such as late at night or when the weather is good and heating and cooling needs are low.

18   It can be daunting to sort out fuel cell efficiency ratings since some include useful heat and others only electric power. To further complicate matters, excess heat from a fuel cell may run a gas turbine, producing still more electricity. See **Appendix A** for fuel cell efficiency ratings and other properties.

19   Stuart Energy of Canada has developed a small hydrogen generator prototype that uses electrolysis. The company has about 50 years' experience building larger industrial units, currently priced at about $650/kW. Their price target for the smaller units is about the same on a per kW basis.

would be sold to employees as automobile fuel. At present, small electrolysis units and re-formers are costly, and issues related to placement, safety, insurance, and so on will have to be confronted for such a scenario to work.

Fuel Storage

Hydrogen may be stored as a gas, a compressed gas, a liquid, or as a metal hydride. Several companies are actively researching these options and the energy consumption, cost, and convenience associated with each of them. The options require energy for compression and other processes. Weight and volume of fuel in relation to the amount of energy stored is also an issue.

Public Acceptance

Initial public acceptance may be limited by the level of comfort and confidence people feel with existing technology, the perceived risk of trying a new technology, and fears of hydrogen's flammability. Having residential electricity generated at home and driving vehicles powered by fuel cells coupled with quiet electric motors may require a leap of faith for many people in industrialized countries. On the other hand, problems such as blackouts and brownouts and concern over global climate change and pollution may become so unacceptable that good alternatives, whether fuel cells or solar and wind generation systems, will be welcomed. In developing countries, where much of the population is not yet accustomed to dependable electricity and automobiles, newer technologies may enjoy ready acceptance if they can compete on a cost basis.

Competition from advanced internal combustion engines and hybrid engine systems presents a challenge for fuel cells, particularly in the near term. These technologies are available now, their efficiency levels are high, and, at least in the case of reciprocating (diesel) engines, their cost is relatively low. Advanced diesel engines may be fueled with natural gas, reducing emissions. They cost about $350 per kilowatt, achieve nearly 50% power generation efficiency and currently dominate the market for smaller (less than 1 megawatt) industrial systems.[20] While fuel cells are quieter, cleaner, and often more efficient, diesel engines may continue to hold the largest market share, at least until fuel cell costs and production volume become more competitive. Micro turbines, a potential competitor, are at about the same stage of development as fuel cells.

---

20   N. Martin et al., Emerging Energy-Efficient Industrial Technologies, Lawrence Berkeley National Laboratory; and R. N. Elliott et al., ACEEE, October 2000, 141. Note: The terms *reciprocating engine* and *diesel engine* appear to be interchangeable in the paper. We chose to use the more common term *diesel engine*.

Resistance from Established Industry

Some utilities are making a transition to distributed power difficult. For example, depending on local regulations, power sold back to the grid from small producers may be subject to restrictions and predatory pricing structures, or it may be refused altogether. High exit fees may be charged to customers who choose to leave the grid and generate their own power. Also, several major coal and petroleum interests, along with other large industrial companies, make up the membership of lobbying organizations that downplay the importance of emissions related to their industries. The reduction of greenhouse gasses and other emissions related to energy generation and transportation is an important factor in the growing interest in fuel cell development. Nevertheless, some utilities and most, if not all, major oil companies and automobile makers are actively developing fuel cells or entering into partnerships with fuel cell developers. Also, some utilities and most major oil companies have significant internal programs to reduce energy use and environmental impacts.[21]

Isn't Hydrogen Extremely Dangerous?

What about the famous Hindenburg (hydrogen-filled air ship) disaster?

As the Los Alamos National Lab put it, "Don't paint your ship with rocket fuel." A NASA scientist used infrared spectroscopy and a scanning electron microscope to examine samples of the fabric that covered the Hindenburg dirigible. It turned out that the cotton fabric had been treated with extremely flammable chemicals including aluminum powder (used in rocket fuel) and cellulose acetate or nitrate (gunpowder). These coatings were almost certainly the cause of the fire.

There are dangers associated with hydrogen storage, transport, and use. Hydrogen is a very small molecule that easily leaks and is highly flammable, but, with proper engineering, hydrogen systems can be as safe as gasoline systems.

---

[21] Resistance to fuel cell technology is by no means universal. Some utilities and most major oil companies are actively researching market opportunities and demonstrating and investing in fuel cell technology. But it is worth noting that many of these companies are trying to cover all bets. They invest in promising energy technologies and other activities aimed at protecting the environment, while at the same time funding organizations like the Global Climate Coalition and the Center for the Study of CO2 and Global Change, both of which promote the views of scientists whose work questions the importance of reducing greenhouse gas emissions.

Types of Fuel Cells

There are several types of fuel cells. They differ primarily in the type of fuel and/or electrolyte[22] used: Alkaline Fuel Cells (AFCs), Polymer Electrolyte Membrane or Proton Exchange Membrane Fuel Cells (PEMFCs), Phosphoric Acid Fuel Cells (PAFCs), Molten Carbonate Fuel Cells (MCFCs), and Solid Oxide Fuel Cells (SOFCs). Solid acid fuel cells have recently emerged from the work of researchers at the California Institute of Technology. Two other kinds of fuel cells, Direct Methanol Fuel Cells (DMFCs) and Regenerative Fuel Cells (RFCs) use a membrane similar to PEMFCs but are often listed as distinct types of fuel cells.

Zinc-air and aluminum oxide cells are less well-known types of electrochemical cells that are sometimes also referred to as fuel cells. Some industry observers regard these cells as more akin to batteries since they use up their internal reactants or require recharging.[23] Operating temperature, bulkiness, and efficiency in energy production are also important distinctions. Some properties of the various types of fuel cells are shown in **Appendix A.**

Potential for the Future

The Rocky Mountain Institute (RMI) takes an optimistic view of the future of electricity from fuel cells, particularly for homes, offices, and automobiles. In fact, RMI researchers envision a seamless network. They suggest that cars could run on hydrogen produced by small electrolytic cells[24] or from natural gas re-formers installed at the home or office. The cars in turn may plug in at work or at home and serve as generators, producing electricity for local use or to be fed back into the grid during the roughly 96 percent of the day and night that a typical car spends parked. In the long run, the electricity to produce hydrogen could come from renewable sources: hydroelectric plants, photovoltaic cells or

---

22 The electrolyte is the substance, such as a polymer membrane or phosphoric acid, through which H+ ions pass as they travel between the negative and positive electrodes in a fuel cell.

23 See David Redstone's comments on zinc-air cells in the Types of Fuel Cells section of the Hydrogen and Fuel Cell Investor Web site, http://www.H2FC.com.

24 Electrolytic cells are essentially fuel cells run in reverse. Electricity is used to split water into hydrogen and oxygen. Hydrogen produced in this way would be expensive unless the electricity powering the fuel cells could be supplied cheaply. Also, emissions from electricity production would be associated with the hydrogen production unless wind power, solar power, or other renewable energy is used.

windmills. Bulk supplies of hydrogen could also come from steam re-forming natural gas at the wellhead[25]. More exotic hydrogen sources include hydrogen-producing algae and direct splitting of water using highly focused sunlight[26]. Remote hydrogen supplies will pose major infrastructure challenges, including pipelines and storage systems.

Conclusions

When compared with existing power sources, fuel cells offer a more reliable, more efficient, and cleaner way of generating electricity, whether in stationary settings, powering vehicles, or providing power for portable electronic devices and generators. The fuel cells that are currently sold for purposes other than demonstration and educational use are placed in settings where relatively high costs are secondary to requirements for uninterrupted delivery of electricity. However, fuel cell developers are quickly reducing costs, increasing efficiency, and overcoming the obstacles that face a transition in the way power is produced and delivered. It is likely that within the next two decades fuel cells will take a measurable share of the energy market, and it is possible that most new automobiles will eventually run on electric motors powered by fuel cells. The implications are significant. Replacing fossil-fueled power plants and vehicle engines with hydrogen-powered fuel cells would eliminate the largest sources of air pollution. This benefit offers public health advantages in industrialized countries as well as in emerging economies faced with rapid urbanization. Fuel cells may eventually contribute to quieter streets and cities, and more efficient, less expensive delivery of power.

Organizations and Web Sites of Interest

The Center for the Study of Carbon Dioxide and Global Change, http://www.co2science.org.
Ballard Power Corporation, http://www.ballard.com.

---

25 Steam re-forming is a way of stripping hydrogen from natural gas, coal gas, or other fossil fuels. Amory Lovins and Brent Williams of the Rocky Mountain Institute note that when this is done at the point of extraction, the "well head," the other product of the separation of natural gas, CO_2, may be pumped back into the well. This sequesters the CO_2 deep underground, rather than adding it to greenhouse gas loadings in the atmosphere, and at the same time forces enough more natural gas out of the well to pay for the reinjection process.

26 A Canadian company, SHEC has produced hydrogen using focused sunlight and researchers at the University of Pennsylvania and elsewhere are experimenting with algal hydrogen production. But these technologies are not yet proven for commercially competitive purposes.

California Air Resources Board, http://www.arb.ca.gov

California Fuel Cell Partnership, http://www.fuelcellpartnership.org

Fuel Cells 2000, http://www.fuelcells.org

Global Climate Change Coalition. http://www.globalclimate.org/.

Los Alamos National Laboratory, http://www.education.lanl.gov/resources/fuelcells

Manhattan Scientifics, Inc., http://www.mhtx.com

Massachusetts Institute of Technology, http://www.mit.edu

Motorola, Inc., http://www.motorola.com

National Aeronautics and Space Administration, http://www.nasa.gov National Fuel Cell Research Center, http://www.nfcrc.uci.edu National Hydrogen Association http://www.ttcorp.com/nha/ Princeton University, http://www.princeton.edu/~ceesdoe Rocky Mountain Institute, http://www.rmi.org

SHEC Labs - Solar Hydrogen Energy Corporation, http://www.solar-h2.com

The American Hydrogen Association, http://www.clean-air.org/

The Fuel Cell Commercialization Group, http://www.ttcorp.com/fccg

The Hydrogen and Fuel Cell Investor, http://www.h2fc.com

The U.S. Fuel Cell Council, http://www.usfcc.com

United Technologies Research Center, http://www.utrc.utc.com

University of Pennsylvania, http://www.upenn.edu

Union of Concerned Scientists, http://www.ucsusa.org

U.S. Department of Energy

Energy Efficiency and Renewable Energy Network, http://www.eren.doe.gov

Federal Energy Technology Center, http://www.fetc.doe.gov

Office of Transportation Technologies, http://www.ott.doe.gov

Appendix A.

Fuel Cell Families

Fuel Cell Type	Fuel	Electrolyte	Operating Temp	Approximate Electricity Generating Efficiency	Applications
PEM-FC	hydrogen/ hydrogen reformed from hydrocarbons	Solid organic polymer membrane	60° to 100° C	50%	Space program, vehicles, small electronics, residential power, stationary power, portable generators
AFC	pure hydrogen and pure oxygen	Aqueous solution of potassium hydroxide	90° to 100° C	70%	Spacecraft, small vehicles, residential power
PAFC	hydrogen/ hydrogen reformed from hydrocarbons	Liquid phosphoric acid	175° to 200° C	40 to 45%	Stationary power generation, Locomotives, large vehicles
MCFC	hydrogen, hydrocarbons	Liquid lithium, sodium, and/or potassium and/or natrium carbonates	600° to 1000° C	50 to 60%	Stationary power generation
SOFC	hydrogen/ hydrogen reformed from hydrocarbons *	Solid zirconium oxide with yttrium added	600° to 1000° C	50 to 60%	Stationary Power, being considered for vehicles
DMFC	methanol	Solid organic polymer membrane	120° to 190° F	about 40%	vehicles, small electronics
RFC	hydrogen				potential space program and other uses

* Solid oxide fuel cells can tolerate impurities in fuel.

Appendix B.

Fuel Cell Benefits and Technical Challenges

Type of Fuel Cell	Benefits	Technical Challenges[1]
PEMFC Proton Exchange Membrane	High power density, quick start and response, low corrosion	High-purity fuel and expensive catalysts[2]
AFC Alkaline	Highly efficient, quick start and response, lower cost catalysts possible	High-purity fuel[3]
PAFC Phosphoric Acid	Commercially available, highly efficient in co-generation	Corrosive electrolyte, expensive catalyst, large size and weight
MCFC Molten Carbonate	High efficiency, fuel choice, lower cost catalysts	High temperature limits cell component longevity. Possible emissions problems with fuels other than hydrogen
SOFC Solid Oxide	High efficiency, fuel choice, lower cost catalysts, simple design	High temperature limits cell component longevity. Possible emissions problems with fuels other than hydrogen
DMFC Direct Methanol	Similar to PEMFC but can run on methanol.	Less efficient than PEMFC, methanol cross-over may cause emissions and corrosion problems[4]
RFC Regenerative	Reverses process to produce hydrogen.	Requires a source of electricity.[5]

[1] Membrane materials need to come down in cost, particularly for SOFCs and MCFCs. Some fuel cell developers claim to have reduced the cost of membranes for PEMFCs and DMFCs.

[2] Several recent breakthroughs are claimed by various fuel cell developers, both in the amount of catalyst needed and in tolerance of fuel impurities.

[3] Astris Energi, of Canada and the Czech Republic, claims to have reduced costs for the catalyst and improved fuel impurity tolerance.

[4] Some fuel cell developers claim to have overcome the problem of methanol crossing the membrane.

[5] RFGs have no emissions when electricity is derived from solar, wind or other non-polluting sources. When hydrogen is produced from grid electricity, the emissions associated with coal, oil, or other fuel used in producing the electricity for the grid must be accounted for.

Appendix C.

Selected Companies Involved in Fuel Cell Production

Representative Companies[1]	Type of Fuel Cell[2]	Current and Projected Products	Partners[3]	Commercial Availability[4]
Astris Energi	AFC	Educational, residential, small vehicle, portable	Czech government (for residential power)	
Avista	PEMFC	Residential, commercial modular		2001
Ballard Power	PEMFC	Vehicle power plants, portable generators, large (sub-100MW) stationary power generation	GPUI, Ebara (Japan), Alstom (France), most large car companies	
DCHT	PEMFC	3W to 10kW		
Ecosoul Electro Chem	RFC PEM, PAFC	Educational, demonstration		
Energy Ventures	DMFC, AFC	Utilities, small vehicles		
Fuel Cell Energy	MCFC	300 kW to 3 MW		
Global Thermoelectric	SOFC	Residential co-generation, auxiliary power for automobiles, small industrial	Enbridge	2002
H Power	PEMFC	1–25 kW residential and commercial, up to 15kW portable and mobile, educational	ECO, DQE, HydroQuébec, Singapore Tech. Kinetics, Sofinov	2001
Hydrocell	PEMFC	Educational and demonstration		now
International Fuel Cells[5]	PAFC, PEMFC	PAFCs for space crafts, residences, and premium industrial. Working on gasoline fueled PEMFC	NASA, VW, BMW, Toshiba, Buderus, Hyundai, Shell	200kW premium power now

136 | Global Sustainability and Innovation

Representative Companies[1]	Type of Fuel Cell[2]	Current and Projected Products	Partners[3]	Commercial Availability[4]
Manhattan Scientific	DMFC	Micro and Mid-range for pagers, phones, power tools, residential, and mobile uses	Novars	
Nuvera	PEMFC DMFC	gasoline powered 1 o 50 kW, residential, transport, premium, commercial	DeNora Group, A.D. Little, Amerada Hess Corporation	2002 to 2005
Plug Power	PEMFC	7kW residential natural gas/propane	GE Microgen, Mechanical Technology	2002
Proton Energy Systems	PEMFC RFC	Stationary generation, vehicles		
SatCon Technology				
Siemens AG	MCFC	Submarine, demonstration, high-end applications		
ZeTek	AFC	2 to 250 kW commercial, residential, marine, automotive	Targeting utilities	2–5 kW stacks available now

[1] This list is by no means comprehensive. The inclusion or omission of companies and/or data does not indicate relative strength or value. All major automobile companies are involved with fuel cell development, either internally or in partnership with other fuel cell companies.

[2] Alkaline Fuel Cell (AFC), Proton Exchange Membrane Fuel Cell (PEMFC), Regenerative Fuel Cell (RFC), Phosphoric Acid Fuel Cell (PAFC), Direct Methanol Fuel Cell (DMFC), Solid Oxide Fuel Cell (SOFC), Molten Carbonate Fuel Cell (MCFC).

[3] This column provides a small sample of the kinds of major contractor agreements, joint ventures, strategic alliances, etc. being formed by fuel cell developers. It is not comprehensive.

[4] Estimated availability varies considerably, particularly for mass production at low cost and for automobile use.

[5] International Fuel Cells (IFC) is a subsidiary of United Technologies Corp. (UTC)

Appendix D.

Fuels for Fuel Cells

FUEL[1]	Production	Storage	Safety	Availability and Distribution	Environmental
Hydrogen	Electrolysis or reformed from natural gas and other fuels	Compressed gas in cylinders, liquid in cryogenic tanks, metal hydrides, carbon nanofibers[2]	Flammable at concentrations of 4% to 75% in air, nearly invisible flame, leak prevention is a challenge, non-toxic, disperses quickly.	May be generated on-site. Distribution infrastructure for centralized production is undeveloped	Zero emissions: electrolysis using solar, wind, hydroelectric, etc., High emissions: electrolysis using coal or oil, etc. Low emissions: using re-formed hydrocarbons such as natural gas[3]
Re-formulated gasoline	Substantial existing production	Conventional tanks	Flammable at concentrations of 1% to 7% in air, potential carcinogen	Large existing infrastructure	Production of gasoline from crude oil and subsequent re-forming of gasoline to make hydrogen results emissions including greenhouse gasses.
Methanol	Abundant natural gas and coal feedstocks. Can be made from biomass[4].	Special tanks. Methanol can be corrosive.	Toxic through skin absorption, no visible flame.	Infrastructure needs development.	High greenhouse gas emissions if made from coal. Low emissions if made from natural gas or biomass.
Natural Gas (NG)	Abundant feedstock.	Compressed gas cylinders.	Flammable at concentrations of 5% to 15% in air, non-carcinogenic, disperses quickly	Infrastructure needs development.	May increase nitrogen oxides, non renewable fossil fuel

Source: This chart was adapted from Fuel Cells - Green Power by Marcia Zalbowitz of Los Alamos National Laboratory with considerable input from Dr. ion Espino of the Department of Chemical Engineering at the University of Virginia

The fuels listed are the most commonly cited fuels for use in fuel cells. Propane, ethanol, biogas, landfill gas, and coal gas have also been used or mention tential fuels by various fuel cell developers. Experimental process.

Different hydrocarbons such as coal gas, gasoline, etc. have a range of emissions during production. CO2 from natural gas re-formed at the wellhead may ected into the well. Experimental process.

Appendix E.

Fuel Cell Financial Information

Representative Companies[1]	Stock Symbol	Share Price 3–22–01	52 Week high	52 Week low	Market Cap. (millions)	Number of Employees[2]
Avista	AVA	16.35	44.50	15.00	749.56	2260
Ballard Power	BLDP	39.94	144.93	32.75	3248.13	238
IMPCO Technologies	IMCO	16.50	52.87	9.75	170.17	581
Fuel Cell Energy	FCEL	49.50	108.75	15.75	633.53	152
H Power	HPOW	6.44	35.94	5.25	295.35	141
Idatech	IDA	34.45	51.81	30.31	1328.20	1720
United Technologies	UTC	67.35	82.50	46.50	33621.17	153800
Plug Power	PLUG	14.50	153.00	9.12	581.78	295
Proton Energy Systems	PRTN	7.94	36.00	5.25	256.39	44
SatCon Technology	SATC	10.19	44.75	8.00	149.52	278
Siemens AG	SI	100.60	113.20	91.89	59568.34	447000

[1] No value judgment is intended by the inclusion or omission of any company in this list. Several of these companies have products and services other than fuel cells. The market capitalization and number of employees does not necessarily indicate the relative importance of any particular company to the development and commercialization of fuel cells.

[2] These figures are approximate. Some are from 1999 and others from 2000.

Appendix F.

Families of Fuel Cells

Alkaline fuel cells (AFCs) have been used for many years on space flights. Several companies, such as Astris Energi and Energy Ventures (both Canadian) and Zetek (England), are developing lower-cost alkaline fuel cells for both small and large-scale applications. Intolerance of fuel impurities and cost are two technical obstacles to these exceptionally efficient fuel cells.

Proton exchange membrane fuel cells (PEMFCs) Many fuel cell makers are now developing PEMFCs. PEM stands for Polymer Electrolyte membrane or Proton Exchange Membrane depending upon whom you ask. Both titles are descriptive of the thin membrane electrolyte with a negatively charged anode and a positively charged cathode. The protons pass through the membrane, but not electrons. PEMFCs operate at a low temperature and can be made small enough for applications such as cell phones and laptops, or they can be arrayed in stacks large enough to provide power for buses and buildings.

Direct methanol fuel cells (DMFCs) are low-temperature fuel cells that run on methanol without re-forming to hydrogen. DMFCs are appealing because methanol can be produced, at a cost similar to gasoline, from natural gas, and gasoline stations would require little modification to dispense methanol. The electrolyte is a membrane, very similar to that used in PEM fuel cells.

Phosphoric acid fuel cells (PAFCs) are already commercially available for customers to whom power reliability is a more important consideration than cost. PAFCs use phosphoric acid as an electrolyte. International Fuel Cell's 200 kW PC25s are currently in use in some hospitals, nursing homes, school computer centers, and other applications where the reliability of electricity supply is exceptionally important. Phosphoric acid fuel cells can also be used in buses and other large vehicles.

Molten carbonate fuel cells (MCFCs) are high-temperature fuel cells that have potential in the stationary power market. They use carbonate as an electrolyte and can be used with hydrogen or a variety of fossil fuels, including gasoline, without re-forming. High-efficiency power generation can be supplemented by cogeneration using waste heat in a gas turbine or directly for district heating or other uses. MCFC (molten carbonate fuel cell) installation costs for recent projects are running about $5,000/kW (Fuel Cell Energy, 2000).

Solid oxide fuel cells (SOFCs) are another type of high temperature fuel cell that may be used for stationary power. They use a solid ceramic oxide electrolyte and can be used with hydrogen or a variety of fossil fuels, including gasoline, without re-forming. Like MCFCs, they are capable of cogeneration. SOFCs are considered a promising technology by many industry observers. A breakthrough in the durability of the ceramic electrolyte would improve the prospects for this technology. "The first SOFC fuel cell/gas turbine hybrid power system (220 kW capacity) is being readied for shipment, installation and operation at the National Fuel Cell Research Center at the University of California, Irvine. The microturbine is said to add an additional 12 percent efficiency."

"Siemens plans on achieving an installation cost of $1,300-$1,500/kW for their SOFC model, after achieving mass production, but the initial costs are much higher (Forbes,

2000). Eventual electricity delivery costs could potentially reach less than $0.05/kWh" (Hydrogen and Fuel Cell Investor, U.S. DOE, 2000).[27]

Solid acid fuel cells are a product of recent research at the California Institute of Technology. They operate at about 160 degrees. These cells could prove more tolerant to carbon monoxide in fuel than low-temperature PEM fuel cells and may start up faster and be less corrosive to internal parts than high-temperature SOFCs and MCFCs.[28]

Regenerative fuel cells (RFCs) generate hydrogen by electrolyzing water using electricity from renewable or traditional grid sources. They then run the reverse reaction, using the hydrogen and producing electricity and water. RFCs could be useful where the electricity to generate the hydrogen in the first place is produced by photovoltaic cells or other renewable sources. NASA and others are researching this type of fuel cell.

Zinc-air, aluminum oxide and certain other electrochemical cells are sometimes referred to as fuel cells. However, some experts regard them as more akin to batteries than fuel cells.[29]

---

27 Martin and Elliott, 145.

28 Peter Weiss, Science News, Vol. 159, (April 21, 2001): 247.

29 According to the Hydrogen and Fuel Cell Investor Internet site: "The devices...have some similarities to fuel cells and claim excellent performance, but they are still really batteries in that they only store energy and require externally generated power to reprocess or recharge their fuel components. True fuel cells generate power, not store it. They rely on a continuous fuel stream and do not consume themselves in use or require replacement or reprocessing of their components to maintain power output." - http://www.h2fc.com.

Discussion Questions

1. Explain in detail how a fuel cell works.

2. What are the different obstacles to large-scale adoption of fuel cell technology? Cite specific examples.

3. What are the different types of fuel cells? Differentiate based on choice of electrolyte and operating temperatures.

4. What type of fuel cell is suitable for specific applications?

In-Class Assignment: Objective Questions

1. Hydrogen can be stored as which of the following?
   a. Gas and a compressed gas only
   b. Gas, compressed gas, liquid only
   c. Compressed gas, metal hydride, and liquid only
   d. Gas, compressed gas, liquid, and metal hydride
2. The operating temperature of the proton exchange membrane fuel cell is which of the following?
   a. 20 to 30 degrees C
   b. 60 to 100 degrees C
   c. 175 to 200 degrees C
   d. None of the above
3. Liquid lithium is used as an electrolyte in which of the following fuel cells?
   a. Molten carbonate fuel cell
   b. Alkaline fuel cell
   c. Solid Oxide fuel cell
   d. None of the above

6 The Need for Nuclear Power

Richard Rhodes and Denis Beller

A Clean Break

The world needs more energy. Energy multiplies human labor, increasing productivity. It builds and lights schools, purifies water, powers farm machinery, drives sewing machines and robot assemblers, stores and moves information. World population is steadily increasing, having passed six billion in 1999. Yet one-third of that number—two billion people—lack access to electricity. Development depends on energy, and the alternative to development is suffering: poverty, disease, and death. Such conditions create instability and the potential for widespread violence. National security therefore requires developed nations to help increase energy production in their more populous developing counterparts. For the sake of safety as well as security, that increased energy supply should come from diverse sources.

"At a global level," the British Royal Society and Royal Academy of Engineering estimate in a 1999 report on nuclear energy and climate change, "we can expect our consumption of energy at least to double in the next 50 years and to grow by a factor of up to five in the next 100 years as the world population increases and as people seek to improve their standards of living." Even with vigorous conservation, world energy production would have to triple by 2050 to support consumption at a mere one-third of today's U.S. per capita rate. The International Energy Agency (IEA) of the Organization for Economic Cooperation and Development (OECD) projects 65 percent growth in world energy demand by 2020, two-thirds of that coming from developing countries. "Given the levels of consumption likely in the future," the Royal Society and Royal Academy caution, "it will be an immense challenge to meet the global demand for energy without unsustainable long-term damage to the environment." That damage includes surface and air pollution and global warming.

Most of the world's energy today comes from petroleum (39.5 percent), coal (24.2 percent), natural gas (22.1 percent), hydroelectric power (6.9 percent), and nuclear power (6.3 percent). Although oil and coal still dominate, their market fraction began declining decades ago. Meanwhile, natural gas and nuclear power have steadily increased their share

Richard Rhodes and Denis Beller, "The Need for Nuclear Power," *Foreign Affairs*, vol. 79, no. 1. Copyright © 2000 by Council on Foreign Relations, Inc. Reprinted with permission.

and should continue to do so. Contrary to the assertions of antinuclear organizations, nuclear power is neither dead nor dying. France generates 79 percent of its electricity with nuclear power; Belgium, 60 percent; Sweden, 42 percent; Switzerland, 39 percent; Spain, 37 percent; Japan, 34 percent; the United Kingdom, 21 percent; and the United States (the largest producer of nuclear energy in the world), 20 percent. South Korea and China have announced ambitious plans to expand their nuclear-power capabilities—in the case of South Korea, by building 16 new plants, increasing capacity by more than 100 percent. With 434 operating reactors worldwide, nuclear power is meeting the annual electrical needs of more than a billion people.

In America and around the globe, nuclear safety and efficiency have improved significantly since 1990. In 1998, unit capacity factor (the fraction of a power plant's capacity that it actually generates) for operating reactors reached record levels. The average U.S. capacity factor in 1998 was 80 percent for about 100 reactors, compared to 58 percent in 1980 and 66 percent in 1990. Despite a reduction in the number of power plants, the U.S. nuclear industry generated nine percent more nuclear electricity in 1999 than in 1998. Average production costs for nuclear energy are now just 1.9 cents per kilowatt-hour (kWh), while electricity produced from gas costs 3.4 cents per kWh. Meanwhile, radiation exposure to workers and waste produced per unit of energy have hit new lows.

Because major, complex technologies take more than half a century to spread around the world, natural gas will share the lead in power generation with nuclear power over the next hundred years. Which of the two will command the greater share remains to be determined. But both are cleaner and more secure than the fuels they have begun to replace, and their ascendance should be endorsed. Even environmentalists should welcome the transition and reconsider their infatuation with renewable energy sources.

Carbon Nations

Among sources of electric-power generation, coal is the worst environmental offender. (Petroleum, today's dominant source of energy, sustains transportation, putting it in a separate category.) Recent studies by the Harvard School of Public Health indicate that pollutants from coal-burning cause about 15,000 premature deaths annually in the United States alone. Used to generate about a quarter of the world's primary energy, coal-burning releases amounts of toxic waste too immense to contain safely. Such waste is either dispersed directly into the air or is solidified and dumped. Some is even mixed into construction materials. Besides emitting noxious chemicals in the form of gases or toxic particles—sulfur and nitrogen oxides (components of acid rain and smog), arsenic, mercury, cadmium, selenium, lead, boron, chromium, copper, fluorine, molybdenum, nickel, vanadium, zinc, carbon monoxide and dioxide, and other greenhouse gases—coal-fired power plants are also the world's major source of radioactive releases into the environment. Uranium and thorium, mildly radioactive elements ubiquitous in the earth's crust, are both released when coal is burned. Radioactive radon gas, produced when uranium in the earth's crust decays and normally confined underground, is released when coal is mined. A 1,000-megawatt-electric (MWe) coal-fired power plant releases about 100 times as much radioactivity into the environment as a comparable nuclear plant. Worldwide releases of uranium and thorium from coal-burning total about 37,300 tonnes (metric tons) annually, with about 7,300 tonnes coming from the United States. Since uranium and thorium are potent nuclear fuels, burning coal also wastes more potential energy than it produces.

Nuclear proliferation is another overlooked potential consequence of coal-burning. The uranium released by a single 1,000-MWe coal plant in a year includes about 74 pounds of uranium-235—enough for at least two atomic bombs. This uranium would have to be enriched before it could be used, which would be complicated and expensive. But plutonium could also be bred from coal-derived uranium. Moreover, "because electric utilities are not high-profile facilities," writes physicist Alex Gabbard of the Oak Ridge National Laboratory, "collection and processing of coal ash for recovery of minerals ... can proceed without attracting outside attention, concern or intervention. Any country with coal-fired plants could collect combustion byproducts and amass sufficient nuclear weapons materials to build up a very powerful arsenal." In the early 1950s, when richer ores were believed to be in short supply, the U.S. Atomic Energy Commission actually investigated using coal as a source of uranium production for nuclear weapons; burning the coal, the AEC concluded, would concentrate the mineral, which could then be extracted from the ash.

Such a scenario may seem far-fetched. But it emphasizes the political disadvantages under which nuclear power labors. Current laws force nuclear utilities, unlike coal plants, to invest in expensive systems that limit the release of radioactivity. Nuclear fuel is not efficiently recycled in the United States because of proliferation fears. These factors have warped the economics of nuclear power development and created a politically difficult waste-disposal problem. If coal utilities were forced to assume similar costs, coal electricity would no longer be cheaper than nuclear.

Decline and Fall of the Renewables

Renewable sources of energy—hydroelectric, solar, wind, geo-thermal, and biomass—have high capital-investment costs and significant, if usually unacknowledged, environmental consequences. Hydropower is not even a true renewable, since dams eventually silt in. Most renewables collect extremely diluted energy, requiring large areas of land and masses of collectors to concentrate. Manufacturing solar collectors, pouring concrete for fields of windmills, and drowning many square miles of land behind dams cause damage and pollution.

Photovoltaic cells used for solar collection are large semiconductors; their manufacture produces highly toxic waste metals and solvents that require special technology for disposal. A 1,000-MWe solar electric plant would generate 6,850 tonnes of hazardous waste from metals-processing alone over a 30-year lifetime. A comparable solar thermal plant (using mirrors focused on a central tower) would require metals for construction that would generate 435,000 tonnes of manufacturing waste, of which 16,300 tonnes would be contaminated with lead and chromium and be considered hazardous.

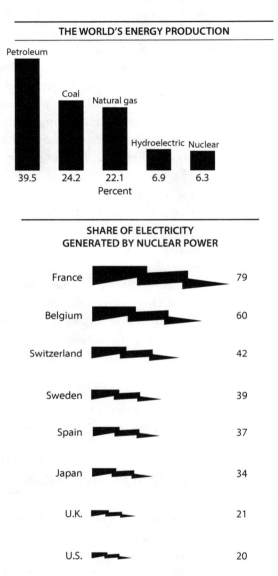

A global solar-energy system would consume at least 20 percent of the world's known iron resources. It would require a century to build and a substantial fraction of annual world iron production to maintain. The energy necessary to manufacture sufficient solar collectors to cover a half-million square miles of the earth's surface and to deliver the electricity through long-distance transmission systems would itself add grievously to the global burden of pollution and greenhouse gas. A global solar-energy system without fossil or nuclear backup would also be dangerously vulnerable to drops in solar radiation from volcanic events such as the 1883 eruption of Krakatoa, which caused widespread crop failure during the "year without a summer" that followed.

Wind farms, besides requiring millions of pounds of concrete and steel to build (and thus creating huge amounts of waste materials), are inefficient, with low (because intermittent) capacity. They also cause visual and noise pollution and are mighty slayers of birds. Several hundred birds of prey, including dozens of golden eagles, are killed every year by a single California wind farm; more eagles have been killed by wind turbines than were lost in the disastrous Exxon *Valdez* oil spill. The National Audubon Society has launched a campaign to save the California condor from a proposed wind farm to be built north of Los Angeles. A wind farm equivalent in output and capacity to a 1,000-MWe fossil-fuel or nuclear plant would occupy 2,000 square miles of land and, even with substantial subsidies and ignoring hidden pollution costs, would produce electricity at double or triple the cost of fossil fuels.

Although at least one-quarter of the world's potential for hydropower has already been developed, hydroelectric power—produced by dams that submerge large areas of land, displace rural populations, change river ecology, kill fish, and risk catastrophic collapse—has understandably lost the backing of environmentalists in recent years. The U.S. Export-Import Bank was responding in part to environmental lobbying when it denied funding to China's 18,000-MWe Three Gorges project.

Meanwhile, geothermal sources—which exploit the internal heat of the earth emerging in geyser areas or under volcanoes—are inherently limited and often coincide with scenic sites (such as Yellowstone National Park) that conservationists understandably want to preserve.

Because of these and other disadvantages, organizations such as the World Energy Council and the IEA predict that hydroelectric generation will continue to account for no more than its present 6.9 percent share of the world's primary energy supply, while all other renewables, even though robustly subsidized, will move from their present 0.5 percent share to claim no more than 5 to 8 percent by 2020. In the United States, which leads the world in renewable energy generation, such production actually declined by 9.4 percent from 1997 to 1998: hydro by 9.2 percent, geothermal by 5.4 percent, wind by 50.5 percent, and solar by 27.7 percent.

Like the dream of controlled thermonuclear fusion, then, the reality of a world run on pristine energy generated from renewables continues to recede, despite expensive, highly subsidized research and development. The 1997 U.S. federal R&D investment per thousand kWh was only 5 cents for nuclear and coal, 58 cents for oil, and 41 cents for gas, but was $4,769 for wind and $17,006 for photovoltaics. This massive public investment in renewables would have been better spent making coal plants and automobiles cleaner. According to Robert Bradley of Houston's Institute for Energy Research, U.S. conservation efforts and nonhydroelectric renewables have benefited from a cumulative 20-year taxpayer investment of some $30-$40 billion—"the largest governmental peacetime energy expenditure in U.S. history." And Bradley estimates that "the $5.8 billion spent by the Department of Energy on wind and solar subsidies" alone could have paid for "replacing between 5,000 and 10,000 MWe of the nation's dirtiest coal capacity with gas-fired combined-cycle units, which would have reduced carbon dioxide emissions by between one-third and two-thirds." Replacing coal with nuclear generation would have reduced overall emissions even more.

Despite the massive investment, conservation and nonhydro renewables remain stubbornly uncompetitive and contribute only marginally to U.S. energy supplies. If the most prosperous nation in the world cannot afford them, who can? Not China, evidently, which expects to generate less than one percent of its commercial energy from nonhydro renewables in 2025. Coal and oil will still account for the bulk of China's energy supply in that year unless developed countries offer incentives to convince the world's most populous nation to change its plans.

Turn Down the Volume

Natural gas has many virtues as a fuel compared to coal or oil, and its share of the world's energy will assuredly grow in the first half of the 21st century. But its supply is limited and unevenly distributed, it is expensive as a power source compared to coal or uranium, and it pollutes the air. A 1,000-MWe natural gas plant releases 5.5 tonnes of sulfur oxides per day, 21 tonnes of nitrogen oxides, 1.6 tonnes of carbon monoxide, and 0.9 tonnes of particulates. In the United States, energy production from natural gas released about 5.5 billion tonnes of waste in 1994. Natural gas fires and explosions are also significant risks. A single mile of gas pipeline three feet in diameter at a pressure of 1,000 pounds per square inch (psi) contains the equivalent of two-thirds of a kiloton of explosive energy; a million miles of such large pipelines lace the earth.

The great advantage of nuclear power is its ability to wrest enormous energy from a small volume of fuel. Nuclear fission, transforming matter directly into energy, is several million times as energetic as chemical burning, which merely breaks chemical bonds. One tonne of nuclear fuel produces energy equivalent to 2 to 3 million tonnes of fossil fuel. Burning 1 kilogram of firewood can generate 1 kilowatt-hour of electricity; 1 kg of coal, 3 kWh; 1 kg of oil, 4 kWh. But 1 kg of uranium fuel in a modern light-water reactor generates 400,000 kWh of electricity, and if that uranium is recycled, 1 kg can generate more than 7,000,000 kWh. These spectacular differences in volume help explain the vast difference in the environmental impacts of nuclear versus fossil fuels. Running a 1,000-MWe power plant for a year requires 2,000 train cars of coal or 10 supertankers of oil but only 12 cubic meters of natural uranium. Out the other end of fossil-fuel plants, even those with pollution-control systems, come thousands of tonnes of noxious gases, particulates, and heavy-metal-bearing (and radioactive) ash, plus solid hazardous waste—up to 500,000 tonnes of sulfur from coal, more than 300,000 tonnes from oil, and 200,000 tonnes from natural gas. In contrast, a 1,000-MWe nuclear plant releases no noxious gases or other pollutants[1] and much less radioactivity per capita than is encountered from airline travel, a home smoke detector, or a television set. It produces about 30 tonnes of high-level waste (spent fuel) and 800 tonnes of low- and intermediate-level waste—about 20 cubic meters in all when compacted (roughly, the volume of two automobiles). All the operating nuclear plants in the world produce some 3,000 cubic meters of waste annually. By comparison, U.S. industry generates annually about 50,000,000 cubic meters of solid toxic waste.

The high-level waste is intensely radioactive, of course (the low-level waste can be less radioactive than coal ash, which is used to make concrete and gypsum—both of which are incorporated into building materials). But thanks to its small volume and the fact that it is not released into the environment, this high-level waste can be meticulously sequestered behind multiple barriers. Waste from coal, dispersed across the landscape in smoke or buried near the surface, remains toxic forever. Radioactive nuclear waste decays steadily, losing 99 percent of its toxicity after 600 years—well within the range of human experience with custody and maintenance, as evidenced by structures such as the Roman Pantheon and Notre Dame Cathedral. Nuclear waste disposal is a political problem in the United States because of widespread fear disproportionate to the reality of risk. But it is not an engineering problem, as advanced projects in France, Sweden, and Japan demonstrate. The World Health Organization has estimated that indoor and outdoor air pollution cause some three million deaths per year. Substituting small, properly contained volumes of nuclear waste for vast, dispersed amounts of toxic wastes from fossil fuels would produce so obvious an improvement in public health that it is astonishing that physicians have not already demanded such a conversion.

The production cost of nuclear electricity generated from existing U.S. plants is already fully competitive with electricity from fossil fuels, although new nuclear power is somewhat more expensive. But this higher price tag is deceptive. Large nuclear power plants require

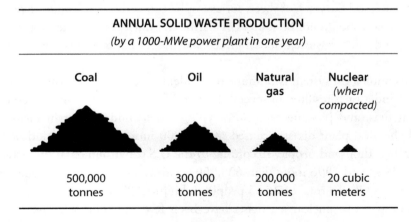

ANNUAL SOLID WASTE PRODUCTION
(by a 1000-MWe power plant in one year)

Coal	Oil	Natural gas	Nuclear *(when compacted)*
500,000 tonnes	300,000 tonnes	200,000 tonnes	20 cubic meters

---

1 Uranium is refined and processed into fuel assemblies today using coal energy, which does of course release pollutants. If nuclear power were made available for process heat or if fuel assemblies were recycled, this source of manufacturing pollution would be eliminated or greatly reduced.

larger capital investments than comparable coal or gas plants only because nuclear utilities are required to build and maintain costly systems to keep their radioactivity from the environment. If fossil-fuel plants were similarly required to sequester the pollutants they generate, they would cost significantly more than nuclear power plants do. The European Union and the International Atomic Energy Agency (IAEA) have determined that "for equivalent amounts of energy generation, coal and oil plants, ... owing to their large emissions and huge fuel and transport requirements, have the highest externality costs as well as equivalent lives lost. The external costs are some ten times higher than for a nuclear power plant and can be a significant fraction of generation costs." In equivalent lives lost per gigawatt generated (that is, loss of life expectancy from exposure to pollutants), coal kills 37 people annually; oil, 32; gas, 2; nuclear, 1. Compared to nuclear power, in other words, fossil fuels (and renewables) have enjoyed a free ride with respect to protection of the environment and public health and safety.

Even the estimate of one life lost to nuclear power is questionable. Such an estimate depends on whether or not, as the long-standing "linear no-threshold" theory (LNT) maintains, exposure to amounts of radiation considerably less than preexisting natural levels increases the risk of cancer. Although LNT dictates elaborate and expensive confinement regimes for nuclear power operations and waste disposal, there is no evidence that low-level radiation exposure increases cancer risk. In fact, there is good evidence that it does not. There is even good evidence that exposure to low doses of radioactivity *improves* health and lengthens life, probably by stimulating the immune system much as vaccines do (the best study, of background radon levels in hundreds of thousands of homes in more than 90 percent of U.S. counties, found lung cancer rates decreasing significantly with increasing radon levels among both smokers and nonsmokers). So low-level radioactivity from nuclear power generation presents at worst a negligible risk. Authorities on coal geology and engineering make the same argument about low-level radioactivity from coal-burning; a U.S. Geological Survey fact sheet, for example, concludes that "radioactive elements in coal and fly ash should not be sources of alarm." Yet nuclear power development has been hobbled, and nuclear waste disposal unnecessarily delayed, by limits not visited upon the coal industry.

No technological system is immune to accident. Recent dam overflows and failures in Italy and India each resulted in several thousand fatalities. Coal-mine accidents, oil- and gas-plant fires, and pipeline explosions typically kill hundreds per incident. The 1984 Bhopal chemical plant disaster caused some 3,000 immediate deaths and poisoned several hundred thousand people. According to the U.S. Environmental Protection Agency, between 1987 and 1996 more than 600,000 accidental releases of toxic chemicals in the United States killed a total of 2,565 people and injured 22,949.

By comparison, nuclear accidents have been few and minimal. The recent, much-reported accident in Japan occurred not at a power plant but at a facility processing fuel for a research reactor. It caused no deaths or injuries to the public. As for the Chernobyl explosion, it resulted from human error in operating a fundamentally faulty reactor design that could not have been licensed in the West. It caused severe human and environmental damage locally, including 31 deaths, most from radiation exposure. Thyroid cancer, which could have been prevented with prompt iodine prophylaxis, has increased in Ukrainian children exposed to fallout. More than 800 cases have been diagnosed and several thousand more are projected; although the disease is treatable, three children have died. LNT-based calculations project 3,420 cancer deaths in Chernobyl-area residents and cleanup crews. The Chernobyl reactor lacked a containment structure, a fundamental safety system that is required on Western reactors. Postaccident calculations indicate that such a structure would have confined the explosion and thus the radioactivity, in which case no injuries or deaths would have occurred.

These numbers, for the worst ever nuclear power accident, are remarkably low compared to major accidents in other industries. More than 40 years of commercial nuclear power operations demonstrate that nuclear power is much safer than fossil-fuel systems in terms of industrial accidents, environmental damage, health effects, and long-term risk.

Ghosts in the Machine

Most of the uranium used in nuclear reactors is inert, a nonfissile product unavailable for use in weapons. Operating reactors, however, breed fissile plutonium that could be used in bombs, and therefore the commercialization of nuclear power has raised concerns about the spread of weapons. In 1977, President Carter deferred indefinitely the recycling of "spent" nuclear fuel, citing proliferation risks. This decision effectively ended nuclear recycling in the United States, even though such recycling reduces the volume and radiotoxicity of nuclear waste and could extend nuclear fuel supplies for thousands of years. Other nations assessed the risks differently and the majority did not follow the U.S. example. France and the United Kingdom currently reprocess spent fuel; Russia is stockpiling fuel and separated plutonium for jump-starting future fast-reactor fuel cycles; Japan has begun using recycled uranium and plutonium mixed-oxide (MOX) fuel in its reactors and recently approved the construction of a new nuclear power plant to use 100-percent MOX fuel by 2007.

Although power-reactor plutonium theoretically can be used to make nuclear explosives, spent fuel is refractory, highly radioactive, and beyond the capacity of terrorists to process. Weapons made from reactor-grade plutonium would be hot, unstable, and of uncertain yield. India has extracted weapons plutonium from a Canadian heavy-water reactor and bars inspection of some dual-purpose reactors it has built. But no plutonium has ever been diverted from British or French reprocessing facilities or fuel shipments for weapons production; IAEA inspections are effective in preventing such diversions. The risk of proliferation, the IAEA has concluded, "is not zero and would not become zero even if nuclear power ceased to exist. It is a continually strengthened nonproliferation regime that will remain the cornerstone of efforts to prevent the spread of nuclear weapons."

Ironically, burying spent fuel without extracting its plutonium through reprocessing would actually increase the long-term risk of nuclear proliferation, since the decay of less-fissile and more-radioactive isotopes in spent fuel after one to three centuries improves the explosive qualities of the plutonium it contains, making it more attractive for weapons use. Besides extending the world's uranium resources almost indefinitely, recycling would make it possible to convert plutonium to useful energy while breaking it down into shorter-lived, nonfissionable, nonthreatening nuclear waste.

Hundreds of tons of weapons-grade plutonium, which cost the nuclear superpowers billions of dollars to produce, have become military surplus in the past decade. Rather than burying some of this strategically worrisome but energetically valuable material—as Washington has proposed—it should be recycled into nuclear fuel. An international system to recycle and manage such fuel would prevent covert proliferation. As envisioned by Edward Arthur, Paul Cunningham, and Richard Wagner of the Los Alamos National Laboratory, such a system would combine internationally monitored retrievable storage, the processing of all separated plutonium into MOX fuel for power reactors, and, in the longer term, advanced integrated materials-processing reactors that would receive, control, and process all fuel discharged from reactors throughout the world, generating electricity and reducing spent fuel to short-lived nuclear waste ready for permanent geological storage.

The New New Thing

A new generation of small, modular power plants—competitive with natural gas and designed for safety, proliferation resistance, and ease of operation—will be necessary to extend the benefits of nuclear power to smaller developing countries that lack a nuclear infrastructure. The Department of Energy has awarded funding to three designs for such "fourth-generation" plants. A South African utility, Eskom, has announced plans to market a modular gas-cooled pebble-bed reactor that does not require emergency core-cooling systems and physically cannot "melt down." Eskom estimates that the reactor will produce electricity at around 1.5 cents per kWh, which is cheaper than electricity from a combined-cycle gas plant. The Massachusetts Institute of Technology and the Idaho National Engineering and Environmental Laboratory are developing a similar design to supply high-temperature heat for industrial processes such as hydrogen generation and desalinization.

Petroleum is used today primarily for transportation, but the internal combustion engine has been refined to its limit. Further reductions in transportation pollution can come only from abandoning petroleum and developing nonpolluting power systems for cars and trucks. Recharging batteries for electric cars will simply transfer pollution from mobile to centralized sources unless the centralized source of electricity is nuclear. Fuel cells, which are now approaching commercialization, may be a better solution. Because fuel cells generate electricity directly from gaseous or liquid fuels, they can be refueled along the way, much as present internal combustion engines are. When operated on pure hydrogen, fuel cells produce only water as a waste product. Since hydrogen can be generated from water using heat or electricity, one can envisage a minimally polluting energy infrastructure, using hydrogen generated by nuclear power for transportation, nuclear electricity and process heat for most other applications, and natural gas and renewable systems as backups. Such a major commitment to nuclear power could not only halt but eventually even reverse the continuing buildup of carbon in the atmosphere. In the meantime, fuel cells using natural gas could significantly reduce air pollution.

Powering the Future

To meet the world's growing need for energy, the Royal Society and Royal Academy report proposes "the formation of an international body for energy research and development, funded by contributions from individual nations on the basis of GDP or total national energy consumption." The body would be "a funding agency supporting research, development and demonstrators elsewhere, not a research center itself." Its budget might build to an annual level of some $25 billion, "roughly one percent of the total global energy budget." If it truly wants to develop efficient and responsible energy supplies, such a body should focus on the nuclear option, on establishing a secure international nuclear-fuel storage and reprocessing system, and on providing expertise for siting, financing, and licensing modular nuclear power systems to developing nations.

According to Arnulf Grübler, Nebojsa Nakicenovic, and David Victor, who study the dynamics of energy technologies, "the share of energy supplied by electricity is growing rapidly in most countries and worldwide." Throughout history, humankind has gradually decarbonized its dominant fuels, moving steadily away from the more polluting, carbon-rich sources. Thus the world has gone from coal (which has one hydrogen atom per carbon atom and was dominant from 1880 to 1950) to oil (with two hydrogens per carbon, dominant from 1950 to today). Natural gas (four hydrogens per carbon) is steadily increasing its market share. But nuclear fission produces no carbon at all.

Physical reality—not arguments about corporate greed, hypothetical risks, radiation exposure, or waste disposal—ought to inform decisions vital to the future of the world. Because diversity and redundancy are important for safety and security, renewable energy sources ought to retain a place in the energy economy of the century to come. But nuclear power should be central. Despite its outstanding record, it has instead been relegated by its opponents to the same twilight zone of contentious ideological conflict as abortion and evolution. It deserves better. Nuclear power is environmentally safe, practical, and affordable. It is not the problem—it is one of the best solutions.

Discussion questions

1. Why is nuclear power preferable to fossil fuels? Cite examples to corroborate your answer.

2. Explain the role played by small modular nuclear power plants as an energy source for the future.

3. What are the disadvantages of using plutonium as a radioactive fuel?

Solar Energy

Travis Bradford

Many energy-industry observers consider solar energy a theoretically elegant but unrealistic solution to the imminent gap between global energy supply and demand. Everyone agrees that clean, limitless, free energy from the sky sounds ideal, but more practical considerations such as relative cost and the sheer scale of the current energy infrastructure seem to doom solar energy to follower status for years to come. Other sources of energy, both conventional and renewable (including wind, geothermal, and biomass), appear to be cheaper, easier to deploy, and better funded and currently enjoy popular support in the media and renewable-energy advocacy circles. In addition, memories of false starts and unfulfilled promises during the twentieth century have tempered general optimism about solar energy's potential. This credibility gap exists not only among members of the conventional energy industry—fuel providers, electric utilities, and all other interested parties—but also among a larger group of environmentalists and solar-energy system installers. Many of these people invested time and money to promote solar energy in response to the first OPEC oil shocks of the 1970s, only to be abandoned after 1982 by the national governments that had supported them. The memory of this disappointment lingers, promoting skepticism that solar could be a viable economic energy solution without substantial government subsidies.

Rapid changes in the photovoltaic industry, technology, and institutional players over the last decade have dramatically altered PV's economic viability, and fundamentally transformed the competitive landscape of the energy industry. Today, solar energy and photovoltaics comprise a global, multibillion-dollar industry providing cost-effective energy to millions of people worldwide in many large and growing markets. As with most technologies, the cost-benefit calculation varies by each potential user and application, making simple generalizations difficult. As a result, the largest remaining obstacle to continued adoption of solar energy is the lack of reliable and current information about its true economic characteristics. This chapter puts this growing global industry in perspective by highlighting its history—its roots, its driving forces and characteristics, the current

Travis Bradford, "Solar Energy," *Solar Revolution: The Economic Transformation of the Global Energy Industry*, pp. 89-111. Copyright © 2008 by MIT Press. Reprinted with permission.

state of its development, and methodologies for estimating how the cost of producing PV will change as the industry matures and grows.

Types of Solar Energy

Typically, an informed discussion about solar energy is limited by various and confusing notions of what the term *solar energy* actually describes. Broadly speaking, *solar energy* could be used to describe any phenomenon that is created by solar sources and harnessed in the form of energy, directly or indirectly—from photosynthesis to photovoltaics. Many of today's environmentalists use the term *solar energy* in its most comprehensive sense to include certain new renewable-energy technologies such as wind power and biomass, arguing that these sources derive energy from the sun, however indirectly. More conservative uses of the term, such as the one that this book employs, discuss direct-only solar sources, whether active, passive, thermal, or electric—that is, sources of energy that can be directly attributed to the light of the sun or the heat that sunlight generates.

This more restrictive classification is useful because a more general characterization of solar energy that includes wind and other technologies tends to obscure various isolated trends within the broader renewable-energy industry. Many renewable-energy technologies sometimes lumped under *solar energy* have very different economic characteristics, making it difficult to draw meaningful conclusions about them. Since the economic drivers discussed in the second half of this book do not apply to all technologies equally, it is helpful to be precise when analyzing specific industrial transformations and the markets in which they will occur.

Understanding direct solar energy requires examining three key continuums in the methods of harnessing it: (1) passive and active, (2) thermal and photovoltaic and (3) concentrating and nonconcentrating. Every solar-energy technology features some combination of these characteristics to harness sunlight. *Passive solar energy* requires a building design that is intended to capture the sun's heat and light. In passive solar design, heat and light are not converted to other forms of energy; they are simply collected. This is done through various design and building methods such as orienting a building toward the sun or including architectural features that absorb solar energy where it is needed and exclude it where it is not useful. The simplest conceptualization of passive solar-energy design for building is in a greenhouse, a design that allows solar light to pass into the interior and then captures the heat it generates inside to maintain year-round growing conditions. Passive solar features—some of which have been used in building design for thousands of years—include site selection and building placement that maximizes synchronized heating and lighting, windows placed in south-facing walls, vents and ducts moved to capture heat through the building, and *trombe walls* (dark, south-facing walls that absorb light and heat), wide eaves, heat-storing slabs, and superinsulation. Passive solar is an elegant way to harness the sun's energy, but it usually has to be designed into the original building plans to be made cost-effective. Once a building design has been finalized with siting, orientation, and structural elements, it is often prohibitively expensive to change or retrofit the facility to capture additional passive solar-energy benefits.

Active solar energy refers to the harnessing the sun's energy to store it or convert it for other applications. These applications include capturing heat for hot water that can be used for cooking, cleaning, heating, or purification; producing industrial heat for melting; or generating electricity directly or through steam turbines. The common characteristic is the active and intentional collection and redirection of the solar energy. These active solar solutions can be broadly grouped as either *thermal* or *photovoltaic* according to the method by which they generate energy for transfer or conversion into other useful forms. Thermal applications include all uses of the sun's energy in heat-driven mechanisms, such as heating water or some other conductive fluid, solar cooking and agricultural drying, or other industrial heat-collection applications—for processes as varied as water treatment or hydrogen generation through water decomposition. The most powerful solar thermal applications are used to superheat water and convert it to steam, which is then used to power a conventional steam engine for thermal electricity generation. Prior to the middle of the twentieth century, all industrial applications of solar energy were thermal in nature, and many of the simplest and most widely used remain so today, including the millions of rooftop solar water-heating systems installed around the world.

Solar photovoltaic is the state of the art in active solar electricity generation. By capturing the photonic energy of light on materials of a specific molecular structure, direct electric current is produced. The photoelectric effect (the description of which won Albert Einstein his Nobel Prize in physics in 1921 and which he believed to be more valuable than his work on the theory of relativity) allows an electric charge to be created on a semiconductive substrate that has been doped with chemical additives to create opposing positive and negative layers.[1] Photons of sunlight striking this surface facilitate an electron moving from the positively charged layer to the negative, creating an electrical current. This shifting of electrons in photovoltaic energy generation occurs without the need for moving parts and in proportion to the amount of light striking the surface. The useful lifetime of a photovoltaic cell is a function of manufacturing methods and the atomic stability of the substrate material, but some PV cells have been in operation for decades in space-based satellite applications, one of harshest possible environments. Land-based applications based on silicon material for PV cells are often warranted by manufacturers for twenty-five years or more, although the expected useful life is much longer.

The final distinction in solar applications is *concentrating* and *non-concentrating*. Concentrating solar applications use mirrors or lenses to focus sunlight. Concentration can significantly increase light intensity in the focus area, similar to the way in which a magnifying glass burns a hole in a leaf. Industrial-scale concentration can be achieved by the *trough method,* in which a long, troughlike parabolic mirror focuses sunlight along the length of a fluid-filled pipe suspended above the mirror. Large-scale concentration can also be created via an array of sun-tracking mirrors arranged to focus sunlight on a central point for thermal or photovoltaic use. Arrays of lenses can concentrate energy on photovoltaic cells, which tend to operate more efficiently (that is, convert more of the sunlight that strikes them into electricity) when the light is brighter. Concentrating systems are, by their nature, more complicated to build and manage than nonconcentrating systems and contain equipment with moving parts that suffer wear and tear as well as problems relating to the significant heat generated by them. Nonconcentrating systems, which allow the sunlight to fall on their energy-gathering parts without concentration by lenses or mirrors, are usually simpler and therefore less expensive maintain; however, they achieve correspondingly lower intensities and temperatures. Nonconcentrating systems include those that use direct sunlight to heat close-set pipes (as in a domestic hot-water system) or open water (such as a swimming pool), as well as most PV panels commonly seen on the roofs of houses and in stand-alone signs and lighting.

Figure 7.1 shows the breakdown of modern active forms of harnessing solar energy, both thermal and photovoltaic, among the various sizes of the generators used. The size classifications (centralized, large distributed, and small distributed) correspond to the different types of users that can use the amounts of power generated (utilities, commercial users, and residential users respectively). These distinctions will be examined in later chapters to discuss the evolving economic decisions that each of these potential users of solar energy face.

	Type of Solar Energy	
System Size	PV	Thermal
Centralized (2 MW–GW)	Concentrating photovoltaic arrays (CPV) / Utility-scale PV	Concentrating solar thermal power (CSP)
Large distributed (20 KW–2 MW)	Commercial-building PV	
On-site distributed (<20 KW)	Residential PV	Home solar hot-water systems

Figure 7.1. Today's mix of active solar-energy technologies by size and type.

A Short History of Solar Energy

Various forms of solar energy have been used since prehistoric times. In fact, passive solar applications have been used in building design and construction for thousands of years. Other early efforts included attempts to harness the power of the sun to wage war. Mythic stories abound regarding the third century BCE scientist and mathematician Archimedes, who, from the safety of the ramparts, defended ancient Syracuse from a Roman military invasion by using an array of solar mirrors to set fire to enemy ships in the harbor. Though the accuracy of these stories is disputed by historians, there have been many recorded attempts to use lenses or mirrors to harness the power of the sun as part of strategic defense and warfare over the last fifteen hundred years. These experiments, none of which seem to have found much practical or long-lived use, included concentrating solar energy to burn, blind, or intimidate the enemy. Other proposed solar-energy applications were more commercial and industrial in nature. Leonardo Da Vinci, inspired by accounts of Archimedes' use of mirrors at Syracuse, designed a gigantic bowl mirror four miles across, to be built and used for large-scale industrial applications, including melting metals. Though the massive mirror was never built, Da Vinci foresaw, by many hundreds of years, the methods by which the sun's power would finally be harnessed for peaceful, commercial applications.

Solar-energy technology saw a burst of new practical applications during the late-nineteenth-century industrial revolution, driven by three solar-energy inventors on different continents. This period in industrial history ushered in a grand expansion in knowledge and invention as entrepreneurs pursued many new energy technologies alternatives to wood and coal. As in all periods of rapid technological growth, an efficient form of commercial Darwinism determined winners and losers. Inventors and entrepreneurial businessmen developed new technologies. Society adopted those that operated and performed faster, better, and cheaper, while all others were put on the shelf until changing relative costs or technological breakthroughs created economic justifications to revisit them.

The first of the solar inventors was William Adams. In Bombay, India, in the 1860s and 1870s, Adams, a former British patent officer and engineer, conducted various solar-energy experiments and created practical devices such as a solar cooker to help ease energy shortfalls and depletion of local wood fuel in colonial India. A solar cooker is effectively a large, bowl-shaped, concave mirror that, when pointed at the sun, creates a cooking hot spot by concentrating the sun's rays at a central focal point. Through trial and error, Adams determined that this technology was an effective means of heating and boiling water, though meat cooked in this way had a disagreeable flavor and smell, a problem Adams solved by using ultraviolet filters to block out the offending part of the solar spectrum. In addition, Adams created a solar-powered boiler for running a steam engine using a bank of flat mirrors to concentrate solar energy on a central vessel. Another of Adams's inventions concentrated solar energy to distill sea water into freshwater for the British enclave at Aden on the Red Sea. His groundbreaking efforts provided tangible evidence of potential solar-energy applications to many across the British empire.

At around the same time period, Augustin Mouchot, a French schoolteacher and inventor, attempted to develop solar-energy generators. He worked on the development of a solar cooker similar to that of Adams but also made and demonstrated a host of practical solar-energy devices, culminating in a large exhibition in 1880 in Paris, where he presented a variety of devices, including his large "Sun Engine," which operated a printing press on which he printed an edition of his newsletter *Journal Soleil*. He also displayed a variety of solar cookers and, much to the amazement and delight of the crowd, a solar machine used to create ice. Mouchot's state-of-the-art devices captured popular imagination, but it was his efforts to develop energy storage that defined his place in history. Mouchot was the first inventor who attempted to use the power of solar radiation to decompose water into its base elements of hydrogen and oxygen and then recombine them to generate electricity, much like the fuel-cell technology of today. While he felt that this was a potentially revolutionary solution to the problem of energy storage and the intermittency of the sun's availability, the increasing scale and cost advantage of coal-based energy prevented Mouchot from developing an economically viable solar-hydrogen energy solution.

John Ericsson, the third solar inventor, was a Swede who moved to the United States in 1839, earning fame and fortune as the designer of the iron-clad Union ship the *Monitor*, which is credited with altering the course of the U.S. Civil War. After the war, Ericsson turned his attention to solar energy and began extensive experiments in the 1870s that continued until his death in 1889. He developed a solar-power engine using hot air to run pistons, an efficient design that limited energy waste. Unfortunately, since the solar-powered machine could perform adequately only in direct summer sun, it remained only marginally valuable unless the problem of energy storage could be overcome. Ericsson experimented with both compressed air and electric batteries in his pursuit of an effective and efficient energy-storage technology, but he was never able to find an economical solution to the problem of storing energy to work these engines in times of low or no sunlight. He did, however, make many significant contributions to the design of solar-energy collectors, including parabolic troughs, many of which are the basis of modern solar thermal designs.

Solar Goes Commercial

Solar energy's first chance for wider commercialization occurred between 1900 and 1915. Using the accumulated knowledge of earlier solar inventors, Aubrey Eneas, a solar entrepreneur in the American Southwest, developed larger solar collectors to power steam engines and pumps for agricultural irrigation water. The dry deserts of this region made an ideal test market with plenty of sun and few alternative fuel sources. Residents of the region were also already comfortable with the practical use of solar energy, with some 30 percent of the homes in Pasadena, California, using solar water heaters to generate hot water by around 1900. Eneas, in an attempt to create commercial-scale solar-energy applications, designed and built a large truncated cone collector to superheat water that powered a steam engine for running irrigation pumps. He managed to convince several customers of the merits of his design and installed a handful of these systems across California and Arizona. Ultimately, however, the construction methods of these large cone collectors proved susceptible to strong and unpredictable weather of the desert region including dust devils, wind storms, and hail. Eneas eventually abandoned his attempts to commercialize concentrating solar energy, believing that they would never be economically viable.

The last major pioneer to attempt to commercialize the power of the sun early in the twentieth century was another American named Frank Shuman. At the time that he began his foray into solar energy, Shuman was already famous for many useful inventions, including shatter-proof wire glass and the safety glass found in today's automobiles. His resulting financial success afforded him the time and money to pursue commercializing solar energy.

Befitting someone who had made his fortune in the glass business, Shuman's first attempt to capture solar energy was through the use of a hotbox, a device similar to a greenhouse in which sunlight enters the box through a glass pane, trapping the heat and causing a dramatic temperature increase inside. By stringing a series of these hotboxes together, Shuman created enough useful heat to run a small boiler for a steam engine. While this design was simpler and less expensive than using mirrors or lenses to concentrate the sun's rays, Shuman ultimately decided to add a row of mirrors to each side to increase the heat generation. After perfecting his techniques in Philadelphia, he found investors and signed a customer in the British protectorate of Egypt for whom he delivered solar pumps for irrigation. Against his better judgment, Shuman was coerced by his partners to abandon the hotbox technology in favor of parabolic troughs similar to John Ericsson's original design. In the end, Shuman's solar-energy irrigation solution was substantially cheaper to operate than any alternative solution but was twice as expensive to set up as the next best solution, the coal-fired engine. Even though the up-front cost differential could be repaid from reduced operating costs in the first two years, Shuman's parabolic troughs remained a marginal choice for investors because there was little long-term financing available at the time. The Great War of 1914 to 1918 forced Shuman to shut down his Egyptian venture, and he returned to America. For over half a century afterward, easy access to coal in the industrial economies of Europe and America along with the increasing availability of petroleum as a cheap power source for motive applications eclipsed nearly every attempt to commercialize the power of the sun.

Solar's Second Chance

Solar energy's next chance for widespread market acceptance came with the development of the first official photovoltaic cell in 1954 by three researchers at Bell Labs. Though the basic photovoltaic effect had been understood in the nineteenth century by Edmund Becquerel, the Bell Lab researchers' work in semiconductors and some fortunate laboratory accidents led to their development of the first working PV module. These modules quickly progressed by the end of the 1950s, stimulating a tremendous amount of excitement among research labs and the governments that funded them about the future of this technology. Continued work and development in the 1960s, much of which was performed to support critical power systems for applications in satellites and space-based vehicles, led to increasing optimism by the international scientific community about the future of PV technologies.

The oil shocks of the 1970s provided a further boost for solar technologies, both PV and thermal. Supply fears and the rapid rise in fuel prices they caused led to strong government promotion of a variety of alternative-energy technologies, including solar energy. In the 1970s, the U.S. government established the Solar Energies Research Institute to help develop solar-energy technologies. President Carter's administration helped bolster the industry by approving a $3 billion program for the development of solar-energy technology and installing a showcase solar water heater at the White House. Excited about the potential for clean, independent energy sources, many American citizens and entrepreneurs began to invest substantial time and money to promote these changes by developing businesses and installing solar water heaters and PV panels. This momentum was cut short with the changeover to Ronald Reagan's administration beginning in 1981. By 1986, the Reagan government had dramatically reduced funding for solar-energy research programs, reduced the federal tax credits for solar water heating, and removed Carter's showcase system from the White House roof. The message to the renewable-and solar-energy communities was clear, and the industry came to a virtual halt not only in America but throughout the world. America represented almost 80 percent of the world market for solar energy at that time, and when research funds in America dried up, the remaining governments of the industrialized world followed in step. A period of decreasing oil prices in the 1980s and early 1990s further diminished the perceived urgency to cultivate renewable-fuel options, leading to another two decades of inertia for the global solar-energy industry. Disparate inventors, research universities, and state energy agencies continued funding research and development in photovoltaic technologies, even as giant energy companies such as Mobil, Shell, and BP purchased the assets and patents of many of the original solar-energy technology companies.

Solar Comes of Age

Over the last ten years and underneath the radar within the broader conventional and renewable-energy industry, solar energy has emerged to make a third attempt at mass commercialization. This time, the opportunity is due almost exclusively to the efforts and programs of the national governments of Japan and Germany, which have led the way in promoting these industries. Although these two countries do not have a naturally high amount of sunlight, their lack of alternative fuel sources has created a dependence on expensive external sources of energy and therefore motivated them to develop less expensive, local, and renewable-energy alternatives. Japan's sunshine program and Germany's 100,000 solar roofs program, which have used various types of subsidies to stimulate robust domestic solar-energy industries, now account for 69 percent of the world market for PV. In these markets and geographies, the current renaissance of interest in solar energy is finding its opportunity, and the cost reductions these markets have experienced have stimulated surprisingly powerful momentum and growth in the last decade. With the global PV market growing from 85 MW in 1995 to over 1.1 GW in 2005 (a 29 percent annual growth rate), the cost of producing PV systems has dropped from $11 per watt to as low as $5 per watt over the same period and continues to fall by 5 to 6 percent per annum.

The Dawn of a Solar Industry

Photovoltaics can be applied cost-effectively at any scale, from handheld gadgets to utility-scale generation. Each application has a unique character, but there are natural groupings. The first and most familiar type is using solar cells without any kind of battery solution and usually at small scale. Applications of this type include solar calculators, irrigation pumps, and freshwater distillers that operate only while the sun is shining. More complex systems include those that store excess photovoltaic electricity in batteries for use at night. For example, homeowners can buy photovoltaic yard lights that charge a battery during the day and glow after sundown until the battery runs out. Area lighting, highway construction signage, and roadside emergency phones are other examples. Together, these two types of PV application—small-scale, single-function systems with or without battery storage—comprise about 15 percent of the PV capacity installed annually worldwide.

Another primary market for solar cells is providing power to homes that do not have access to electricity in other forms. The economics of these off-grid applications are compelling and are discussed more fully in the following chapters but solar photovoltaics are often a cost-effective solution in the absence of alternative off-grid energy solutions. Unlike gasoline-powered generators, which are the traditional alternative for off-grid power, solar cells require no fuel deliveries, operate silently, and never refuse to start. The system equipment includes the solar modules themselves as well as some form of mounting that can be either stationary or can track the sun on one or two axes to maximize the sun collected. Solar cells also typically require a lead-acid battery storage solution to provide power in hours when the sun is unavailable. A specific form of lead-acid battery called a *deep-cycle battery* usually works best for these applications because they last longer than traditional twelve-volt car batteries. These off-grid solar applications can be found in industrialized countries where there homes and businesses are located outside the range of the existing grid but more commonly are found in the developing world where access to reliable and consistent grid power is not available. Combined, these off-grid applications represent some 18 percent of the total PV installed worldwide on an annual basis.

The bulk of the remaining solar market represents the growth of *grid-tied systems* for residential and commercial customers. These systems use sets of PV panels ranging from a few hundred watts to a few megawatts of peak capacity and are located on rooftops of homes and buildings. During the day, as the energy collected by these PV systems exceeds the energy needed for the local homes or business, the system feeds the excess back into the utility grid at effectively the same rate that the customer would pay for that electricity, a concept commonly referred to in the United States as *net metering*. This system obviates the need for a localized battery solution for these installations because the utility grid is used instead as the equivalent of a huge storage battery. Once a building owner meets the connection requirements of the local utility, whereby the utility confirms the safety of the equipment and the correct connection parameters, a customer's meter can flow in both directions—positive at night when the home or business is using more power than it is generating and negative in the sunny part of the day when the opposite is true. This system does not require a battery but needs a device called a *grid-tied inverter*. These inverters collect the direct current (DC) of solar PV and convert it into the alternating-current (AC) power of grid electricity. The remaining solar modules and associated mounting and wiring are essentially the same as any other off-grid application.

This grid-connected market is the engine driving the solar industry's commercial-scale growth and transformation. While total solar applications have collectively grown 29 percent per annum over the decade through 2005, the grid-connected segment has experienced a growth rate of over 50 percent per annum over that same period. Figure 7.2 shows how much the growth of this segment has contributed to the total growth of solar photovoltaics. This sustained growth in the use of grid-tied systems will continue to propel the PV industry in the coming decades as grid-tied PV economics and the technology's innate reliability increasingly provide incentives for customers to adopt such systems.

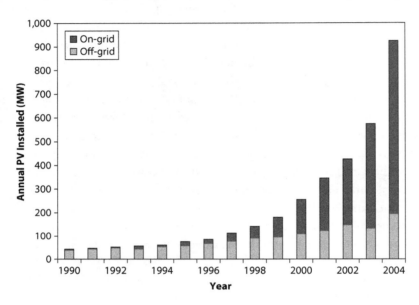

Figure 7.2. Growth in annual installations of grid-connected and off-grid photovoltaic cells, 1990 to 2004 (MW of peak capacity).

Source: Solarbuzz (2005).

Finally, an application known as *centralized systems,* while small today at only 2 percent of the total PV market, should become increasingly relevant in the future as more large energy users and utilities adopt PV. Using this application, industrial customers or utilities can take advantage of good solar characteristics in a given location to generate utility-scale power in large fields of ground-mounted solar arrays. To date, this application for PV is not widely cost-effective in the face of industrial-scale alternatives, though the next chapter shows where it is becoming so. Centralized systems have the potential to contribute significant amounts of electricity, as one-third of the earth's surface is covered by sun-rich deserts, creating a potentially vast amount of energy resource. Some 4 percent of which (just over 1 percent or the total land area of the world) would meet the entire world's energy needs from these sources even at today's efficiency levels.

Geographic Markets

The driver of the solar electricity industry has been grid-tied applications but primarily from only a few geographic markets. As recently as 1998, the United States was the world leader in PV installations, but concerted government programs in both Germany and Japan have enabled these countries to surpass the United States in terms of PV capacity installed. Figure 7.3 shows just how quickly the markets have taken off since 2000 and projected growth in 2005 and 2006. However, U.S. markets are beginning to develop further as many state-level governments institute programs to develop renewable-energy industries or markets.

The Japanese photovoltaic market is currently the largest globally, with over 1 GW of peak capacity installed, of which 277 MW were installed in 2004 alone, 36 percent higher than the prior year. Over the last decade, this rapid growth resulted primarily from the Japanese government's residential PV-dissemination program, which created targeted incentives for installation of solar photovoltaic modules on residential rooftops. Chapter 9 details this successful government program to foster technology growth. The most important fact about this program is that it has helped lower the cost of installation of grid-tied residential PV in Japan by half since its implementation in 1996. As of 2004, although the subsidy was reduced to around 7 percent of the system price from an initial level of 50 percent, demand continued to soar, signaling a belief by homeowners that the solution is now nearly at parity with retail electricity rates of twenty-one cents per kWh. The government is forecasting continued robust growth even without the aid of ongoing subsidy programs and hopes to be adding well over a GW per year by the end of the decade.

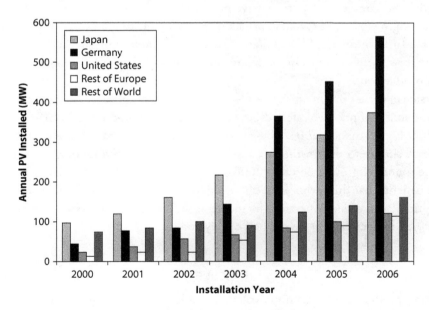

Figure 7.3. Growth in annual installations of photovoltaic cells in the major geographic markets, 2000 to 2004 with estimates for 2005 and 2006.

Source: Solarbuzz (2005).

The European market for photovoltaics, while significantly smaller than its wind-power market, has also seen dramatic growth over the last decade with 50 percent annual increases in domestic production in both 2003 and 2004. Performance-based incentives to pay system owners a premium for generating clean PV electricity instituted in Germany in 1999 and renewed in 2004 have fostered PV growth to 366 MW in 2004, an amazing growth of 150 percent over the prior year, primarily in the grid-tied market. Today, Germany represents 80 percent of the solar photovoltaics installed in Europe, though this percentage will decrease in the next few years as other countries such as Spain and Portugal begin to increase their use of PV. However, supply is not projected to keep up with demand because many manufacturers have already committed over a year's worth of production to customers, with Germany currently importing some half of its PV modules installed. Market expectations across Europe are for continued robust growth for the rest of the decade with some forecasts as high as 40 percent annual growth over this period, which would result in over 1.3 GW installed per year in Germany by 2010, out of a total of 1.7 GW for all of Europe.

The United States' historical leadership in PV technologies has been usurped by Japan and Germany, but the United States still represents the third-largest world market. In 2004, U.S. users installed 84 MW of peak generating capacity. Its growth rate has not been as spectacular as those of Japan and Germany, but it has still averaged 25 percent annually over the last decade. With the recent growth in many states' subsidy payments for installing PV systems, reimbursing up to half or more of the system cost, grid-tied residential and commercial systems have grown much faster. California has experienced the greatest growth in photovoltaic installations, today making up over 80 percent of the U.S. total. In addition, the United States exports a net amount of 58 MW of solar cells to make up for supply shortfalls in Germany and the rest of the world, totaling 139 MW of PV cell production.

Outside of these three main markets, which together comprise 89 percent of the world's total production of photovoltaic cells, the rest of the world produces 124 MW of product, 38 percent higher than in 2003. These markets have experienced volume growth over the last decade similar to that in the Japanese and German grid-tied markets, averaging over 50 percent annual growth between 2000 and 2004, albeit from a low base. Despite the relative benefits that this technology can bring to the developing world, there are limits to its adoption, such as lack of system financing as well as sales and service channels, which are explored in detail in the second half of this book.

Producers

In general, producers of these photovoltaic modules are located in the major markets in which they sell their products. Table 7.1 shows the top ten producers worldwide and their annual production for the last four years. Given the relative size of the Japanese market, it is no surprise that four of the ten largest PV producers are Japanese firms and represent some of Japan's largest and most powerful industrial firms such as Sharp, Kyocera, Mitsubishi, and Sanyo. These companies are dominant electronics firms with market penetration and expertise in cost reduction via large-scale production development, and they have been primarily responsible for the reductions in system prices for these markets since 1994. In Europe, PV production is led by domestic producers such as Germany's Q-cells and Schott Solar and Isofoton of Spain and is also bolstered by BP Solar (British) and Shell Solar (Dutch). America's PV production is dominated by these two oil-company owned solar divisions, and General Electric entered the fray in 2004 by purchasing America's largest independent PV producer, Astropower, out of bankruptcy. Each of these major producers is looking at substantial additions to capacity in the next twelve months, with some planning to double their line capacity. Many PV producers outside of these major markets are also planning production increases, including Motech of Taiwan in the top ten global producers. Countries like China and India are leading the way in setting ambitious growth targets for solar photovoltaic installations.

Table 7.1.

The top ten global solar-cell producers in 2004 and their past production from 2001. (Production in peak MW.)

	2001	2002	2003	2004
Sharp (Japan)	75	123	198	324
Kyocera (Japan)	54	60	72	105
BP Solar (United States)	54.2	73.8	70.2	85
Mitsubishi (Japan)	14	24	40	75
Q-cells (Germany)	N.A.	N.A.	28	75
Shell Solar (Germany)	39	57.5	73	72
Sanyo (Japan)	19	35	35	65
Schott Solar (Germany)	23	29.5	42	63
Isofoton (Spain)	18	27.4	35.2	53.3
Motech (Taiwan)	N.A.	N.A.	N.A.	35
Total	**296.2**	**430.2**	**593.4**	**952.3**

Note: N.A. = Data are not available. Source: Maycock (2005).

Technologies

There is a constant and growing effort to find the most inexpensive and reliable technologies for producing these solar cells to continue to develop the photovoltaic market worldwide. While the dominant technology today is silicon based, many advanced silicon and nonsilicon variants are being explored to meet these needs.

Monocrystalline silicon cells are roughly similar to those PV cells originally created in 1954 at Bell Labs. Today, they are generally formed from ingots of pure silicon that are sliced into thin wafers and then chemically treated and etched to operate as solar cells in a process similar to that used to produce chips in the microprocessor industry. Their advantage is that they possess the highest levels of conversion efficiency for turning sunlight into energy. Their disadvantage is that they are the most costly to produce of all possible choices because they are fabricated using energy- and capital-intensive methods derived from similar processes in their microprocessor counterparts and often at a quality standard much higher than necessary for use in current photovoltaic applications. The alternative silicon-based cells are *polycrystalline* and are produced using slightly different manufacturing methods, creating a less efficient but also less costly end product. The major PV producers use variants of these technologies, with the American producers favoring the monocrystalline and the Japanese favoring polycrystalline. Combined, these basic silicon solutions make up 85 percent of the solar cells produced today.

The remaining market for PV is comprised of a second-generation technology called *thin-film PV* that eliminates the need to have freely supported solar cells in favor of depositing the photovoltaic layers directly on a supporting substrate. This allows for a further reduction in cost and a more creative configuration of cells such as embedding the cells within building materials, rolled sheets, and roof tiles. The modules developed using this process are usually of lower efficiency and lower cost but have continued to improve in both respects as they are commercialized. With so many interesting applications being pursued, including those that eliminate the use of silicon entirely, most of the major industry research associations expect thin-film PV to be a major contributor to long-term growth of the industry as cost effectiveness improves. Ultimately, the progress of thin-film PV will be a function of how quickly technologies in silicon-based crystalline solutions reduce manufacturing costs and the third-generation technologies discussed below are deployed.

Third-generation (or 3G) technologies—so termed by Martin Green of University of New South Wales, who is one of the world's leading researchers on solar cells—are poised to make a much larger long-term impact in solar cells but probably not until the second decade of this century when these technologies will finally hit the mass market. The technologies being explored include photosynthetic chemical processes similar to those that occur in plants and trees. There are also a number of exciting technologies for producing spherical solar cells—that is, small bubbles that reduce the amount of silicon needed, resulting in thin, flexible panels that can be used in, for example, microelectronics applications. In addition, about half a dozen companies are attempting to figure out ways to use printing technology to print solar cells directly onto a substrate, a development that could dramatically reduce the cost of manufacturing. None of these technologies is yet widely available, but they have the potential of bringing the cost of solar energy well below today's price once the technical hurdles of mass producing them are overcome.

PV Supply Chain

The modern PV industry is a $10 billion dollar industry worldwide and comprises many types of manufacturers, installers, and service providers. Figure 7.4 shows the PV production chain for the support services required to develop and deploy PV systems.

The solar-component manufacturers comprise the cell producers discussed above as well as the raw silicon and ingot producers, production-line equipment manufacturers, and providers of the other metals and raw materials needed to make PV cells. Product manufacturers (often but not always the same companies that make the PV cells) combine the cells with glass, frames, and electronic busing to create finished modules. These are then aggregated by distributors or installers with the balance of system components, including inverters, batteries when necessary, and wiring to be installed in end-user applications.

Supporting this process are the research and development by companies and research institutes; the banks and capital markets that provide the capital to create the manufacturing and finance the end-user installation of the systems; and the designers, architects, and engineers who enable integration of these systems in homes, buildings, and new centralized plants. The many phases of production require various technical skills and combine to make PV production a labor-intensive process among electricity technologies, even at an equivalent level of electricity production. As the industry grows, new jobs at all stages of the PV supply chain will need to be filled in proportion to the overall industry growth rate, with many of these at the local design and installation levels.

Figure 7.4. Photovoltaic industry production chain and various support services.

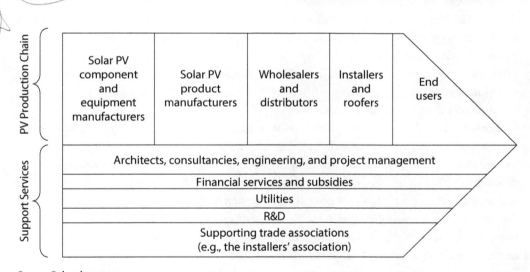

Source: Solarplaza.com.

Trends and Projections in the Cost of PV

Typically, emerging technologies in any industry start off as expensive, complex, and inefficient to produce. The most promising ones—those with the potential to fill a previously unmet need or to provide an existing product or service more cheaply or effectively—begin to generate growing interest, and more versions and refinements are introduced. In this research and development phase, industry works out methods by which the technology can be produced and begins to envision its final form and application. As users are convinced of the technology's viability and long-term economic value, industry producers begin to produce limited quantities of standardized products to meet specific market applications. Mass production usually begins in limited lot sizes, but as market demand increases, producers usually find faster, better, and cheaper ways to produce their products. Sometimes just making more units can reduce the average cost of each unit because of bulk pricing on raw materials and a fuller utilization of people and manufacturing facilities, known as *economies of scale*. Institutional experience becomes embedded, too, as more efficient production methods are systematically discovered and reduce manufacturing costs.

The current cost of any technology, then, is partly a function of the history and scale of the production of that technology. A useful depiction of this process is the *experience curve*. This curve, often represented graphically, describes the relationship of the technology's current cost to the cumulative quantity produced over its history. In the case of an electricity-generating technology such as PV, cumulative quantity produced is measured in peak capacity, usually watts, produced. Therefore, experience shows up as decreasing cost as cumulative production increases.

In the case of photovoltaic modules, the cost to produce them in the late 1970s was around $25 per watt but has since dropped to less than $3.50 per kW, an 86 percent reduction, while the cumulated production has grown a thousandfold, or roughly ten doublings over that time. Calculating experience curve for grid-tied solar photovoltaic modules over the period from 1980 to 2003 implies a learning rate of around 18 percent, meaning that for every doubling in the installed volume of the technology over time the per-unit cost drops another 18 percent.

Forecasting future learning rates is as treacherous as forecasting anything else, but a few general parameters are helpful. Starting with a historical rate is fine, but that rate may change in the future, either increasing or decreasing. Typically, a technology experiences a slowing of its learning rate as it grows and becomes established. Because the most obvious and easy gains in cost efficiency are usually the first to be captured, driving costs out of the production process gets progressively harder as it matures. Scale of production helps further, but these gains also level off as an industry reaches critical mass. As a result, most technologies have a natural limit beyond which cost improvements slow to nearly nothing. Each industry is different, but at some point, usually well before market saturation, a technology stops gaining cost advantages from increases in scale. That being said, occasional breakthroughs in technology can occur that can greatly accelerate the learning rate for a period of time.

Once an experience curve is projected, the expected market growth of the technology in question can be layered in, and an expected annual cost reduction for the technology can be calculated. In the photovoltaic case, an 18 percent learning rate and a 30 percent annual market growth rate would translate into a cost decline of 5 percent to 6 percent a year—numbers that correspond with recent historical experience for both growth and cost reduction in the global PV module market.

Using assumptions for experience rates and total market growth, it is possible to build a forecast model of the expected future cost of PV electricity. Figure 7.5 shows a forecast model through 2040 for three locations with different levels of available sun and the expected worldwide growth of the underlying market. As the next chapters explain, potential residential, commercial, and utility adopters of PV face different economic choices and different methods of financing the cost of their system. The underlying market growth rates span all of these sectors as it is cumulative global growth that determines expected system prices. The projected PV electricity costs in this forecast were developed assuming the economics faced by residential customers in the United States and the type of supplemental mortgage financing that they would most likely use when installing PV on their home.

Figure 7.5. Forecast of the global annual market for PV, expected cost per watt of installed PV systems, and resulting cost per kilowatt hour of PV electricity (unsubsidized) for three sun scenarios through 2040.

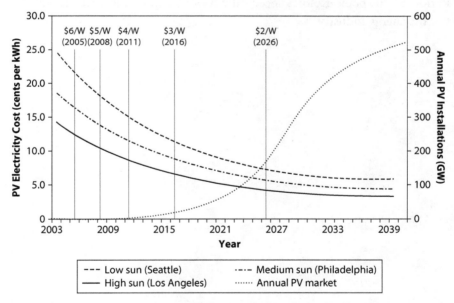

The projection model in Figure 7.6 illustrates that, even under conservative assumptions, the evolving economics of PV will make these systems progressively cheaper over time, which will make PV electricity increasingly cost-effective in locations with a wide disparity of solar resources and allow PV to grow in dominance within the global energy and electricity mix. The cost levels of PV systems (in dollars per watt) are used for more specific forecasting later in the book and examined for reasonableness and limitations.

When asked about their views on various energy technologies, over 90 percent of people believe that solar energy is a desirable solution, making it the most popular of all energy technologies. Although most people have never seen it deployed successfully, solar energy is popular because it is conceptually simple—almost deceptively so. Solar energy is safe and clean and has no moving parts, making it reliable and long-lived, and the sun as a source of energy cannot be bought, sold, or metered. As a result, solar energy offers nations and individuals unprecedented and unlimited control over their own vital source of energy. For these potential benefits alone, many people genuinely want to see solar energy become a widely deployed energy alternative.

Despite wide pessimism about the cost and capacity of PV technology, a dramatic growth in the rise of solar photovoltaic electricity over the last ten years has occurred primarily as a result of Japanese and German government policies that promote the development of grid-tied PV systems. As the next two chapters show, the increasing scale of global PV deployment has decreased the cost of grid-tied systems to the point where PV is cost effective in many large and growing industrial nations. As PV continues to be adopted in these markets and global production volume increases, predictable cost reductions will make photovoltaics more affordable for both industrial and developing-country consumers worldwide—leading to greater sales, larger volumes, and further cost reductions. The next section of this book explores specifically how and when these trends will shape the future of the energy industry.

Discussion Questions

1. What are different types of solar energy? Explain citing specific examples.

2. Explain the stratification of the PV market.

3. Explain the critical role played by photovoltaics in the solar energy market.

4. Describe in detail the various components of the PV supply chain.

In-Class Assignment: Objective Questions

1. Which of the following is a disadvantage of solar power?
    a. It is a clean source of energy.
    b. It is a safe source of energy.
    c. Efficiency of solar cells is about 10%-20%.
2. Which of the following scientists invented the first solar-powered steam engine?
    a. James Watt
    b. Thomas Edison
    c. Marconi
    d. Auguste Mouchout
    e. Enzo Ferrari
3. The manufacture of solar cells produces highly toxic waste metals and solvents that require special technology for disposal.
    a. True
    b. False

Wind Power

Andrea Larson and Stephen Keach

First, there is the power of the Wind, constantly exerted over the globe. ... Here is an almost incalculable power at our disposal, yet how trifling the use we make of it! It only serves to turn a few mills, blow a few vessels across the ocean, and a few trivial ends besides. What a poor compliment do we pay to our indefatigable and energetic servant!

—Henry David Thoreau[1]

Vestas Wind Systems A/S

---

1 Henry *David* Thoreau (1817–1862), U.S. philosopher, author, naturalist. "Paradise (To Be) Regained" (1843) in *The Writings of Henry David Thoreau*, vol. 4 (New York: Houghton Mifflin, 1906), 286–287.

Andrea Larson, Stephen Keach, and Giles Jackson, "Wind Power (Technical Note)," pp. 1-22. Copyright © 2002 by Darden Business Publishing. Reprinted with permission.

Introduction

Wind power has a simple elegance. Perhaps Thoreau, who loved simplicity and nature, would have been gratified by the growing use of modern wind turbines. Indeed, there is little reason not to be drawn to the possibility of an endless source of inexpensive, nonpolluting power.

The world market for power generation is changing dramatically, due to utility deregulation and privatization, availability of alternate technologies, environmental pressures, and growing demand for electricity. As a consequence, technological advances, incentives,[2] and environmental concerns are making renewable energy sources[3] like wind, biomass, solar, and geo-thermal power increasingly attractive throughout the world. While each of these options may have economic advantages in particular applications, these technologies have not generally reduced costs to the point necessary to transcend their current niche markets. This is not the case for wind power. Pollution-free and economically competitive with fossil fuel sources, wind power has become the fastest-growing source of new electricity generation worldwide. A combination of increased demand, steadily decreasing cost, and concern for the environment has expanded the market for wind power. Wind, in combination with other power sources that do not require fossil fuels (oil, coal, and natural gas), holds the potential for working in tandem with improved energy efficiency to pave the way for a cleaner energy future that depends less on imported fuels.

New wind turbines are quieter, more efficient, and less expensive (per unit of energy production capacity) than previous generations, and they can be built and installed more quickly than fossil fuel plants. As the technology has become increasingly sophisticated, companies have improved materials performance, variable speed turbines, torque control capability, blade pitch changes, and the system's ability to withstand hurricane-strength winds. These improvements have dramatically improved power output.

Today, wind turbines as small as 250 Watts are used for home or business electricity generation, and to pump water or charge batteries. Utility-scale wind turbines generate

---

2   See *Incentives* below. Renewable energy enthusiasts are quick to point out that fossil fuel power is also heavily subsidized, including less direct subsidies, like the benefits of mining laws and defense department protection for its supply chain (Middle East oil). The industry also fails to ameliorate the high environmental and public health costs imposed on current and future generations.

3   Non-fossil fuel energy sources, with the exception of nuclear power are known as renewable energy resources. Nuclear power is not usually thought of as a "renewable" energy source, presumably because there is no sustainable program for avoiding or adequately dealing with the radioactive wastes that it generates.

from 0.5 to 2.5 megawatts (MW)[4] of electricity apiece with rotors 50 to 80 meters in diameter and towers 80 meters high.[5] They are placed in arrays called wind farms that feed electricity into a power grid.

Wind power generates electricity for:

- distribution grids (wind farms)
- supplemental electricity supply on-site—reducing dependence on the grid
- power in remote locations
- battery recharging
- pumping water

Factors driving interest and investment in wind power include:

- deregulation of the electric power industry in the United States
- increased demand for electricity and insufficient supply worldwide
- high cost and other drawbacks of traditional fossil fuel electric generation plants
- increased interest in less polluting energy sources
- commitment to carbon dioxide and other greenhouse gas emission reductions (particularly in Europe)
- preference for national sources vs. dependence on imported fuels
- growing interest in distributed power vs. centralized, large-scale power plants
- improved economics of wind power vs. alternatives
- improved wind turbine design, materials, and operation (quieter, more efficient, easier to install and maintain)
- the movement toward customer choice and "green product" offerings by utilities
- the appeal of zero-emission technologies in light of regulations

The History of Wind Power

Variations in the earth's rotation, atmospheric temperature, and topography produce winds. Plant cover, water bodies, and other qualities of the earth's surface also influence the wind flow patterns. The first windmills pumped water in Asia thousands of years ago. From the earliest sailboats to windmills for pumping water and grinding grain in Asia and later in Europe, wind power has been harnessed for human needs. Small wind turbines have generated electricity in the rural United States since 1900 and in

---

4   A megawatt (MW) is one million watts or one thousand kilowatts (kW). A typical home requires between 1 and 10 kilowatts of power capacity.

5   The Danish Wind Turbine Manufacturers Association (F.A.Q.s) www.windpower.org.

Europe since 1910, but low-cost grid-supplied electricity, generated from fossil fuels, edged out most wind power by mid-century. Wind turbine research and use in remote settings continued, but it took high oil prices in the 1970s to spark a renewed interest in wind power. Since that time, wind turbines have become increasingly efficient and cost-effective.

Wind Energy's Role in the Larger Energy Picture

In 2001, the vast majority of primary energy worldwide, about 85 percent, was generated from burning fossil fuels such as coal, oil, and natural gas. Most of the remaining 15 percent came from nuclear and hydroelectric power. In this energy picture, electricity accounted for about 50 percent of world energy production, with demand growing faster for electricity than for any other type of energy (see Appendix 6).

> Over the outlook period [2000 through 2020], nearly 3,000 [Gigawatts] GW of new generating capacity are projected to be installed around the world. More than half of the new capacity will be in developing countries, much of it in developing Asia. The total cost of these plants is estimated at nearly $3 trillion at today's prices, not including the cost of expanding the transmission and distribution network. The developing countries will need to invest around $1.7 trillion in new generation plant.
> —International Energy Agency.[6]

In Organization for Economic Cooperation and Development (OECD) countries, 60 percent of electricity generation came from combustible fuels, 24 percent from nuclear power, and 16 percent from hydroelectric and other sources.[7] Combined-cycle natural gas power plants[8] emerged in the 1990s as the electrical generation technology of choice in the United States. This reflects the broader shift worldwide to gas as the preferred fossil fuel.

---

6 World Energy Outlook 2000, International Energy Agency (February, 2001), p. 25.

7 The Organization for Economic Cooperation and Development (OECD) includes many of the world's developed nations, including the United States, Japan, and the E.U. These percentages are based on data from the International Energy Agency, Monthly Electric Survey – June 2001.

8 Combined-cycle plants run a steam turbine by heating water with the exhaust gasses from the gas turbine. Both cycles generate electricity. Many people look at these plants as a transition technology providing an economical way to produce electricity with lower emissions and higher efficiency than coal-fired plants, while renewable technologies struggle to become competitive.

While wind power remains a small portion of world energy capacity, it is the fastest-growing segment and has the potential to claim a much larger share of the market. Economic incentives and technological breakthroughs propelled wind power to an average annual growth of 24 percent during the 1990s.[9] By the end of 2000, wind energy was responsible for 17,300 MW of electricity and about $4 billion in sales. In 2001, Denmark already produced 12–13 percent of its electricity from wind (with a goal of 50 percent by 2030) and Germany more than 2 percent.[10] More than 5000 MW of new installations were projected for 2001.[11] In 2001, Merrill Lynch predicted wind power would achieve a 20 percent compound annual growth rate to 2010 and would reach nearly 4 percent share of electricity generation worldwide by 2020.[12]

Wind Power Generation Capacity

In 2001, Germany was the world's largest wind energy generator (6,113 MW), followed by the United States (2,554 MW), Denmark (2,300 MW), and Spain (2,235 MW). Several other European countries as well as Canada, Japan, and Australia have small but significant wind power installations. The European Wind Energy Association (EWEA) set a goal of 60,000 MW of installed wind energy generating capacity for 2010.[13] The European Union (EU) has targeted renewables to supply 22 percent of electricity (and 12 percent of total energy) by 2010.

Reported wind power capacity in developing countries:

- India 1,176 MW
- China 265 MW
- Morocco 50 MW (with 200 additional MW proposed or under development)

---

9   Seth Dunn, "Micropower: The Next Electrical Era," Worldwatch Paper 151, July 2000 (estimate based on BTM Consult, International Wind Energy Development: World Market Update (Copenhagen: various years).

10   Flavin, Christopher, "Rich Planet Poor Planet," *State of the World 2001*, World Watch Institute, p. 16.

11   2000 Global Wind Energy Market Report, Wind Energy and Energy Policy, www.awea.org.

12   Merrill Lynch, Wind Power Update, 12 July 2001.

13   Global Wind Energy Market Report, Wind Energy and Energy Policy, American Wind Energy Association (AWEA), May 2001.

- Egypt 30 MW
- Turkey has contracts for several hundred megawatts.
- Argentina has contracts and government approval for about 3,000 MW.
- Costa Rica has a 20 MW facility (see quotation below).

The utility-scale wind farms in Guanacaste, Costa Rica, are operating with … one of the best performance records in the hemisphere. At 5.5–7 cents/kWh in the region, wind energy is more than competitive with petroleum-fired plants, which get paid 13 cents/kWh in some cases in those countries. Small wind systems are economical in many locations and help provide power autonomy and security in a region that lacks rural electrification and is often hit by natural disasters. In spite of this evidence, investments in renewable energy are only a trickle compared to those in fossil fuels.[14]

Figure 8.1.

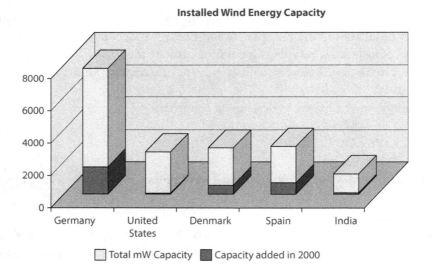

According to the American Wind Energy Association's 2000 Global Wind Energy Market Report,[15] 3,800 MW of utility-scale wind energy was added in 2000, mostly in Europe. Wind energy sales in 2000 were roughly $4 billion. This increase brought global wind energy generating capacity to 17,300 MW. The AWEA predicted new installations

---

14   Ibid.

15   Ibid.

of over 5,000 MW in 2001 and worldwide installed capacity in excess of 20,000 MW by the end of 2001. Continued stable and rapid growth was anticipated for Europe, and the United States was entering a period of expansion.

Environmental Benefits

Apart from material and energy waste from production, transportation, and installation, wind turbines are a nonpolluting source of electricity. During operation, wind turbines are emission-free and eliminate the need for fossil fuel electricity generation and related pollution.

> Unlike conventional power plants, wind plants emit no air pollutants or greenhouse gases. In 1990, California's wind power plants offset the emission of more than 2.5 billion pounds of carbon dioxide and 15 million pounds of other pollutants that otherwise would have been produced. It would take a forest of 90 million to 175 million trees to provide the same air quality.[16]

Once the technology became competitive with conventional energy sources on a cost basis, wind power's environmental attributes began to tip the scale, and new projects increased at a rapid pace. With the Kyoto Protocol on Global Climate Change in 1997, concern over potentially expensive compliance with regulations supporting the treaty caused many governments and international companies to look for ways to lower emissions. Wind power reduced the threat of climate change by eliminating the need for fossil fuel plants and consequent greenhouse gas emissions. A one MW turbine in a typical location avoids 2,000 tons of carbon dioxide that would have been produced by one megawatt of output from a coal-fired plant.[17] In addition to their contribution to global climate change, fossil fuel power plants cause air pollution linked to acid rain and several other ecological and public health problems. Their cooling-systems intakes kill millions of fish and the warm water discharges damage ecosystems. When health and environmental impacts of other power generation methods are built into the cost assessment, wind emerges as one of the least expensive energy sources.

Other benefits to wind energy exist. In some rural areas, particularly in developing countries, the cost and transportation of diesel or other fossil fuels for combustion-based power generation may be more expensive than wind power. In developed countries as well, fossil fuel supplies and prices vary with political and market forces. The growing demand

---

16 U.S. Department of Energy: Wind Energy Program (Quick facts), http://www.eren.doe.gov/wind/web.html.

17 The Danish Wind Turbine Manufacturers Association, www.windpower.org.

for electricity may cause more frequent shortages and unacceptable price fluctuations. A more diverse portfolio of energy resources could decrease the frequency of power interruptions and reduce the potential for political manipulation. Unlike fossil fuels, wind will not be depleted. Wind power also provides more jobs per unit of energy than other forms of energy generation[18] and offers rural landowners and farmers an alternative source of income.

Industry Analysis and Market Update

Initially fragmented, the world industry became relatively concentrated in the late 1990s, with six European firms accounting for 87 percent of the market. In 2001, the wind energy market was considered young, fast-moving, and innovative. There were about 40 wind turbine manufacturers in the world, with Denmark controlling about half of the market.[19] The Danish company Vestas led the industry with 29 percent of the world market.[20] Of the 16 turbine manufacturers listed in the American Wind Energy Association members directory,[21] seven companies made only large systems (greater than 50 kW), five made only small systems (less than 50 kW), three made both sizes, and six made systems for water pumping. Only a few of the wind turbine manufacturers listed by AWEA were publicly traded. Enron Wind Corporation, (a subsidiary of Enron Corporation) was the only American wind turbine maker readily identifiable as a publicly traded company. Mitsubishi Heavy Industries, a Japanese firm, and three Danish manufacturers, NEG Micon, Nordex, and Vestas, rounded out the publicly traded companies on the AWEA list. The AWEA directory also listed 53 wind power-plant project developers. About 25 developers were also project operators, and about 30 were also independent power producers.

The market for wind power is changing so fast that it is hard to make a statement that will be relevant for more than a few months. Government renewable energy production incentives, fossil fuel price fluctuations, technological improvements, public sentiment about environmental issues, energy utility deregulation, the shift to distributed power, government and public interest in reducing reliance on outside energy sources, and the

---

18   http://www.awea.org.

19   The Danish Wind Turbine Manufacturers Association, www.windpower.org.

20   HSBC Investment Bank plc, "Engineering the Future of Power," 2 February 2001.

21   The directory is not comprehensive, but does include many international manufacturers and developers; http://www.awea.org/directory.

growing demand for electricity are among the forces in the changing power market. However, some facts and trends seem likely to carry into the long term:

1. Wind power is the fastest-growing form of energy production worldwide. A global compound annual growth rate in wind turbine sales of 30 percent is expected by 2004.[22]
2. Wind power is economically competitive with any other form of energy generation. Subsidies, important in the early development, have become less important as turbine efficiencies have improved and as economies of scale are achieved. From the mid 70s to about 1980, at a time when coal plants generated electricity for about $0.02 per kWh, utility-scale wind power typically cost more than $0.30 per kWh. Today, some wind farms, in regions with strong dependable winds, produce electricity for less than $0.03 per kWh. An energy crisis early in 2001 sent the price of wholesale electricity in California over $0.30 per kWh on the spot market. As a consequence of the shifting economics, wind energy developers are rushing to lease choice sites and initiate new projects.
3. For developed countries, in Europe and elsewhere, which are struggling with global climate change and compliance with the Kyoto Protocol,[23] wind power's environmental attributes have added to its attractiveness. Although the United States officially rejected the protocol in the spring of 2001, significant consumer interest has created a niche for renewable energy sources, and many states and companies are independently seeking reductions in greenhouse gas emissions. Wind power is also an attractive option for technology transfer from developed countries to less developed countries. Despite the fact that developing countries account for 38 percent of global CO_2 emissions,[24] the Kyoto Protocol does not require them to attain the emissions-reduction targets expected of developed countries. However, the protocol does allow developed countries to offset greenhouse gas emissions by assisting developing countries in reducing their contributions.
4. Wind power generation technology is maturing in efficiency and reliability, broadening the range of potential applications for small and large systems worldwide and adding to investor interest.
5. The banking community is beginning to view wind power generation as a viable technology, worthy of lending rates more comparable to other forms of power generation.

---

22 HSBC Investment Bank plc, "Engineering the Future of Power," 2 February 2001, p.1.

23 The 1997 Kyoto Protocol sets specific greenhouse gas emission reduction targets for signatory nations. For more information on global climate change see the Darden technote on atmospheric change.

24 World Energy Outlook 2000, International Energy Agency, February, 2001, p. 27.

6. Wind power is well suited to distributed generation. New energy storage methods, software, and electronic controls are smoothing out the variability and fluctuations in energy delivery and enabling power generators to better match electricity supply with demand. Ubiquitous dependence on computers and electronics has created a demand for reliability and quality in electric power that may lead to a new model for power generation and transmission known as distributed generation. Distributed generation means the generation of electricity by small facilities close to the point of use. It may gradually replace the current model of centralized power plants with extended transmission grids. In many cases it offers end users both a supplemental power source and an alternative to electricity from the utility grid. For developing economies, distributed power offers the benefits of electricity without the construction costs of building a national distribution grid.

Site Selection

Location is exceedingly important for wind power generation. Higher wind speed delivers more power and more electricity to sell with the same equipment costs as a site with lower wind speed. Wind power generation potential accelerates as approximately the cube of wind speed. Wind speed increases with height from the ground. Wind power also varies with air density changes due to temperature and altitude. Colder wind generally carries more power than warmer wind at the same speed. The United States and much of the world has been mapped for wind speed and stability. Potential sites are rated on a scale of 1 to 7, with 7 assigned to sites with the highest speed, most dependable winds.

Figure 8.2. Wind Speed Classification

Wind Class	Average Wind Speed In mph @ 98 ft	Average Wind Speed In mph @ 164 ft
1	0.00 – 11.4	0.00 – 12.5
2	11.4 – 13.2	12.5 – 14.3
3	13.2 – 14.6	14.3 – 15.7
4	14.6 – 15.7	15.7 – 16.8
5	15.7 – 16.6	16.8 – 17.9
6	16.6 – 18.4	17.9 – 19.7
7	> 18.4	> 19.7

Data Source: Union of Concerned Scientists[25]

---

25   Michael W. Tennis, et al., "Assessing Wind Resources," *UCS Reports* (Union of Concerned Scientists), December 1999, http://www.ucsusa.org/energy/WRA.pdf, p.8.

Figure 8.3. Annual Average Wind Resource Estimates in the Contiguous United States[26]

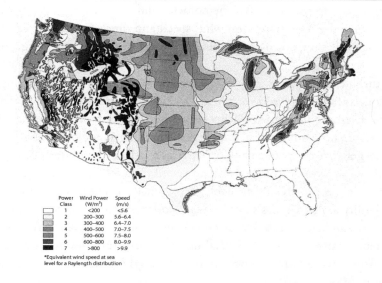

U.S. Department of Energy, Pacific Northwest Laboratory.

Manufacturers have developed models suited for the physical characteristics of a variety of sites including:

- Inland with modest wind speed
- Inland and coastal with high wind speed
- Steep terrain
- Areas with height restrictions
- Offshore

Mechanics of Wind Turbines[27]

Viewed from the outside, a typical large wind turbine consists of two or three blades, attached to a hub on the front of a housing known as the nacelle. Inside the nacelle are a gearbox, an electrical generator, a controller, hydraulic and mechanical braking systems, and a cooling system. The rotor blades for a typical 600 kW turbine are about 20 meters long and shaped like an airplane's wing. The rotors capture wind power and turn the hub which spins a low-speed shaft, typically at about 19 to 30 revolutions per minute (rpm). The low-speed shaft powers a gearbox that spins a high-speed shaft at about 1,500 rpm

---

26   http://rredc.nrel.gov/wind/pubs/atlas/maps/chap2/2-06m.html.
27   For a more thorough discussion of wind turbine mechanics, visit the Danish Wind Industry Association's web site, http://www.windpower.org/tour/wtrb/comp/index.htm.

that turns the generator. An electronic controller uses a computer to optimize the overall performance, avoiding overheating and controlling the yaw mechanism that keeps the turbine oriented into the wind. An anemometer and a wind vane serve as sensors for the controller, providing information on wind speed and direction. In case of system failure, the controller will shut down the turbine and send a message to the turbine operator's computer.

Small turbines range from 250 W models, primarily used for recharging batteries, to 50 kW models that can provide a primary or supplemental electricity source for a small business or a remote village. A tail fin keeps the turbine facing into the wind. Smaller turbines that are not connected to a large distribution grid are often used in conjunction with a diesel generator or with a bank of batteries to assure an uninterrupted power supply. Large, utility-scale, wind turbines are usually rated from 500 kW to 1.5 MW, although 2.5 MW turbines are commercially available. Most small turbines are mounted on towers or rooftops. Utility-scale turbines sit on either tubular or lattice towers ranging from about 40 to 80 meters in height. Tubular towers cost more but may have internal ladders that are safer for maintenance personnel. Several hundred large turbines may be placed on land-based or offshore wind farms connected to a power grid, generating enough electricity for a large community.

Improvements in Technology

Technological breakthroughs in turbine, rotor, and tower design over the past two decades have raised efficiency and reliability while lowering costs. Where 100 kW to 300 kW was typical for large utility-class turbines in the 1980s, many newer turbines generate 1 MW more quietly, efficiently, and less expensively per kilowatt hour than earlier models. New turbines operate with greater than 98 percent reliability. Specific areas of improvement include: taller, lower-cost towers, more efficient rotors, variable speed turbines that allow more efficient capture of wind energy at a variety of speeds, and electronic controllers that optimize every aspect of wind turbine performance. Offshore project developers are experimenting with floating arrays of wind turbines and seabed footings near coastal, high-wind areas.

Energy storage devices that are rapidly approaching commercialization may help improve the dependability of power delivery from wind turbines. Excess electricity generated during off-peak periods can be used in producing hydrogen through electrolysis of water. The hydrogen may be stored for later use in a fuel cell to generate electricity, when needed during periods of low wind or peak energy use. Low-friction, high-speed flywheels kinetically store energy that can spin a generator and produce electricity on demand. Excess wind power may also be used to pump water into an elevated reservoir and later release it to spin a hydroelectric turbine.

Small Wind Turbines

The Bergey BWC EXCEL 10 kW wind turbine, shown here, is intended to supply enough electricity to cover most of the needs of an all-electric home in an area with an average wind speed of 12 miles per hour. It is designed to interface with a local electric grid, charge batteries, or pump water with an electric pump.

"World Wind Watch—It's A Sign Of The Times." True to this oft-used company advertising jingle, when a major U.S retailer begins selling wind turbines via its website, something is definitely changing. Admired and trendsetting Target is now offering one-kilowatt Bergey turbines through its Target.com on-line store. The Bergey XL.1 turbine is being offered as environmentally clean and back-up power for homes and small businesses. The eight-foot diameter turbine operates in a wind speed range of 7–24 miles per hour with maximum power output at higher wind speeds. Target is offering the XL.1 for $1,700 and 32 or 40 foot towers for $490 and $690 respectively.

Before you jump into wind energy for your home, perform a wind and electricity analysis of your property. Make sure a small wind turbine will meet your expectations or recoup your investment in it. Most U.S. homes need more than 1 kilowatt of power. Check local building codes and consult with your neighbors too. Visit Target at http://www.target.com/ (click Home Furnishings—Energy Savers for the turbine).[28]

Photo: Bergey Windpower Co., Inc.

---

28 World Wind Watch, "It's a Sign of the Times," *Green Energy News,* http://www.Energies@nrglink.com, July 29, 2001.

The American Wind Energy Association membership directory lists 9 small wind turbine manufacturers.[29] Small wind turbines, typically from 500 Watts to 50 kW, have been used for many years on farms and in other rural settings as a supplemental or stand-alone power source. Small-scale wind power technology has improved considerably over the past few decades and is gaining ground in remote settings in both developed and developing countries.

The Cost and Pricing of Wind Power

In addition to the turbine itself, a wind power generation system may include towers, wiring, storage systems (usually a bank of batteries for small systems), inverters (used to transform direct current generated by the turbine into alternating current that may be used or fed back into the utility grid), and installation.

Small systems start at about $3,030 for a Bergey XL1 one-kilowatt system, including a wind turbine, 30-foot tower, inverter, and battery bank. This system is intended for use in remote settings and often incorporates a supplemental gas, propane, or diesel generator for low wind periods (the backup generator is not included in the above price). Added expenses for taxes, permitting, wiring, foundation, and tower erection typically cost from $500 to $1500. A slightly larger system of, say, 10 kilowatts, could cost $25,000 to $30,000 fully installed with inverter and connected to the local utility grid. Utility-scale, 1MW-class turbines are priced at about $850,000 per unit, installed. Site leasing and development (including assessments, permits, agreements, etc.) and balance of plant equipment bring costs to roughly $1 million per MW of capacity. A study reported in the journal, *Science* (24 August 2001), pointed out that wind energy costs less than coal power. The authors note that a new coal plant produces energy at 3.5 to 4 cents per kWh but has significant environmental and human health impacts that add between 2 and 4.3 cents per kWh to coal power's true cost. Black lung disease alone kills 2,000 U.S. miners each year and has cost the federal government $35 billion since 1973. Acid deposition, smog, visibility degradation, and global warming are among the most troubling environmental problems associated with coal plants. Asthma, respiratory, and cardiovascular disease, and increased mortality rates round out the list of human health issues related to fossil fuel combustion. The newest wind turbines produce electricity for between 3 and 4 cents per kWh, similar to the price of coal-generated electricity. According to some analysts, the United States could meet its Kyoto Protocol greenhouse gas emissions reduction goals by replacing 59 percent of coal-generated electricity with wind power. The infrastructure transformation would require about 214,000 to 236,000 new wind turbines placed on land or the ocean. The estimated annual amortized capital and operating costs of $149,000 to $183,000 per

---

29   http://www.awea.org/directory/wtgmfgr.html.

turbine would be fully recovered from electricity sales to customers over a twenty-year period with no added cost to taxpayers.[30]

The cost of wind power is a function of several factors:

- Site purchase or lease
- Wind power monitoring and assessment
- Permitting
- Site development
- Environmental assessment
- Taxes
- Equipment and installation costs, including turbines, rotors, towers, wiring, substation with power inverter, etc.
- Maintenance costs
- Connection to the utility grid
- Financing costs.

Figure 8.4. The Cost of Wind Energy Generation

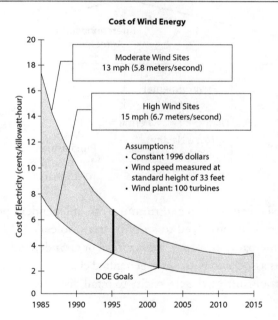

Source: United States Department of Energy[31]

---

30   Mark Z. Jacobson and Gilbert M. Masters, "Exploiting Wind Versus Coal," *Science* (August 24, 2001): p.1438, http://www.sciencemag.org/.

31   http://www.eren.doe.gov/wind/wttr.html.

The price of wind energy is usually expressed in dollars per kilowatt hour (kWh)[32] and incorporates the total cost, government incentives, and the providers' required rate of return. Price will also reflect the speed and stability of the wind (the number of hours of wind per day and days per year, during peak and non-peak periods), and the efficiency of the wind generation equipment. The project developer: identifies a potential site, leases the land,[33] assesses wind potential, arranges for an environmental review, acquires a permit, negotiates an interconnection agreement, arranges financing, installs equipment, and either sells to an IPP or is an IPP. Risks in the project decrease and its value increases as it approaches installation and grid connection.

Figure 5 depicts a schedule and value interpretation for a wind energy project.

Figure 8.5. Risk/Value Analysis In Wind Power Development[34]

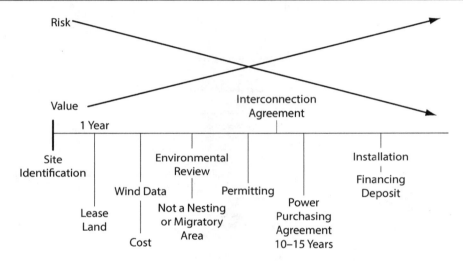

Once a wind power facility is up and running, the independent power producer (IPP) typically sells the electricity produced to a utility at a wholesale price. The utility may be investor-owned, municipality-owned, a cooperative, or an independent power marketer in need of the tax credits associated with wind-generated electricity. The utility is then responsible for selling and transmitting the electricity to retail distributors. Retail distributors sell

---

32 A kilowatt hour (kWh) is the power of 1,000 watts over an hour's time. One kilowatt hour is the energy required to keep a 100 Watt bulb burning for 10 hours.

33 Land is usually leased, rather than purchased, out of concern for good community relations.

34 This diagram was developed from a sketch drawn by Greenlight Energy, Inc., CEO, Stanislav A. Reisky de Dubnik, during an interview.

to individual and commercial end users. "Green attributes"—such as credits for reducing carbon or other emissions—may be marketed as well.

Stages in Electricity Pricing per Kilowatt Hour:
  IPP wholesale to utility   4 cents
  Utility transmits   0.5 cents
  Distributor distributes   1 cent
  Retail sale   1 cent
  Retail price   6.5 cents per kWh
  Green attributes   ?

Technical Challenges—Transmission Line Capacity

Some wind power advocates have argued that transmission infrastructure is more important to the future of wind power than improved turbine design. Where wind power is captured with the intent of sale to a distant market, existing transmission lines may not be up to the task. Minnesota's deputy commissioner of energy, Linda Taylor, noted a project in Buffalo Ridge, Minnesota, that is currently designed to produce 450 MW of wind power by the end of 2002 and could be expanded to produce 3,000 to 4,000 MW. However, the additional power would overload the transmission lines. Bringing the transmission capacity to the necessary level could cost up to $1 million per mile.[35]

In the future, rural wind-generated electricity may be used to produce hydrogen from water through electrolysis. The hydrogen would then be piped or shipped for use in fuel cells that generate electricity where needed for residences, vehicles, and other uses.[36]

Financial Challenges—Terms of Sale of Electricity to Utilities and Bank Financing

Negotiating with utilities to buy their electricity at a reasonable price and finding acceptable financing rates are major challenges for new wind-generation facilities. Once a land lease, environmental review, wind assessment, permits, and interconnection agreements are completed, developers must negotiate a power purchasing agreement at a high enough rate to generate a profit. While a power purchasing agreement is usually for a term of 10 to 15 years and makes cash flow predictable, finding bank financing at reasonable rates for turbine and balance of plant installation costs can still be a

---

35   Janet Roloff, "Power Harvests," *Science News* (July 21, 2001): p 46.

36   For more information on fuel cells see the Darden technote on fuel cells, UVA-E-0215.

substantial hurdle. Banks that are accustomed to dealing with fossil fuel power plants and unfamiliar with wind power may assign an undue level of risk and correspondingly higher-interest, shorter-term loans.

Researchers at Lawrence Berkeley Laboratory estimated that wind power could be generated at 3.5 cents per kWh rather than nearly 5 cents per kWh if a plant were financed by an investor-owned utility rather than by a wind power developer. The difference lies in longer-term debt at a lower cost for the utility (assuming a 20-year loan at 7.5 percent for the utility versus a 12-year loan at 9.5 percent for the developer). The developer might also be subject to a debt service coverage ratio, requiring annual cash generation in excess of loan payments.[37]

Distributed Power

Many electric power industry watchers predict that deregulation, technological changes, and market forces will push the electric power industry into a "distributed power" system with many local, small generators selling energy for grid distribution and/or local use. The current model for distribution typically consists of one or a few large central plants feeding a distribution grid. If utilities accept a network of small local plants as a new paradigm for electrical power generation and distribution, wind energy producers will benefit. An opportunity will open up for independent energy producers to deliver extra power to the grid during periods of peak energy use and premium prices. Some wind farms might also sell directly to end users and enjoy some of the profit currently going to utilities, distributors, and retailers.

For smaller, independent wind power generators, economic viability is often dependent on state and local regulations and the policies of local utilities. Many states in the United States offer net metering, allowing independent energy generators to feed electricity back into the grid at full retail value up to 100 percent of consumption during a billing period. U.S. state and utility policies vary regarding power in excess of consumption. Excess power may be granted to the utility at no cost, purchased by the utility at avoided cost, or purchased at an average retail rate. In some cases a utility may even charge an exit fee to customers wishing to produce their own electricity. Small wind turbines are a likely future source for distributed generation, providing power to utility grids during periods of peak demand and avoiding costly upgrades in distribution equipment.[38]

---

37   Ryan Wiser and Edward Kahn, "Alternative Windpower Ownership Structures: Financing Terms and Project Costs," report number LBNL-38921, published 1996 by the Lawrence Berkeley National Lab (LBNL), Berkeley, CA.

38   http://www.nrel.gov/wind/smalltur.html.

The U.S. Department of Energy's National Renewable Energy Laboratory (NREL) operates a program designed to help U.S. industry develop hybrid alternating-current electric power systems for remote communities in developing countries. These systems, known as "village power," are intended for non-grid connected applications from 1 to 50 kW. In areas with good wind resources, a single wind turbine can provide electricity at lower cost than diesel generation for a home, school, clinic, water pump, or small industry. "Mini-grid" village hybridpower systems, which provide electricity to small communities and use multiple wind turbines, are often more economical than transmission line extension for communities in remote, windy regions.

Incentives[39]

A range of incentives have been put in place to encourage development of wind energy. In the United States, the Public Utilities Regulatory Policies Act (PURPA) in 1978 required utilities to buy electricity from renewable sources whenever it was cheaper than a utility's internal generation costs. It also requires utilities to allow customers to offset energy they use with energy they produce up to the total amount of energy that they consume.

Germany has a generous program known as the Electricity Feed Law, which requires electric utilities to purchase renewable energy at a fixed percentage of the retail electricity price for the life of the plant. Denmark's Windmill Law and India's state utility purchase agreements also guarantee that wind-generated electricity will be purchased well above wholesale market prices.

Property tax reductions and investment tax credits are available to wind developers in several U.S. states.

Renewable energy production tax credits (PTC) are the primary government incentive in the United States. The program offers a tax credit of 1.7 cents per kWh for wind-generated electricity credit through December, 2001.

Accelerated tax depreciation allows investors to fully depreciate the taxable value of wind power property over a short time period (5 years in the United States, 1 year in India) rather than the normal 15 to 20 year depreciation period (for US non-renewable power projects).

India allows electricity generators 5 years of income without tax.

The Renewable Energy Production Incentive (REPI)[40] provides a cash incentive of up to 1.5 cents per kWh to U.S. state and local government and non-profit organizations that own wind facilities that started operation between October 1, 1993, and September 30,

---

39 This section is primarily based on Louise Guey-Lee, "Wind Energy Developments: Incentives" in *Selected Countries,* US DOE, EIA, 1997.

40 U.S. Energy Policy Act of 1992, Section 1212.

2003. Actual payments depend on annual allocations and may be pro-rated if the program is over-subscribed.

Grants offered by U.S. federal and state governments include programs for technological innovation and development, resource assessments, and feasibility studies.

U.S. national labs assist with wind technology research and development. The German and Danish governments also provide research and development support.

Germany's "250-megawatt program" offered commercial-scale wind turbine operators grants of about 3 to 4 cents per kWh. Another grant program funds up to 70 percent of project costs for joint ventures between German companies and their partners in the Southern Hemisphere.

State government-subsidized loans provide below-market rates in Minnesota and other states.

Germany offers wind turbine operators fixed-rate loans at 1 to 2 percent below capital market rates.

Denmark's International Development Agency provides export assistance in the form of grants and low-interest, long-term loans to developing countries that purchase or set up licensing agreements for Danish wind equipment. (Denmark is the world's leading producer of wind turbines.)

Several U.S. states offer "net metering:" allowing a wind power producer to sell electricity back to the utility, effectively running the meter in reverse, up to the amount of power consumed during a billing period. The customer then pays only for consumption in excess of his own electricity generation during the billing period. Any excess power the customer generated beyond his own needs is either given to the utility or sold back, often at a reduced rate.

State and federal governments may offer assistance with site assessment, transmission issues, bird population studies, zoning negotiations, and permitting.

Several states have Renewable Portfolio Standards (RPS) requiring energy retailers to use renewable resources for a minimum percentage of electricity sold.

Germany exempts wind turbines from open land building restrictions.

The United States DOE's 1999 program, Wind Powering America, provided states with assistance in raising public awareness, creating a wind development strategy, and implementing action plans. The program's goals include:

- provide at least 5 percent of the nation's electricity with wind by 2020
- install more than 5000 megawatts by 2005
- have more than 10,000 megawatts on-line by 2020
- double the number of states that have more than 20 megawatts of wind capacity to 16 by 2005 and triple the number to 24 by 2010
- increase wind's contribution to federal electricity use by 5 percent by 2010

The benefits they envision for this program include:

- adding $60 billion in capital investment in rural America over 20 years. Reaching $8 billion in annual investment by 2020
- providing $1.2 billion in new income for American farmers, Native Americans, and rural landowners over 20 years
- creating 80,000 permanent jobs by 2020
- displacing 35 million tons of atmospheric carbon by 2020

Concerns About Wind Power

Environmental and community groups historically have expressed three areas of concern about wind power:

- noise
- view-shed (visual) impacts
- bird mortality

Recent improvements in turbine and rotor design have eliminated much of the vibration and most of the noise from wind power structures, making this issue less of a concern. Visual impact is a more intractable issue. A potential solution for aesthetic degradation of the landscape is placement of wind farms in rural areas where relatively few people will see them and find them offensive. Adequate sites may exist in offshore and inland areas that are not considered scenic tourist destinations. In some areas, such as farm land, the local community consensus may be that wind turbines are not unattractive and/or that wind power benefits outweigh their aesthetic costs. Concern over aesthetic impact is one of the factors driving the industry to develop offshore.

At the Altamont Pass wind farm in California, more than 30 threatened golden eagles and 75 other raptors died or were injured from electrocutions and collisions during a three-year period. Such concerns, particularly for migrating birds, are mirrored in Europe. An August 2001 paper from the (U.S.) National Wind Coordinating Committee estimated that bird collisions with wind generation facilities in the United States killed about 10,000 to 40,000 birds annually, a number that is likely to decrease as new generation wind towers and turbines replace older structures. Despite the significance of these losses, wind generation proved to be a relatively small contributor to collisions that cause avian mortality. A variety of other man-made structures, including vehicles, buildings and windows, power lines, and communications towers caused from 100 million to over 1 billion avian deaths annually.[41] Changes in turbines and towers, coupled with careful siting assessments have greatly decreased bird fatalities. Some newer generation large turbines have

---

41   Erikson, Wallace P., et at, "Avian Collisions with Wind Turbines: A Summary of Existing Studies and Comparisons to Other Sources of Avian Collision Mortality in the United States," a source document

been redesigned to make them less attractive and less dangerous for birds. Tubular towers and enclosed generators have reduced or eliminated perching and nesting sites in some designs, and taller towers with larger rotors may help to reduce fatal collisions as well. An environmental assessment including avian migratory patterns, flyways, and nesting habits is considered a prerequisite before final siting decisions are made.

Wind Power in the Future

Photo: Warren Getz/National Renewable Energy Laboratory[42]

Significant wind power potential exists in nearly every country in the world and the world market outlook for wind energy is favorable. Technological advances have made wind power competitive. Growing demand for electricity and government policies encouraging renewable energy technologies are moving wind energy to center stage. In September 2000, the European Wind Energy Association (EWEA) set a target of 150,000 megawatts of installed capacity by 2020. A few years earlier, a goal of 100,000 MW had been set for 2030, based on a 25 percent to 30 percent annual growth rate. The organization set the new target in light of an actual growth rate of about 40 percent in installed wind

---

of the NWCC (National Wind Coordinating Committee), August 2001, http://www.nationalwind.org/pubs/avian_collisions.pdf.

42   http://www.nrel.gov/data/pix/Jpegs/00042.jpg.

capacity throughout the past 6 years.[43] Marine wind farms have begun to capture the energy potential of powerful offshore winds of Denmark, Sweden, the Netherlands, and the United Kingdom. Offshore wind projects are also planned for the United States, China, and several other countries.[44] Technological progress— particularly the development of high-capacity (1 MW and higher) turbines, direct drive generators, and variable speed systems—has contributed to the high growth rate.

The Union of Concerned Scientists noted that (in 1999), "levels of [renewable energy resource] development represented only a tiny fraction of what could be developed. ... Winds in the United States contain energy equivalent to 40 times the amount of energy the nation uses."[45]

Randall Swisher, executive director of the AWEA in Washington, D.C., notes that U.S. wind-power potential "is comparable to or larger than Saudi Arabia's energy resources." The untapped wind power in Texas, North Dakota, and Kansas could in theory provide for all of the current electricity needs of the United States. Within 100 miles of every major U.S. metropolitan area there is an area with a commercially exploitable wind resource. According to Lester Brown of Earth Policy Institute, global wind electricity generation is close to four times what it was five years ago and U.S. capacity is expected to grow 60 percent in 2001 alone.[46]

Sources for Additional Information

U.S. Department of Energy —Wind Energy Program (quick facts):
http://www.eren.doe.gov/wind/web.html
The American Wind Energy Association—http://www.awea.org/
BTM Consult ApS—http://www.btm.dk/
Energy Efficiency and Renewable Energy Network (EREN)—http://www.eren.doe.gov/
European Renewable Energy EXchange—http://www.nrel.gov/
European Wind Energy Association—http://www.ewea.org/src/links.htm
National Renewable Energy Laboratory—http://www.nrel.gov/
National Wind Technology Center—http://www.nrel.gov/wind/
Offshore Wind Energy Network—http://www.owen.eru.rl.ac.uk/
The Danish Wind Turbine Manufacturers Association—www.windpower.org
Wind Energy Research Group—http://www.uni-muenster.de/Energie/wind/Welcomee.html

---

43 European Wind Energy Association, 2000, www.ewea.org.
44 Offshore Wind Energy Network, http://www.owen.eru.rl.ac.uk.
45 "Powerful Solutions: Seven Ways to Switch America to Renewable Electricity," Union of Concerned Scientists, 1999, www.ucsusa.org.
46 Janet Roloff, "Power Harvests," *Science News* (July 21, 2001): p 45.

Appendix 8.1. 1999 World Energy Consumption[47]

- "Other" includes all other energy sectors, such as hydroelectric, wind power, biomass combustion, and other renewable energy resources.

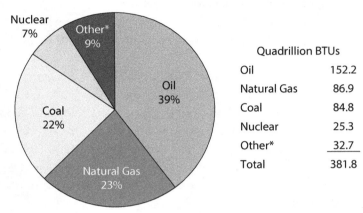

1999 World Energy Consumption[1]

	Quadrillion BTUs
Oil	152.2
Natural Gas	86.9
Coal	84.8
Nuclear	25.3
Other*	32.7
Total	381.8

This chart includes energy used for heating, powering vehicles, and other nonelectric uses. Electricity generation uses less oil and more coal than the figures shown above.

Appendix 8.2. World Electricity Generation by Source, 1999

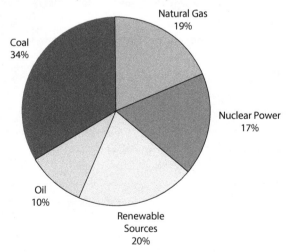

Data Source: U.S. Department of Energy, EIA International Energy Outlook 2001.[48]

---

47 Data from International Energy Annual 1999, Energy Information Administration (EIA) DOE/EIA-0219(99), January, 2001, http://www.eia.doe.gov/oiaf/ieo/tbla1_a8.html#a8.

48 http://www.eia.doe.gov/oiaf/ieo/electricity.html.

Appendix 8.3. OECD Electricity Supply, 2000

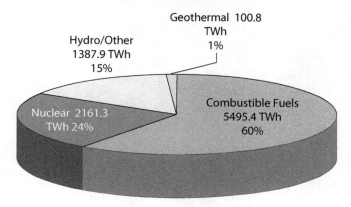

Data on electricity supply in Organization for Economic Cooperation and Development (OECD) countries from the International Energy Agency.[49]

Appendix 8.4. Sources of U.S. Electrical Power, 2000

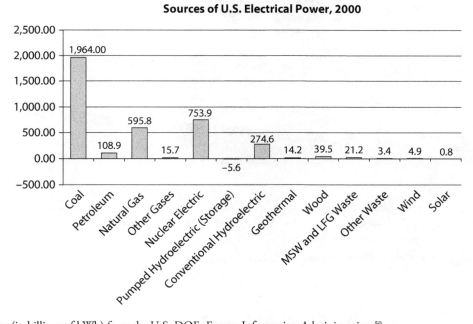

Data (in billions of kWh) from the U.S. DOE, Energy Information Administration.[50]

---

49 International Energy Agency, Monthly Electricity Survey, June 2001, http:www.iea.org/stats/files/mes.html (TWh stands for terawatt hours).

50 http://www.eia.doe.gov/emeu/aer/txt/tab0802.htm.

Appendix 8.5. Global Annual Growth Rates (%), 1990–98

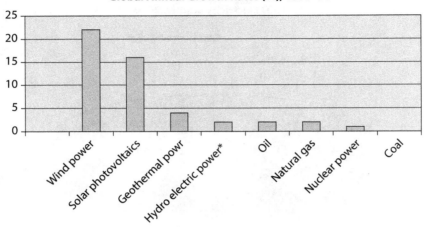

Hydroelectric—1990–97 Data from State of the World, 2000, Worldwatch Institute.

Appendix 8.6. World Electricity Consumption Projections 1990–2020

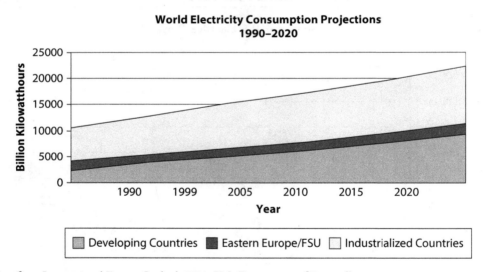

Data from International Energy Outlook 2001, U.S. Department of Energy.[51]

---

51 Energy Information Administration (EIA), http:/www.eia.doe.gov/oiaf/ieo/tbl_20.html.

Discussion Questions

1. Explain the different factors driving interest and investment in wind power.

2. What are the environmental benefits of adopting wind power?

3. What are the different factors affecting site selection when establishing a wind power facility?

4. How does a wind turbine work? Explain in detail by specifying the function of each component.

5. Explain the various factors that affect the cost of a wind turbine.

6. What are the different incentives provided by the U.S. government to encourage increased adoption of wind power?

In-Class Assignment: Objective Questions

1. Winds are neither local nor regional.
    a. True
    b. False
2. Wind power generating capacity is still under the 30,000MW (mega Watt) mark.
    a. True
    b. False
3. 3. The reliability of modern wind turbines is about
    a. 75 %
    b. 50 %
    c. 98 %
    d. 85 %

Wind Power and Biofuels:
A Green Dilemma for Wildlife Conservation

Gregory D. Johnson and Scott E. Stephens

Renewable or green energy is defined as energy generated from natural processes that are replenished over time; it includes electricity and heat generated from solar, wind, hydropower, biomass, geothermal resources, and biofuels derived from renewable resources. In 2006, around 18 percent of global energy use was derived from renewable sources (REN21 2008). The Obama Administration has made the development of renewable energy a top priority for economic expansion, to reduce U.S. dependence on fossil fuels and to lower greenhouse gas emissions contributing to climate change. More than twenty states have enacted laws requiring that a portion of the electricity supply come from renewable energy (American Wind Energy Association 2006). The U.S. Department of Energy (2008a) reports that it is technically feasible to generate 20 percent of the nation's electricity by 2030 from wind energy, and there is a goal of replacing 30 percent of transportation fuel consumption with renewable fuels by the year 2030. Although developing renewable energy sources is generally considered environmentally friendly, impacts on wildlife and their habitats can be associated with many forms of renewable energy (McDonald et al. 2009). With the expected increase in renewable energy, careful planning is needed to avoid conflicts between the development of green energy and concerns with wildlife impacts.

With the exception of hydropower, the two primary sources of commercial renewable energy in North America are electricity produced from wind (7 percent of U.S. renewable energy consumption in 2008) and biofuels (19 percent of U.S. renewable energy consumption in 2008) from food and forage crops. For comparison, U.S. solar (including photovoltaic) energy made up only 1 percent of renewable energy consumption in 2008. Technological advances are under way in the solar sector that may soon improve efficiencies and reduce costs, but for the foreseeable future, wind and biofuels will probably remain the dominant sources of renewable energy (other than hydropower) in North America.

Existing native grasslands, restored grasslands, and shrublands of the western United States support a great diversity of wildlife and provide critical ecological goods and services, such as maintenance of water quality and sequestration of atmospheric carbon. Bird

Gregory D. Johnson and Scott E. Stephens, "Wind Power and Biofuels: A Green Dilemma for Wildlife Conservation," *Energy Development and Wildlife Conservation in Western North America*, pp. 131-155. Copyright © 2011 by Island Press. Reprinted with permission.

diversity is especially high in unique regions such as the glaciated Prairie Pothole Region (PPR) in the midcontinent of North America, and expanses of sagebrush shrublands in the Rocky Mountains provide critical habitat for migrating ungulates and upland bird species. The way in which new demand for biofuels and wind plays out across the West will largely determine whether grassland- and shrubland-dependent populations of wildlife benefit or suffer.

In this chapter, we summarize what is known about how wind and biofuel energy production affect wildlife and their habitats across western North America, and discuss the extent of land potentially affected by these two sources of energy development if the United States and Canada are to meet their stated renewable energy goals. We close with recommendations to reduce wildlife impacts through proper siting, preassessment impact studies, and mitigation.

Wind Energy and Wildlife Conservation

Commercial wind energy facilities have been constructed in thirty-five U.S. states (American Wind Energy Association [AWEA] 2009), and total wind power capacity in the United States increased from 10 megawatts of nameplate capacity in 1981 to 29,440 megawatts as of June 2009 (AWEA 2009), enough to provide electricity to 8 million average households. Despite rapid growth, wind energy amounted to less than 1 percent of U.S. electricity generation in 2006. The wind industry is planning to generate 6 percent of the country's electricity supply by 2020 (AWEA 2006). Canada has 2,854 megawatts of installed wind capacity, meeting about 1 percent of Canada's demand (Canadian Wind Energy Association 2009).

Wind energy has the potential to reduce environmental impacts caused by fossil fuels because wind power does not generate atmospheric contaminants or thermal pollution (National Academy of Sciences [NAS] 2008). Therefore, wind-generated electricity does not have many of the negative environmental impacts associated with other energy sources, such as air and water pollution, or greenhouse gas emissions associated with climate change (Arnett et al. 2007). Wind power is a domestic source of energy that can be produced without water consumption, mining, drilling, refining, waste storage, or many of the other problems that accompany traditional forms of energy generation (Federal Advisory Committee 2009).

Although wind energy has many environmental benefits, wind energy development has caused the deaths of birds and bats that collide with turbines and has resulted in indirect impacts to wildlife through behavioral displacement and habitat loss (Arnett et al. 2007; NAS 2008). We define western North America as the Canadian provinces of Manitoba, Saskatchewan, Alberta, and British Columbia and the states of North and South Dakota, Nebraska, Kansas, Oklahoma, Texas, Montana, Wyoming, Colorado, New Mexico, Idaho, Utah, Arizona, Nevada, Washington, Oregon, and California. Although we limit our review to these states and provinces, some relevant data collected in other states (e.g.,

Minnesota) were included. Of the seventeen western U.S. states, only Nevada and Arizona did not have any installed wind energy as of June 27, 2009 (AWEA 2009). The remaining fifteen states had 19,951 megawatts of installed capacity, which represents 68 percent of all installed wind energy in the United States. Texas, with 8,361 megawatts, is by far the largest, followed by California, with 2,787 megawatts. The amount of installed capacity in the other western states ranges from 20 megawatts in Utah to 1,575 megawatts in Washington (AWEA 2009). Of the western Canadian provinces, Alberta has the largest installed capacity (524 megawatts), followed by Saskatchewan (171 megawatts) and Manitoba (104 megawatts) (Canadian Wind Energy Association 2009). Unlike several wind energy facilities in the eastern United States, all wind energy facilities in western North America are located in nonforested habitats, including agricultural fields, grasslands, and shrub-steppe.

Collision Mortality

Concerns about avian collision mortality at wind energy facilities originated when high raptor fatality rates were first reported at the Altamont Pass Wind Resource Area (APWRA) in California (Orloff and Flannery 1992). An estimated 881–1,300 raptors are killed annually in the APWRA, which equates to 1.5–2.2 raptor fatalities per megawatt per year, the most common being golden eagles (*Aquila chrysaetos*), red-tailed hawks (*Buteo jamaicensis*), American kestrels (*Falco sparverius*), and burrowing owls (*Athene cunicularia*) (Smallwood and Thelander 2004). The APWRA consists primarily of small, older-generation turbines, many of which have lattice support towers, and many of the electrical lines are above ground, providing opportunities for raptors to perch throughout the facility. There are currently more than 5,000 turbines of various types and sizes with an installed capacity of 550 megawatts. Most of the turbines range in size from 40 to 300 kilowatts, the most common size being 100 kilowatts (Arnett et al. 2007). The two other large, older-generation wind energy facilities in California (San Gorgonio and Tehachapi) have not experienced the level of raptor fatalities observed at the APWRA (Anderson et al. 2004, 2005). Differences in raptor fatality rates between these three sites appear to be related to raptor densities.

More recent wind energy developments consist of much larger turbines, ranging in size from 0.66 to 2.5 megawatts, with tubular steel towers and three-blade rotors; most electrical lines are buried. Raptor fatality rates at facilities with modern turbines in western North America have generally been much lower than at the APWRA. At eighteen modern facilities in western North America where raptor fatality estimates are available, raptor fatality rates have ranged from 0 to 1.79 per megawatt per year and averaged 0.19 per megawatt per year (table 9.1).

The two facilities with the highest raptor fatality rates (1.79 and 0.53 per megawatt per year) are in California. Of the sixteen facilities located outside California where raptor fatality rates were reported, raptor fatality rates have ranged from 0 to 0.15 and averaged 0.07 per megawatt per year, or approximately seven raptors for each 100 megawatts of development. These facilities include nine located in Washington and Oregon, three in

Table 9.1. Avian and bat fatality rates at modern wind energy facilities in western North America.

Study Area	Raptor Fatalities (per MW/year)	Bird Fatalities (per MW/year)	Bat Fatalities (per MW/year)	Total MW	Reference
Diablo Winds, CA	1.79	5.67	0.78	20.0	Smallwood and Karas (2009)
Judith Gap, MT	0.09	3.01	8.93	135.0	TRC Environmental Corporation (2008)
Blue Canyon II, OK	—	0.38	3.71	151.2	Burba et al. (2008)
Nine Canyon, WA	0.05	2.76	2.47	48.0	Erickson et al. (2003)
Foote Creek Rim, WY	0.05	2.50	2.23	41.4	Young et al. (2003)
High Winds, CA	—	1.36	2.02	162.0	Kerlinger et al. (2006)
Big Horn, WA	0.15	2.54	1.90	199.5	Kronner et al. (2008)
Combine Hills, OR	0.00	2.56	1.88	41.0	Young et al. (2006)
Stateline, WA/OR	0.09	2.92	1.70	300.0	Erickson et al. (2004)
NPPD Ainsworth, NE	0.06	1.63	1.16	59.4	Derby et al. (2007)
Vansycle, OR	0.00	0.95	1.12	24.9	Erickson et al. (2000)
Klondike, OR	0.00	0.95	0.77	24.0	Johnson et al. (2003)
Hopkins Ridge, WA	0.14	1.23	0.63	150.0	Young et al. (2007)
Oklahoma Wind Energy Center, OK	—	0.08	0.53	102.0	Piorkowski (2006)

Table 9.1. Continued

Study Area	Raptor Fatalities (per MW/year)	Bird Fatalities (per MW/year)	Bat Fatalities (per MW/year)	Total MW	Reference
Klondike II, OR	0.11	3.14	0.41	75.0	Northwest Wildlife Consultants, Incorporated, and Western EcoSystems Technology, Incorporated (2007)
Wildhorse, WA	0.09	1.55	0.39	229.0	Erickson et al. (2008)
Buffalo Gap, TX	0.10	1.32	0.10	134.0	Tierney (2007)
SMUD, CA	0.53	0.99	0.07	15.0	URS et al. (2005)
Castle River, AB	0.01	0.29	0.84	40.0	Brown and Hamilton (2002)
McBride Lake, AB	0.09	0.55	0.71	173.0	Brown and Hamilton (2006)
Summerview, AB	0.11	1.06	12.41	70.0	Baerwald (2008), Baerwald et al. (2009)

Alberta, and one each in Montana, Wyoming, Nebraska, and Texas. Although raptor fatality rates are generally low at most modern wind energy facilities, the number of raptor fatalities is still much higher than that of passerines (i.e., songbirds) relative to the number of individuals exposed to collisions (NAS 2008).

Mortality estimates for all bird species combined are publicly available for twenty-one wind energy facilities in western North America, including those mentioned earlier plus two facilities in Oklahoma and one additional California facility. Bird fatality rates have ranged from 0.08 to 5.67 per megawatt per year and averaged 1.78 per megawatt per year (table 9.1). Avian mortality in western North America is lower than the U.S. national average. Using mortality data from a 10-year period from wind energy facilities throughout the entire United States, the average number of bird collision fatalities is 3.1 per megawatt per year (National Wind Coordinating Committee 2004).

Based on data from twenty-one fatality monitoring studies conducted in western North America at modern wind energy facilities (table 9.1), where 1,247 avian fatalities representing 128 species were reported, raptor fatalities made up 19.4 percent of the identified wind energy facility–related fatalities. The most common raptor fatalities were American kestrel (eighty-two fatalities), red-tailed hawk (forty-six), turkey vulture (*Cathartes aura*; forty-two), and burrowing owl (thirteen). Passerines were the most common collision victims, making up 59.3 percent of the fatalities, with horned lark (*Eremophila alpestris*; 272 fatalities), golden-crowned kinglet (*Regulus satrapa*; forty-seven), and western meadowlark (*Sturnella neglecta*; forty-five) experiencing the highest numbers of fatalities. Upland gamebirds were the third most common group found, making up 9.6 percent of the fatalities. Ring-necked pheasant (*Phasianus colchicus*; forty-five fatalities), gray partridge (*Perdix perdix*; thirty-eight), and chukar (*Alectoris chukar*; eighteen) were the most common fatalities found. Mourning doves (*Zenaida macroura*; twenty-nine fatalities) and rock pigeons (*Columba livia*; seventeen) made up 3.8 percent. Waterbirds such as American coot (*Fulica americana*; ten fatalities) and western grebe (*Aechmophorus occidentalis*; seven) were uncommon, representing 4.0 percent of all fatalities. Waterfowl, primarily mallard (*Anas platyrhynchos*; nine fatalities), were also infrequently found (1.9 percent of all fatalities). Only three shorebirds (0.2 percent of all fatalities) were found. Other groups, such as nighthawks, woodpeckers, and swifts, combined accounted for 1.9 percent of all fatalities. Birds that could not be identified to any avian group also made up 1.9 percent of reported fatalities.

Bat collision mortality at wind energy facilities appears to occur worldwide, as it has been documented in Australia, Germany, Sweden, Spain, and Canada (Johnson 2005). The highest bat fatality rates have occurred at facilities located on forested ridges in the eastern United States (14.9–53.3 per megawatt per year), although high bat fatality rates have also occurred in the Northeast and Upper Midwest and in southern Alberta, suggesting that wind energy facilities located in nonforested areas may also have high bat fatality rates (Arnett et al. 2007). Bat fatality estimates are available for twenty-one wind energy facilities located throughout western North America, where bat fatalities have ranged from 0.07 per megawatt per year at a wind energy facility in California to 12.41 per megawatt per year over a 3-year period at a facility in Alberta, and averaged 2.13 fatalities per megawatt per year (table 9.1), which is slightly higher than avian mortality at wind energy facilities in western North America.

Although it has been assumed that most bat fatalities at wind energy facilities were caused by blunt trauma, based on necropsy results of bats found in Alberta, it was determined that 90 percent of bat fatalities had internal lung hemorrhaging consistent with barotrauma, and it was hypothesized that direct contact with turbine blades accounted for only half of the fatalities (Baerwald et al. 2008). The barotraumas were assumed to be caused by rapid air pressure reduction near moving turbine blades.

Most of the mortality throughout North America occurred among migratory tree-roosting bats, namely the eastern red bat (*Lasiurus borealis*), hoary bat (*Lasiurus cinereus*), and silver-haired bat (*Lasionycteris noctivagans*) (Johnson 2005; Arnett et al. 2008). Of 2,343 bat fatalities reported from studies conducted in western North America, of which 2,285 were identified, hoary bat made up 55.9 percent, silver-haired bat made up 33.1 percent, and Brazilian free-tailed bat (*Tadarida brasiliensis*) made up 6.8 percent. Species that made up less than 2 percent each of the identified fatalities included little brown bat (*Myotis lucifugus*), big brown bat (*Eptesicus fuscus*), eastern red bat, western red bat (*Lasiurus blossevillii*), and evening bat (*Nycticeius humeralis*). Approximately 90 percent of bat fatalities occur from mid-July through the end of September, with more than 50 percent occurring in August (Johnson 2005). At most sites, mortality during the spring migration and breeding season is much lower. However, as wind energy expands into the southwestern United States, where large populations of Brazilian free-tailed bat occur, impacts may occur in breeding bat populations. Studies at facilities located in Oklahoma and California have found that this species made up 41.3 percent and 85.6 percent of the bat fatalities, respectively (Arnett et al. 2008), although total bat mortality at both of these facilities was low (2.02 per megawatt per year in California and 0.53 per megawatt per year in Oklahoma; table 9.1). Many of the free-tailed bat fatalities in Oklahoma involved breeding bats, rather than migrants, unlike most facilities in western North America (Johnson 2005; Arnett et al. 2008). Most bat mortality occurs during low wind speeds at night (Arnett et al. 2008). At wind energy facilities where bat mortality is high, curtailment of turbines during these low-wind situations has been shown to reduce bat mortality by 50–70 percent at a site in Alberta (Baerwald et al. 2009) and by 53–87 percent at a site in Pennsylvania (Arnett et al. 2009).

Although collision mortality is well documented at most wind energy facilities, population-level effects have not been detected, although few studies have addressed this issue. Available data from wind energy facilities suggest that fatalities of passerines from turbine strikes generally are not significant at the population level, although exceptions could occur if facilities are sited in areas where migrating birds or rare species are concentrated (Arnett et al. 2007). In such situations, Desholm (2009) developed a framework for ranking bird species in terms of the relative sensitivity to turbine collisions based on relative abundance and demographic sensitivity (e.g., survival rates). The framework allows preconstruction identification of the species that are at high risk of being adversely affected. When this framework was applied to an offshore wind project in Denmark, results suggested that raptors and waterbirds had the highest risk of being affected, and passerines showed low risk.

Johnson and Erickson (2008) examined the potential for population-level impacts caused by avian collision mortality associated with 6,700 megawatts of existing and proposed wind energy development in the Columbia Plateau Ecoregion of eastern Oregon and Washington. The number and species composition of bird collision fatalities were estimated based on results of eleven existing mortality studies in the ecoregion. Estimated breeding population sizes were available for most birds in the ecoregion based on Breeding Bird Survey data. Predicted mortality rates for avian groups and species of concern were compared with published annual mortality rates. Because the additional wind energy–associated mortality was found to make up only a small fraction of existing mortality rates, it was concluded that population-level impacts would not be expected for the ecoregion as a whole, but that local impacts to some species could occur. In the only study to quantitatively assess potential population-level impacts, Hunt (2002) conducted a 4-year radio telemetry study of golden eagles at the APWRA and found that the resident golden eagle population appeared to be self-sustaining despite sustaining high levels of fatalities, but the effect of these fatalities on eagle populations wintering within and adjacent to the APWRA was unknown. Additional research conducted in 2005 by Hunt and Hunt (2006) found that all fifty-eight territories occupied by golden eagle pairs in the APWRA in 2000 remained active in 2005.

Bats are long-lived species with low reproductive rates. Therefore, their populations are much slower to recover from large fatality events than other species, such as most birds, that have much higher reproductive rates (Kunz et al. 2007b). Because migratory tree bats are primarily solitary tree dwellers that do not hibernate, it has not been possible to develop any suitable field methods to estimate their population sizes (Carter et al. 2003). As a result, impacts on these bat species caused by wind energy development cannot be put into perspective from a population impact standpoint. Based on their estimates of cumulative bat impacts from wind energy development in the eastern United States, Kunz et al. (2007b) concluded that wind energy development could have a substantial impact on bat populations, especially given that evidence suggests the eastern red bat, and perhaps other species, are in decline throughout much of their range (Carter et al. 2003). Although bat mortality at most wind energy facilities in western North America is lower than in other portions of the United States (Johnson 2005; Arnett et al. 2008), the potential for causing significant population-level impacts cannot be determined without estimates of population sizes. To help solve this problem, population genetic analyses of DNA sequence and microsatellite data are being conducted to provide population size estimates, to determine whether populations are growing or declining, and to see whether these populations consist of one large population or several discrete subpopulations that use spatially segregated migration routes (Amy L. Russell, Grand Valley State University, personal communication).

Indirect Effects

In addition to direct effects through collision mortality, wind energy development results in direct loss of habitat where infrastructure is placed and indirect loss of habitat through behavioral avoidance and habitat fragmentation. Direct loss of habitat associated with wind energy development is minor for most species compared with most other forms of energy development. Although wind energy facilities can cover substantial areas, the permanent footprint of wind energy facilities, such as turbines, access roads, maintenance buildings, substations, and overhead transmission lines, generally occupies only 5–10 percent of the entire development area (Bureau of Land Management 2005). Estimates of temporary construction impacts range from 0.2 to 1.0 hectares (0.5–2.5 acres) per turbine (AWEA 2009). However, behavioral avoidance may render much larger areas unsuitable or less suitable for some species of wildlife, depending on how far each species is displaced from wind energy facilities. Based on some studies in Europe, displacement effects associated with wind energy were thought to have a greater impact on birds than collision mortality (Gill et al. 1996). The greatest concern with displacement impacts for wind energy facilities in western North America has occurred where these facilities are constructed in native habitats such as grasslands or shrublands (Leddy et al. 1999; Mabey and Paul 2007).

Most studies on raptor displacement at wind energy facilities indicate that effects appear to be negligible. A before–after control–impact (BACI) study of avian use at the Buffalo Ridge wind energy facility in Minnesota found evidence of northern harriers (*Circus cyaneus*) avoiding turbines on both a small scale (less than 100 meters [328 feet] from turbines) and a larger scale (105–5,364 meters [345–17,598 feet]) in the year after construction (Johnson et al. 2000a). Two years after construction, however, no large-scale displacement of northern harriers was detected.

The only published report of avoidance of wind turbines by nesting raptors occurred at the Buffalo Ridge facility, where raptor nest density on 261.6 square kilometers (101 square miles) of land surrounding the facility was 5.94 nests per 101.0 square kilometers (39 square miles), yet no nests were present in the 31.1-square-kilometer (12-square-mile) facility itself, even though habitat was similar (Usgaard et al. 1997). At a wind energy facility in eastern Washington, extensive monitoring using helicopter flights and ground observations revealed that raptors still nested in the study area at approximately the same levels after construction, and several nests were located within 0.8 kilometer (0.5 mile) of turbines (Erickson et al. 2004). Howell and Noone (1992) found similar numbers of raptor nests before and after construction of Phase 1 of the Montezuma Hills wind energy facility in California, and anecdotal evidence indicates that raptor use of the APWRA in California may have increased since the installation of wind turbines (Orloff and Flannery 1992; AWEA 1995). At the Foote Creek Rim wind energy facility in southern Wyoming, one pair of red-tailed hawks nested within 0.5 kilometer (0.3 mile) of the turbine strings, and seven red-tailed hawk nests, one great horned owl (*Bubo virginianus*) nest, and one golden eagle nest located within 1.6 kilometers (1 mile) of the wind energy facility successfully fledged young (Johnson et al. 2000b; Western EcoSystems Technology, Inc., unpublished data). The golden eagle pair successfully nested 0.3 kilometer (0.5 mile) from the facility for three different years after the project became operational.

Studies in western North America concerning displacement of nonraptor species have concentrated on grassland passerines and waterfowl. Wind energy facility construction appears to cause small-scale local displacement of some grassland passerines, probably because the birds avoid turbine noise and maintenance activities. Construction also reduces habitat effectiveness because of the presence of access roads and large gravel pads surrounding turbines (Leddy 1996; Johnson et al. 2000a). Leddy et al. (1999) surveyed bird densities in Conservation Reserve Program (CRP) grasslands at the Buffalo Ridge wind energy facility in Minnesota and found that mean densities of ten grassland bird species were four times higher at areas located 180 meters (591 feet) from turbines than they were at grasslands nearer turbines. Johnson et al. (2000a) found reduced use of habitat within 100 meters of turbines by seven of twenty-two grassland-breeding birds after construction of the Buffalo Ridge facility. At the State-line wind energy facility in Oregon and Washington, use of areas less than 50 meters from turbines by grasshopper sparrow (*Ammodramus savannarum*) was reduced by approximately 60 percent, with no reduction in use more than 50 meters from turbines (Erickson et al. 2004). At the Combine Hills facility in Oregon, use of areas within 150 meters of turbines by western meadowlark was reduced by 86 percent, compared with a 12.6 percent reduction in use of reference areas over the same time period (Young et al. 2005a). However, horned larks showed significant increases in use of areas near turbines at both these facilities, possibly because the cleared turbine pads and access roads provided habitat preferred by this species.

Shaffer and Johnson (2008) examined displacement of grassland birds at two wind energy facilities in the northern Great Plains. Intensive transect surveys were conducted in grid cells that contained turbines and in reference areas. The study focused on five species at two study sites, one in South Dakota and one in North Dakota. Based on this analysis, killdeer (*Charadrius vociferous*), western meadowlark, and chestnut-collared long-spur (*Calcarius ornatus*) did not show any avoidance of wind turbines. However, grasshopper sparrow and clay-colored sparrow (*Spizella pallida*) showed avoidance out to 200 meters (656 feet).

At the Buffalo Ridge facility, the abundance of several bird types, including shorebirds and waterfowl, was found to be significantly lower at survey plots with turbines than at reference plots without turbines (Johnson et al. 2000a). The report concluded that the area of reduced use was limited primarily to areas within 100 meters of the turbines. These results are similar to those of Osborn et al. (1998), who reported that birds at Buffalo Ridge avoided flying in areas with turbines.

Results of a long-term mountain plover (*Charadrius montanus*) monitoring study at the Foote Creek Rim wind energy facility in Wyoming suggest that construction of the facility resulted in some displacement of mountain plovers. The mountain plover population was reduced during construction but has slowly increased since, although not to the same level as it was before construction. It is not known whether the initial decline was due to the presence of the wind energy facility or to regional declines in mountain plover populations. The subsequent increase may also be influenced by regional changes in mountain plover abundance. Nevertheless, some mountain plovers have apparently become habituated to the turbines, as several mountain plover nests have been located within 75 meters (246 feet) of turbines, and many of the nests were successful (Young et al. 2005a).

Breeding puddle ducks (mallard, blue-winged-teal [*Anas discors*], gadwall [*A. strepera*], northern pintail [*A. acuta*], and northern shoveler [*A. clypeata*]) were counted on wetland complexes within two wind energy facilities and on similar reference areas in North and South Dakota during the 2008 and 2009 breeding seasons (Ducks Unlimited Inc. and

U.S. Fish and Wildlife Service [USFWS], unpublished data). Based on results of the surveys, breeding puddle duck abundance was not lower than expected in areas of wind energy development in 2008, but the 2009 data suggested lower pair densities in the wind developed sites. The study is continuing through 2010 to further assess the response of breeding ducks to wind energy development.

Wind Energy and Prairie Grouse

Much debate has occurred recently regarding the potential impacts of wind energy facilities on prairie grouse (*Centrocercus* and *Tympanuchus* spp.). It is currently unknown how prairie grouse, which are accustomed to a low vegetation canopy, would respond to numerous wind turbines hundreds of meters taller than the surrounding landscape. Some scientists speculate that such a skyline may displace prairie grouse hundreds of meters or even kilometers from their normal range (Manes et al. 2002; USFWS 2003; NWCC 2004). If birds are displaced, it is unknown whether, in time, local populations may become acclimated to elevated structures and return to the area. The USFWS argued that because prairie grouse evolved in habitats with little vertical structure, the placement of tall human-made structures, such as wind turbines, in occupied prairie grouse habitat may result in a decrease in habitat suitability (USFWS 2004). Several studies have shown that prairie grouse avoid other anthropogenic features, such as roads, power lines, oil and gas wells, and buildings (Robel et al. 2004; Holloran 2005; Pruett et al. 2009a, 2009b). Much of the infrastructure associated with wind energy facilities, such as power lines and roads, is common to most forms of energy development, and it is assumed that impacts would be similar. Nevertheless, there are substantial differences between wind energy facilities and most other forms of energy development, particularly related to human activity. Although results of these studies suggest that the potential exists for wind turbines to displace prairie grouse from occupied habitat, well-designed studies examining the potential impacts of wind turbines on prairie grouse are lacking. Ongoing telemetry research being conducted by Kansas State University to examine the response of greater prairie chickens to wind energy development in Kansas, and a similar study being conducted by Western EcoSystems Technology, Inc. (Johnson et al. 2009a) on greater sage-grouse (*Centrocercus urophasianus*) response to wind energy development in Wyoming, will help address this knowledge gap.

Other than these two ongoing telemetry studies, we are aware of only three publicly available studies that examined the response of prairie grouse species to wind energy development. The Nebraska Game and Parks Commission (2009) monitored greater prairie chicken (*Typanuchus cupido*) and sharp-tailed grouse (*T. phasianellus*) leks after construction of the thirty-six-turbine Ainsworth wind energy facility in Brown County, Nebraska. Surveys for leks were conducted 4 years after construction (2006–2009) within a 1.6- to 3.2-kilometer (1- to 2-mile) radius of the facility, an area that covered approximately 65 square kilometers (25 square miles). The numbers of leks of both species combined in the study area were thirteen, twelve, nine, and twelve in the first 4 years after construction. The number of greater prairie chickens counted on leks increased from seventy to ninety-five during the 4-year period, whereas the number of sharp-tailed grouse decreased from sixty-six to fifty-six. No preconstruction data were available on prairie grouse leks near the site; however, densities of lekking grouse on the study area at the Ainsworth facility were within the range of expected grouse densities in similar habitats in Brown County and the adjacent Rock County (Nebraska Game and Parks Commission 2009). The leks ranged from 0.7 to 2.7 kilometers (0.4 to 1.7 miles) from the nearest turbine, with an average distance of 1.4 kilometers (0.9 miles).

At a three-turbine wind energy facility in Minnesota, six active greater prairie chicken leks were located within 3.2 kilometers of turbines, with the nearest lek within 1 kilometer (0.6 mile) of the nearest turbine (USFWS 2004). During subsequent research at this facility based on forty nest locations, it was found that nesting hens were not avoiding turbines. Based on extensive research of the prairie chicken population in the vicinity of this wind energy facility from 1997 to 2009, it was concluded that the distribution and location of leks and especially nests were determined by the presence of adequate habitat in the form of residual grass cover, not the presence of vertical structures such as trees, woodlots, power lines, and wind turbines (Toepfer and Vodehnal 2009).

Greater prairie chicken lek surveys were conducted at the Elk River wind energy facility in Butler County, Kansas, within the southern Flint Hills, beginning 3 years before and continuing 4 years after construction (Johnson et al. 2009a). The facility consists of 100 1.5-megawatt turbines. During the year immediately before construction of the project (2005), ten leks were present on the project area, with 103 birds on all leks combined. By 2009, 4 years after construction, only one of these ten leks remained active, with three birds on the lek. The ten leks were located 88–1,470 meters from the nearest turbine, with a mean distance of 587 meters; eight of the ten leks were located within 0.8 kilometer (0.5 mile) of the nearest turbine. Although this decline may be attributable to development of the wind energy facility, greater prairie chicken populations have declined significantly in the Flint Hills due to the practice of annual spring burning. During the same time frame that leks were monitored at the Elk River facility, the estimated average number of greater prairie chickens in the southern Flint Hills declined by 65 percent from 2003 to 2009. In Butler County, the estimated number of birds declined by 67 percent from 2003 to 2009 (Kansas Department of Wildlife and Parks, unpublished data). This regional decline is attributed primarily to the practice of annual spring burning and heavy cattle stocking rates, which remove nesting and brood-rearing cover for prairie chickens (Robbins et al. 2002). Therefore, it seems unlikely that the decline of prairie chickens on the Elk River site is due entirely to the presence of wind turbines, although the presence of turbines may have contributed to the decline (Johnson et al. 2009a).

Another grouse with a lek mating system, the black grouse (*Lyrurus tetrix*), was found to be negatively affected by wind power development in Austria (Zeiler and Grünschachner-Berger 2009). The number of displaying males in the wind power development area increased from twenty-three to forty-one during the 3-year period immediately before construction but then declined to nine males 4 years after construction. In addition to the decline in displaying males, the remaining birds shifted their distribution away from the turbines. One lek located within 200 meters of the nearest turbine declined from twelve birds 1 year before construction to no birds 4 years after construction.

Although the data collected in the United States on response of prairie grouse to wind energy development indicate that prairie grouse may continue to use habitats near wind energy facilities, research conducted on greater sage-grouse response to oil and gas development has found that population declines due to oil and gas development may not occur until 4 years after construction (Naugle et al. 2011). Therefore, data spanning two or more grouse generations will be needed to adequately assess the impacts of wind energy development on prairie grouse.

There is little information regarding wind energy facility operation effects on big game. At the Foote Creek Rim wind energy facility, pronghorn antelope (*Antilocapra americana*) observed during raptor use surveys were recorded year-round (Johnson et al. 2000b). The mean number of pronghorn antelope observed at the six survey points was 1.07 per survey before construction of the wind energy facility and 1.59 and 1.14 per survey the 2 years immediately after construction, indicating no reduction in use of the immediate area. During a study of interactions of a transplanted elk (*Cervus elaphus*) herd with operating wind energy facilities in Oklahoma, no evidence was found that operating wind turbines have a measurable impact on elk use of the surrounding area (Walter et al. 2009). Current telemetry studies being conducted to assess the response of elk to wind energy development in Wyoming and Oregon and pronghorn antelope response in Wyoming will help address potential impacts on big game.

Planning for Wind Energy

Meeting the goal of 20 percent wind energy by 2030 would require use of an area of approximately 50,000 square kilometers (19,300 square miles) for land-based wind energy facilities and more than 11,000 square kilometers (4,245 square miles) for offshore facilities (NAS 2008), although the direct loss of land would make up only 5–10 percent of this amount (Bureau of Land Management 2005). McDonald et al. (2009) examined energy development impacts on habitat under various cap-and-trade scenarios and ranked the expected wind energy development footprint behind only that of biofuels, with 72.1 square kilometers (28 square miles) needed for each terawatt-hour produced per year by 2030. As wind energy expands across the region, the potential for cumulative impacts increases. Major expansion of renewable energy is still in its infancy, so the opportunity to avoid earlier mistakes associated with siting of oil and gas developments, as well as other forms of development in western North America, is still possible. Many government agencies and several nongovernment organizations (NGOs) have developed resource maps and recommendations to help guide future wind energy development to reduce impacts on wildlife and sensitive habitats at state (e.g., Martin et al. 2009) and national levels, such as the mapping effort being conducted jointly by the American Wind and Wildlife Institute and The Nature Conservancy (American Wind and Wildlife Institute 2009). The USFWS, along with numerous stakeholders, including the wind energy industry, academia, and conservation organizations, through the Federal Advisory Council, are developing guidelines for siting wind energy facilities, conducting preconstruction wildlife surveys, and evaluating the impacts of the facilities.

Wind energy development mitigation measures have been developed to prevent or reduce impacts and to compensate for impacts through offsite habitat mitigation when warranted (Johnson et al. 2007). Because environmental impacts determined on a project-by-project basis can underestimate the cumulative impacts of many projects, Kiesecker et al. (2009) have developed a framework to show how combining landscape-level conservation planning with application of a mitigation hierarchy can be used to determine where impacts on biodiversity can be offset and where they should be avoided or reduced (chap. 9). Use of these planning tools, guidelines, and appropriate mitigation strategies will facilitate sustainable development of wind energy in western North America, while reducing associated impacts on wildlife resources.

While wind farms are cropping up throughout the high plains and the far West, biofuels are farmed predominantly farther east (e.g., North Dakota, South Dakota, and Kansas). This geographic separation nearly ensures that all western communities are touched by some, if not many, forms of renewable energy production.

Biofuels: Boon or Bane for Grassland Wildlife?

As a rule, change comes slowly to the agricultural landscape of the West. Change can be risky, and farmers are risk averse by nature. Modern agribusiness is characterized by huge capital investment, large volumes of money moving into and then back out of checkbooks, and—in the end— thin profit margins. There is little room for experimentation and a failed crop year. But a daring new attitude is spreading across farm country, and it is setting the stage for dramatic changes in land use. Perhaps this is best characterized as the twenty-first-century gold rush of ethanol production.

Companies have been fermenting corn into ethanol for decades, and many of us have filled our vehicles with a 10 percent blend of their product. But until recently, few of us have given serious thought to the need to wean the United States off foreign oil. And only a few botanists were familiar with a plant called switchgrass that, given time and new technology, might be an important ingredient in the recipe for energy independence.

How quickly things change. Reducing our dependence on foreign oil and reducing greenhouse gas emissions are now cornerstones of federal policy. And the obscure plant called switchgrass emerged in the public consciousness when former President Bush referenced it in his 2006 State of the Union address. From all indications, biofuels are here to stay.

Energy from Biomass

Biofuels are energy sources derived from plants or other living organisms (i.e., biomass). Converting plant matter to biofuel can be as simple as picking the stems and leaves from a field, bundling them up, and transporting them to a power plant, where they are burned—together with coal and other fuels—to generate electricity. But most often, the term *biofuel* refers to a liquid transportation fuel derived from plant material. Ethanol and biodiesel are the most common examples.

Energy technology is evolving quickly, and the list of biofuels and useful coproducts continues to grow. Today, one biofuel—ethanol—is the focus of attention. Fermenting ethanol from corn or other grains is a proven technology, and engines have already been adapted to burn gasoline–ethanol blends. Currently, there are 170 corn ethanol plants in the United States, and 24 more are being built or are planned. At a conversion rate of 10.6 liters (2.8 gallons) of ethanol per bushel of corn, producing the 39.9 billion liters (10.5 billion gallons) of ethanol flowing from today's plants takes nearly 3.8 billion bushels. Corn acreage in the United States rose by 5.5 million hectares (13.5 million acres) in 2007 to nearly 37.6 million hectares (93 million acres) (National Agricultural Statistics Service 2007), a level of production that has not been seen since 1944 (fig. 9.1).

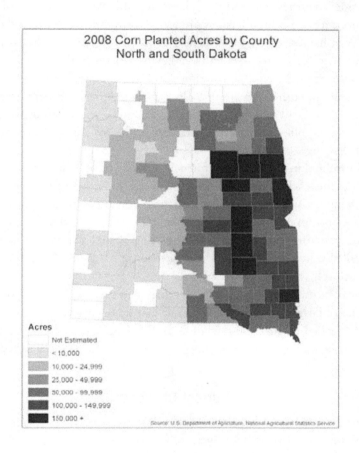

Figure 9.1. Distribution of corn-planted acres by county in North Dakota and South Dakota, 2008. As new varieties of corn are developed, the distribution of corn acres is likely to shift northwest from Minnesota and Iowa into central North and South Dakota. Additional tillage for corn will place our highest wetland density landscapes at increased risk of wetland drainage (fig. 9.2). (From U.S. Department of Agriculture, National Agricultural Statistics Service.)

This is probably not an anomaly but rather a trend, because if all the new plants are built, corn ethanol production will reach 47.6 billion liters (12.6 billion gallons) and probably continue to increase. In December 2007, former President Bush signed the Energy Independence and Security Act of 2007, which established a new Renewable Fuels Standard of 136.8 billion liters (36 billion gallons) per year by 2022. Fifty-seven billion liters (15 billion gallons) are expected to come from corn ethanol. And the U.S. Environmental Protection Agency (2009) is considering increasing the allowable ethanol blend content from the current 10 percent to 15 percent. All of this is moving the United States toward the "30 by 30" goal of replacing 30 percent of transportation fuel consumption with renewable fuels by the year 2030.

But consider that these old and new plants, together with plant expansions, will have a combined need for 4.5 billion bushels of corn. That number represents 37 percent of the 2008 U.S. corn harvest (fig. 8.1). This new demand will cut into the corn supply that is already being used for animal feed and as a key ingredient in thousands of food products, most notably processed food. Speculation continues to mount that someday soon there may not be enough corn to go around. Recent passage of the Energy Independence and Security Act of 2007 will ease that speculation, but it is unclear for how long. There is little doubt that corn supplies will be stretched thin, and a new dimension has been added to the ethanol challenge.

Ethanol production goals in the United States cannot be achieved solely with corn and other starch grains. Achieving these goals will require the use of new sources of biomass, such as switchgrass, wheat straw, and corn stover (i.e., cornstalks, cobs, and leaves). In each case, the portion of the plant digested to make ethanol is the cellulose—the material that provides structural support for plant growth. Ethanol that results from the digestion process is called cellulosic ethanol. It is created using specialized enzymes and a series of complex processes.

In the near future, corn and cellulosic ethanol feedstocks may well be competing for the same acres of farmland. This includes land that is now growing wheat, beef, or—in the case of land enrolled in the CRP—wildlife. In the grasslands of the West, big changes are coming. The key question is the net effect on wetlands, grasslands, and wildlife.

Current Challenges

Most experts believe that commercial-scale cellulosic ethanol processing plants will not come online for several years. However, some progress is being made, with twenty-three cellulosic plants under development or construction in the United States as of early 2009 (Renewable Fuels Association 2009). Until then, corn will remain the primary product market and demand for corn ethanol will increase, as indicated by the passage of a new Renewable Fuels Standard. Regrettably, when it comes to wildlife habitats, it is hard to find a silver lining in a forecast that calls for growing more corn. In fact, the demand for more corn is likely to increase conversion pressure on existing grasslands and wetlands throughout the PPR (fig. 9.2).

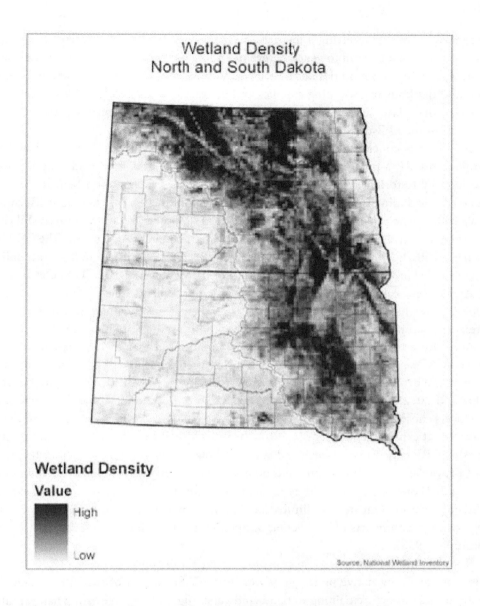

Figure 9.2. The highest wetland and waterfowl densities in the United States occur in the central Dakotas, an area that currently contains few harvest acres of corn (fig. 9.1). But as new varieties of corn are developed, the distribution of corn acres is likely to shift northwest from Minnesota and Iowa into the heart of the Prairie Pothole Region (wetland density map courtesy of U.S. Fish and Wildlife Service National Wetland Inventory).

Typical of any supply-and-demand relationship, when the demand for corn is high, so is its price. If corn growers receive a high price for their product, they can then afford to invest more in land and land rental (fig. 9.1). Simple supply-and-demand economics force those currently receiving a low return on their land investment to reconsider how their land can best be used to generate revenue. More area planted to corn means less area planted to other commodities, such as wheat and soybeans. Lower production of these commodities leads to dwindling supplies and increased prices. It is a vicious cycle and a situation that places an enormous amount of conversion pressure on the grasslands and wetlands of the PPR that has not been seen for decades.

An obvious concern associated with increased corn production is the fate of land enrolled in CRP, a program that was created in the 1985 Farm Bill to idle highly erodible land by restoring it to grassland. As the many other conservation benefits of CRP emerged—including the annual addition of 2.1 million ducks to the fall flight (Reynolds et al. 2006)—the program became recognized as the most significant and successful conservation initiative ever implemented by the U.S. Department of Agriculture. The CRP has been particularly important for wetlands in the PPR (fig. 9.2), with more than 1.2 million acres enrolled in various wetland practices in North and South Dakota (U.S. Department of Agriculture, Farm Services Agency, Economic and Policy Analysis Staff, September 2007). Now, those in search of more corn ground are viewing land enrolled in CRP as a reserve of cropland waiting to be brought back into production. The U.S. Department of Agriculture supported this position when it announced that there would be no new general CRP signups in 2007 or 2008—all for the stated purpose of providing more area to meet the demand for corn. Unfortunately, much area in the PPR will expire between 2007 and 2012, threatening hundreds of thousands of hectares of wetlands. Agricultural producers chose not to re-enroll 271,000 hectares (670,000 acres) across North and South Dakota that expired September 30, 2007. Two-thirds or 1.4 million hectares (3.4 million acres) of CRP grasslands in North and South Dakota will expire by 2012. This is bad news for PPR wetlands. Wetland regulations do not apply to wetlands restored as a requirement of the CRP contract, and therefore they are eligible for drainage once the contract expires. Wetlands that do remain have limited habitat value when surrounded by corn and are degraded by the impacts of sedimentation, pesticides, herbicides, and fertilizer (Gleason and Euliss 1998).

The demand for corn has also put pressure on the 9 million hectares (22.3 million acres) of remaining native prairie grassland in the U.S. portion of the PPR, which also encompasses our best remaining wetland landscapes (fig. 9.2). Most existing native prairie is not suitable for growing a high-yield corn crop. In general, the soils are poor and the climate is often extremely dynamic, two factors that generally lead to poor corn yields. However, some newly broken prairie may be suitable for wheat and other crops that will be displaced by corn on existing cropland. And the demand for corn and cropland in general is driving up the cost of all land dramatically. These increased land values also affect the region's other producers: cattle ranchers.

The livestock industry is native prairie's reason for being in an economic sense. If that industry were to disappear today, no obvious economic return could then be realized from native prairie grasslands and wetlands. Yet history tells us that some creative mind will find a way to make money from the land, but most uses will not be as compatible as ranching is with maintaining the wetlands and grasslands.

Even though cattle prices have been good for the last few years, the economics of ranching hinge on being able to buy or rent pasture at an affordable price. Higher land prices and rental rates squeeze the rancher's bottom line. Further compounding the problem are the high costs ranchers incur when they take their livestock to the feeder, the last step in the cattle-rearing process, when the animals are fattened on an increasingly valuable commodity: corn. Cattle feeders simply pass on their higher corn prices to ranchers. If a combination of high pasture values, high feeder costs, and a collapse in cattle prices occurs, many ranchers in the PPR may be driven out of business, which could spell disaster for grasslands and wetlands.

Expansion of corn for ethanol may also have unintended and unforeseen consequences for the other critical element in the wildlife equation: wetlands. Millions of wetlands in the PPR are currently protected by a strong disincentive in the federal Farm Bill called Swampbuster. Very simply, if farmers choose to drain a wetland, they disqualify themselves from all Farm Bill programs, including commodity payments that compensate them for low yields and low crop prices. To date, this disincentive has been very effective at protecting wetlands from drainage. But its effectiveness hinges on low or volatile commodity prices and the need for farmers to manage their risk in the face of these circumstances. If prices for corn and other commodities increase and remain stable, some farmers may not need or expect to receive commodity payments. That scenario effectively removes the protection afforded wetlands, and widespread wetland drainage could be the result.

These additional pressures on grasslands and wetlands in the PPR also reinforce the need for stronger protection of existing grassland. Conservation organizations have been strong advocates for a new protection measure called Sodsaver in the new Farm Bill. The Sodsaver provision would remove eligibility for crop insurance and disaster assistance on any areas without a cropping history. Federal crop insurance and disaster payments have been important incentives to convert existing grassland to cropland (Governmental Accountability Office 2007). Protection of native grasslands also prevents degradation of the adjacent wetlands.

Future Opportunities

If the wetlands and grasslands of the PPR can weather the corn ethanol storm, positive trends may be on the horizon. If implemented in a thoughtful and environmentally friendly way, the cellulosic ethanol industry could be an asset to the PPR. The key lies in the nature of the feedstock, what land uses it displaces, and how it is harvested. A perennial grass crop such as switchgrass clearly holds the potential for benefits to wetlands and wildlife.

Switchgrass, like other perennial plants being considered as biofuels, is a tall, dense grass that makes the most of the sun's energy and the soil's nutrients. The economics of cellulosic ethanol are all about the tons of biomass that can be grown on an acre of ground.

Currently, switchgrass varieties being grown in the PPR yield 10–15 tons per hectare (2–3 tons per acre) annually. However, new varieties being developed through advanced genetic technology may eventually yield as much as 50 tons per hectare (10 tons per acre) annually. From a wildlife perspective, grasses used to produce cellulosic ethanol are harvested after the growing season, in fall or winter, which is beneficial to grassland-nesting birds.

Switchgrass and other perennial grasses can also play a key role in wetland conservation, especially if they are planted on cropland that used to be cultivated every year. The addition of grassland around these wetlands will greatly improve water quality, wetland function, and wetland value for wildlife. Annually tilled crops degrade wetlands and provide poor habitat for breeding grassland birds. However, the key issue is whether switchgrass replaces existing cropland or replaces other grassland that has higher value for wildlife. Switchgrass could have negative impacts on wildlife and grassland resources, especially if native grasslands or CRP are converted to monotypic stands of switchgrass (Murray and Best 2003).

One important economic consideration for cellulosic feedstocks such as switchgrass is the transportation distance from field to processing plant. At current switchgrass yields, the maximum transport distance is about 80.5 kilometers (50 miles) (Hettenhaus 2004). So policies intended to encourage the cellulosic ethanol industry should recognize the need to cluster dedicated biomass crops near proposed or existing plants. Conservation groups have suggested that existing acres of CRP in the PPR are not a viable feedstock because they are randomly scattered across the landscape. It is estimated that most of the biomass would be too far from a processing plant to be used for fuel production.

Domestic energy needs can be met while maintaining the important natural resource values and ecological goods and services provided by grassland and wetland habitats. Detrimental effects on grassland and wetland systems will be reduced in the near term if corn acreage is expanded into existing cropland and the crops displaced do not expand into native prairie or lands now enrolled in CRP. However, recent expiration of hundreds of thousands of acres of CRP and ongoing conversion of native prairie (Stephens et al. 2008) make it clear that expansion of corn acreage is having an impact on other crops and on producer decisions. On the other hand, the production of biofuels from switchgrass or other perennial grasses could provide substantial benefits for both energy and natural resources if they replace existing cropland and are properly managed.

Conclusion

Few emerging industries have as much potential as renewable energy to shape the landscape and the habitat on which wildlife depends. Conservation, industry, and government leaders must recognize this potential and become deeply engaged in energy issues to identify opportunities to benefit our natural resources, to minimize potential adverse impacts, and to mitigate the impacts that are unavoidable. The way forward is clear enough. Where feasible and cost-effective, we need to embed renewable energy production in already disturbed areas, through solar panels on the rooftops of our cities, geothermal heating of buildings, wind turbines on agricultural and mined lands, and switchgrass and other perennial grasses planted on existing croplands. When impacts do occur, we need to apply offsite mitigation planning tools to identify the species and systems most affected and fund conservation projects that offset these effects. These changes, coupled with the widespread adoption of smart grids that conserve and monitor consumer energy use and emerging clean energy technology advances on the horizon, may finally advance renewable energy production well above the current level of 7 percent of U.S. domestic energy production without compromising our wildlife and ecosystems in the process. With these changes, we can solve the green dilemma.

Discussion Questions

1. Explain the role played by wind energy in conserving wildlife.

2. How is the prairie grouse effected by wind energy?

3. How is grassland wildlife affected by biofuels?

4. How is energy obtained from biomass sources?

5. What is the unique position of cellulosic ethanol within the biofuels spectrum?

In-Class Assignment: Objective Questions

1. The two facilities with the highest raptor mortality rates are in California.
    a. True
    b. False
2. Raptor fatality rates have averaged _____ out of the sixteen facilities located outside California.
    a. 0.0 per MW per year.
    b. 2 per MW per year
    c. 0.07 per MW per year.
    d. None of the above.
3. Barotrauma is assumed to be caused by which of the following:
    a. High RPMs of the wind turbines.
    b. Air pressure increase near turbines.
    c. Rapid air pressure reduction near moving turbine blades.
    d. None of the above.
4. Companies have been fermenting crops into ethanol for decades.
    a. True
    b. False
5. The current conversion rate of ethanol is which of the following?
    a. 5 liters of ethanol per bushel of corn.
    b. 20 liters of ethanol per bushel of corn.
    c. 10.6 liters of ethanol per bushel of corn.
    d. None of the above.

10   Climate Change[1]

Andrea Larson and Mark Meier

The thickness of the air, compared to the size of the Earth, is something like the thickness of a coat of shellac on a schoolroom globe. Many astronauts have reported seeing that delicate, thin, blue aura at the horizon of the daylit hemisphere and immediately, unbidden, began contemplating its fragility and vulnerability. They have reason to worry.

—Carl Sagan[2]

Introduction

Since the beginning of their history, humans have altered their environment. Only recently, however, have we realized how human activities influence Earth's terrestrial, hydrological, and atmospheric systems to the extent that these systems may no longer maintain the stable climate we have assumed as the basis of our existence. The science of climate change developed rapidly in the late 20th century as researchers established a correlation between increasing atmospheric concentrations of certain gases, human activities emitting those gases, and a rapid increase in global temperatures. Many, but by no means all, international

---

1   This technical note on climate change updates and incorporates work from three technical notes by Stephen Keach and Andrea Larson, all published by Darden Business Publishing in 2002: "The Challenge of Climate Change: Overview" (UVA-ENT-0036), "The Science of Climate Change" (UVA-ENT-0037), and "Government and Corporate Response to Climate Change" (UVA-ENT-0038). It replaces that series.

2   Carl Sagan, *Billions and Billions: Thoughts on Life and Death at the Brink of the Millennium* (New York: Ballantine, 1997).

Andrea Larson and Mark Meier, "Climate Change," pp. 1-14. Copyright © 2010 by Darden Business Publishing. Reprinted with permission.

policy makers spurred research forward, as it became apparent that impacts ranging from melting polar icecaps to acidified oceans and extreme weather were attributed to anthropogenic (human-caused) influences on climate. Global businesses, many of which initially balked at potential economic disruption from changes in the use of fossil fuel and other business practices, have largely acceded to the need for change. Nonetheless, the overall response to the challenge has been slow and not without resistance, thereby increasing the potential opportunities and urgency.

The Science of Global Climate Change

In the early 1820s, Joseph Fourier, the French pioneer in the mathematics of heat diffusion, became interested in why some heat from the sun was retained by the earth and its atmosphere rather than being reflected back into space. Fourier conceived of the atmosphere as a bell jar with the atmospheric gases retaining heat and thereby acting as the containing vessel. In 1896, Swedish Nobel laureate and physicist Svante August Arrhenius published a paper in which he calculated how carbon dioxide (CO_2) could affect the temperature of the earth. He and early atmospheric scientists recognized that normal CO_2 levels in the atmosphere contributed to making the earth habitable. Scientists also have known for some time that air pollution alters weather. For example, certain industrial air pollutants can significantly increase rainfall downwind of their source. As intensive agriculture and industrial activity have expanded very rapidly around the world since 1850, a growing body of scientific evidence has accumulated to suggest that humans influence global climate.

The earth's climate has always varied, which initially raised doubts about the significance of human influences on climate or suggested our impact may have been positive. Successive ice ages, after all, likely were triggered by subtle changes in the earth's orbit or atmosphere and would presumably recur. Indeed, changes in one earth system, such as solar energy reaching the earth's surface, can alter other systems, such as ocean circulation, through various feedback loops. The dinosaurs are thought to have gone extinct when a meteor struck the earth, causing tsunamis, earthquakes, fires, and palls of ash and dust that would have hindered photosynthesis and lowered oxygen levels and temperatures. Aside from acute catastrophes, however, climate has changed slowly, on the scale of tens of thousands to millions of years. Paleoclimatological data also suggest a strong correlation between atmospheric carbon dioxide levels and surface temperatures over the past 400,000 years and indicate that the last 20 years have been the warmest of the previous 1,000.[3]

In the last decades of the 20th century, scientists voiced concern over a rapid increase in "greenhouse gases." Greenhouse gases (GHGs) were named for their role in retaining heat in earth's atmosphere, causing a greenhouse effect similar to that in Fourier's bell

---

3 NOAA Paleoclimatology, "A Paleo Perspective on Global Warming," July 13, 2009, http://www.ncdc.noaa.gov/paleo/globalwarming/home.html (accessed August 19, 2010).

jar. Increases in the atmospheric concentration of these gases, which could be measured directly in modern times and from ice core samples, were correlated with a significant warming of the earth's surface, monitored using meteorological stations, satellites, and other means. The gases currently of most concern include CO_2, nitrous oxide (N_2O), methane (CH_4), and chlorofluorocarbons (CFCs). Carbon dioxide, largely a product of burning fossil fuels and deforestation, is by far the most prevalent greenhouse gas, albeit not the most potent. Methane, produced by livestock and decomposition in landfills and sewage treatment plants, contributes per unit 12 times as much to global warming as does carbon dioxide. Nitrous oxide, created largely by fertilizers and coal or gasoline combustion, is 120 times as potent. CFCs, wholly synthetic in origin, have largely been phased out by the 1987 Montreal Protocol because they degraded the ozone layer that protected earth from ultraviolet radiation. The successor hydrochlorofluorocarbons, however, remain greenhouse gases with potencies one to two orders of magnitude greater than CO_2.

In response to such findings, the United Nations and other international organizations gathered in Geneva to convene the First World Climate Conference in 1979. In 1988, a year after the Brundtland Commission called for sustainable development, the World Meteorological Organization and the United Nations Environment Programme (UNEP) created the Intergovernmental Panel on Climate Change (IPCC). The IPCC gathered 2,500 scientific experts from 130 countries to assess the scientific, technical, and socio-economic aspects of climate change, its risks, and possible mitigation.[4] The IPPC's First Assessment Report, published in 1990, concluded that the average global temperature was indeed rising and that human activity was responsible. This report laid the groundwork for negotiation of the Kyoto Protocol, an international treaty to reduce GHG emissions that met with limited success. Subsequent IPCC reports and myriad other studies indicated that climate change was occurring faster and with worse consequences than initially anticipated.

---

4 The IPCC comprises three working groups and a task force. Working Group I assesses the scientific aspects of the climate system and climate change. Working Group II addresses the vulnerability of socio-economic and natural systems to climate change, negative and positive consequences of climate change, and options for adapting to those consequences. Working Group III assesses options for limiting greenhouse gas emissions and otherwise mitigating climate change. The Task Force on National Greenhouse Gas Inventories implemented the National Greenhouse Gas Inventories Program. Each report has been written by several hundred scientists and other experts from academic, scientific, and other institutions, both private and public, and has been reviewed by hundreds of independent experts. These experts were neither employed nor compensated by the IPCC nor by the United Nations system for this work.

Effects and Predictions

The Fourth IPCC Assessment Report in 2007 summarized much of the current knowledge about global climate change, which includes actual historical measurements as well as predictions based on increasingly detailed models.[5] These findings represent general scientific consensus and typically have 90% or greater statistical confidence.

- The global average surface temperature increased 0.74°C ± 0.18°C (1.3°F ± 0.32°F) from 1906 to 2005, with temperatures in the upper latitudes (nearer the poles) and over land increasing even more. In the same period, natural solar and volcanic activity would have decreased global temperatures in the absence of human activity. Depending on future GHG emissions, the average global temperature is expected to rise an additional 0.5°C to 4°C by 2100, which could put over 30% of species at risk for extinction.
- Eleven of the twelve years from 1995 to 2006 were among the twelve warmest since 1850, when sufficient records were first kept. August 2009 had the highest ocean temperatures ever recorded for August and the second hottest land temperatures, and early signs indicated 2010, the time of this writing, began as another unusually warm year.[6]
- Precipitation patterns have changed since 1900, with certain areas of northern Europe and eastern North and South America becoming significantly wetter, while the Mediterranean, central Africa, and parts of Asia have become significantly drier. Record snowfalls in Washington, D.C., in the winter of 2009–10 reflected this trend as warmer, wetter air dumped nearly one meter of snow on the U.S. capital in two storms.[7]
- Coral reefs, crucial sources of marine species diversity, are dying, due in part to their sensitivity to increasing ocean temperatures and ocean acidity. Oceans acidify as they absorb additional carbon dioxide; lower pH numbers indicate more acidic conditions. Ocean pH decreased 0.1 points from 1750 to 2000 and is expected to decrease an additional 0.14 to 0.35 pH by 2100. (A pH difference of one is the difference between lemon juice and battery acid.)

---

5 Available from http://www.ipcc.ch/publications_and_data/ar4/syr/en/contents.html (accessed August 19, 2010). A fifth assessment report was begun in January 2010. Unless otherwise footnoted, all statistics in this list are from the Fourth IPCC Assessment. Recent (2010) independent reviews of the Fourth IPCC report are discussed on the IPCC home page, www.ipcc.ch/ (accessed February 4, 2011).

6 Data more current than the fourth IPCC report are available from NASA and NOAA, among other sources, at http://data.giss.nasa.gov/gistemp/ and http://www.noaanews.noaa.gov/stories2009/20090916_globalstats.html (accessed August 19, 2010).

7 Bryan Walsh, "Another Blizzard," *Time*, February 10, 2010.

- Glaciers and mountain snow packs, crucial sources of drinking water for many people, have been retreating for the past century. From 1978 to 2006, Arctic ice coverage declined between 6% and 10%, with declines in summer coverage of 15% to 30%.
- Seas have risen 20 to 40 centimeters over the past century as glaciers melt and water expands from elevated temperatures. Sea levels rose at a rate of 1.8 (±0.5) millimeters per year from 1961 to 2003. From 1993 to 2003 alone, that rate was dramatically higher: 3.1 (±0.7) millimeters per year. An additional rise in sea level of 0.4 meters to 3.7 meters (1.3 to 12.1 feet) is expected by 2100. The former amount would threaten many coastal ecosystems and communities;[8] the latter would be enough to submerge completely the archipelago nation of the Maldives.
- Trees are moving northward into the tundra. A thawing permafrost, meanwhile, would release enough methane to catastrophically accelerate global warming.[9] Other species, too, are migrating or threatened, such as the polar bear. The population of polar bears is expected to decline two-thirds by 2050 as their icepack habitats disintegrate under current trends.[10] Warmer waters will also increase the range of cholera and other diseases and pests.[11]
- At the same time that humans have increased production of greenhouse gases, they have decreased the ability of the earth's ecosystems to reabsorb those gases. Deforestation and conversion of land from vegetation to built structures reduces the size of so-called carbon sinks. Moreover, conventional building materials such as pavement contribute to local areas of increased temperature, called heat islands, which can be 12°C (22°F) hotter than surrounding areas. These elevated local temperatures further exacerbate the problems of climate change for communities.[12]

By impairing natural systems, climate change impairs social systems. Dramatic shifts in climate alter population distribution, natural resource availability, health parameters, and political dynamics. Droughts and rising seas that inundate populous coastal areas will force migration on a large scale. Unusually severe and extreme weather patterns have already increased economic costs and death tolls from hurricanes, floods, heat waves, and other natural disasters. Melting Arctic ice packs have led countries in a scramble to dominate

---

8   U.S. Environmental Protection Agency, *Coastal Elevations and Sensitivity to Sea-Level Rise: A Focus on the Mid-Atlantic Region*, January 2009, http://www.epa.gov/climatechange/effects/coastal/sap4-1.html (accessed August 19, 2010).

9   National Science Foundation, "Methane Releases From Arctic Shelf May Be Much Larger and Faster Than Anticipated," press release, March 4, 2010.

10 U.S. Geological Survey, "New Polar Bear Finding," press release, October 2, 2009.

11 World Health Organization, "Cholera," http://www.who.int/mediacentre/factsheets/fs107/en/index.html (accessed August 19, 2010).

12 U.S. Environmental Protection Agency, "Urban Heat Island," February 23, 2010, http://www.epa.gov/heatisland/ (August 19, 2010).

possible new shipping routes. When the chairman of the Norwegian Nobel Committee awarded the 2007 Nobel Peace Prize to the IPCC and Al Gore, he said, "A goal in our modern world must be to maintain 'human security' in the broadest sense." Similarly, albeit with different interests in mind, the United States' 2008 National Intelligence Assessment, which analyzes emerging threats to national security, focused specifically on climate change.

Scientists have tried to define acceptable atmospheric concentrations of CO_2 or temperature rises that would still avert the worst consequences of global warming while accepting we will likely not entirely undo our changes. NASA scientists and others have focused on the target of 350 parts per million (ppm) of carbon dioxide in the atmosphere.[13] Their paleoclimataological data suggest that a doubling of carbon dioxide in the atmosphere, which is well within some IPCC scenarios for 2100, would likely increase the global temperature 6°C (11°F). Atmospheric carbon dioxide levels, however, passed 350 ppm in 1990 and reached 388 ppm by early 2010. This concentration will continue to rise rapidly as emissions accumulate in the atmosphere. Even if the CO_2 concentration stabilizes, temperatures will continue to rise for some centuries, much the way a pan on a stove keeps absorbing heat even if the flame is lowered. Hence, some scientists have begun to suggest that anything less than zero net emissions by 2050 will be too little, too late; policymakers have yet to embrace such aggressive action.[14]

International and U.S. Policy Response

The primary international policy response to climate change was the United Nations Framework Convention on Climate Change (UNFCCC). The convention was adopted in May 1992 and became the first binding international legal instrument dealing directly with climate change. It was presented for signature at the Earth Summit in Rio de Janeiro and went into force in March 1994 with signatures from 166 countries. By 2010 the convention had been accepted by 192 countries. UNFCCC signatories met in 1997 in Kyoto and agreed to a schedule of reduction targets known as the Kyoto Protocol. Industrialized countries committed to reducing emissions of specific greenhouse gases, averaged over 2008–12, to 5% below 1990 levels. The European Union committed to an 8% reduction and the United States to 7%. Other industrialized countries agreed to lesser reductions or to hold their emissions constant, while developing countries made no commitments but hoped to industrialize more cleanly than their predecessors. Partly to help developing countries, the Kyoto Protocol also created a market for trading GHG emission allowances.

---

13   James Hansen et al., "Target Atmospheric CO_2: Where Should Humanity Aim?" *The Open Atmospheric Science Journal* 2 (2008): 217–31.
14   H. Damon Matthes and Ken Caldeira, "Stabilizing Climate Requires Near-Zero Emissions," *Geophysical Research Letters* 35 (2008).

If one country developed a carbon sink, such as by planting a forest, another country could buy the amount of carbon sequestered and use it to negate the equivalent amount of its own emissions.

The Kyoto Protocol has suffered from a lack of political will behind enforcement in the United States and abroad. The United States signed it, but the Senate never ratified it. U.S. President George W. Bush backed away from the emission reduction targets and eventually rejected them entirely. Already by the time he took office in 2001, a 7% reduction from 1990 levels for the United States would have translated into a 30% reduction from 2001 levels. U.S. GHG emissions, instead of declining, rose 17.2% from 1990 to 2007, the last year measured.[15] Almost all other Kyoto signatories will also fail to meet their goals. The EU, in contrast, in 2010 was on track to meet or exceed its Kyoto targets. GHG pollution allowances for major stationary sources have been traded through the EU Emissions Trading System since 2005. The consensus in Europe is that the Kyoto Protocol is necessary and action is required to reduce greenhouse gases.

The Kyoto Protocol expires in 2012, so meetings have begun to negotiate new goals. In December 2007, UNFCCC countries met in Bali to begin to discuss a successor treaty. The conference made little headway, and countries met again in December 2009 in Copenhagen. That conference again failed to generate legally binding reduction goals, but the countries confirmed the dangers of climate change and agreed to strive to limit temperature increases to no more than 2°C total. Moreover, in exchange for industrializing countries' pledge to set emissions reductions targets by 2020, industrialized countries agreed to provide $130 billion in financing by 2020 in addition to technology and support for forestry.[16] The next major meeting is planned in Mexico.

Individual countries and U.S. states and agencies have acted, nonetheless, in the absence of broader leadership. In 2007, EU countries set their own future emissions reduction goals, the so-called 20-20-20 strategy of reducing emissions 20% from 1990 levels by 2020 while reducing energy demand 20% through efficiency and generating 20% of energy from renewable resources.[17] In the Northeast U.S., 10 states collaborated to form the Regional Greenhouse Gas Initiative (RGGI), which caps and gradually lowers GHG emissions from power plants by 10% from 2009 to 2018. A similar program, the Western Climate Initiative, is being prepared by several Western U.S. states and Canadian provinces, and California's AB 32 "Global Warming Solutions Act" set a state GHG emissions limit for 2020. Likewise, the federal government under President Barack Obama committed to reducing its emissions, while the U.S. Environmental Protection Agency, in response to a 2007 lawsuit led by the state of

---

15 U.S. Environmental Protection Agency, *2009 Greenhouse Gas Inventory Report*, April 2009.

16 http://unfccc.int/meetings/cop_15/items/5257.php (accessed August 19, 2010).

17 European Union, "The EU Climate and Energy Package," press release, February 23, 2010.

Massachusetts, prepared to regulate GHGs under the Clean Air Act. Members of Congress, however, have threatened to curtail the EPA's power to do so, either by altering the procedures for New Source Review that would require carbon controls or by legislatively decreeing that global warming does not endanger human health.[18] One bill, in contrast, to combat climate change would have reduced U.S. emissions 80% from 2005 levels by 2050. It passed the House of Representatives in 2009 but failed to make it to a Senate vote.

Corporate Response and Opportunity

Certain industries are more vulnerable than others to the economic impacts of climate change. Industries that are highly dependent on fossil fuels, such as oil and gas companies, automobile manufacturers, airlines, and power plant operators, as well as cement producers that emit significant amounts of CO_2, are closely watching legislation related to greenhouse gas emissions. The reinsurance industry, which over the past several years has taken large financial losses due to extreme weather events, is deeply concerned about global climate change and growing liabilities for its impacts.

Given the potential costs of ignoring climate change, the costs of addressing it appear rather minimal. In 2006, the UK Treasury released the *Stern Review on the Economics of Climate Change*. The report estimated that the most immediate effects of global warming would cause damages of "at least 5% of global GDP each year, now and forever. If a wider range of risks and impacts is taken into account, the estimates of damage could rise to 20% of GDP or more." Actions to mitigate the change, in contrast, would cost only about 1% of global GDP between 2010 and 2030.[19]

Corporate reactions have ranged from taking action now to reduce or eliminate emissions of greenhouse gases to actively opposing new policies that might require changes in products or processes. Anticipatory firms are developing scenarios for potential threats and opportunities related to those policies, public opinion, and resource constraints. Among those companies actively pursuing a reduction in GHGs, some cite financial gains for their actions. In 2005, WalMart and GE both committed to major sustainability efforts, as have many smaller corporations. Excessive greenhouse gas emissions may reflect inefficient energy use or loss of valuable assets, such as when natural gas escapes during production or use. The Carbon Disclosure Project emerged in 2000 as a private organization to track GHG emissions for companies that volunteered to disclose their data. By 2010,

---

18 Environmental News Service, "Coal State Senators Battle EPA to Control Greenhouse Gases," February 23, 2010; Juliet Eilperin and David A. Fahrenthold, "Lawmakers Move to Restrain EPA on Climate Change," *Washington Post*, March 5, 2010.

19 HM Treasury, "Executive Summary," *Stern Review on the Economics of Climate Change*, 2006, 1.

over 1,500 companies belonged to the organization, and institutional investors used these and other data to screen companies for corporate social responsibility. Out of concern for good corporate citizenship and in anticipation of potential future regulation, greenhouse gas emissions trading has become a growing market involving many large corporations. The emissions trading process involves credits for renewable energy generation, carbon sequestration, and low-emission agricultural and industrial practices that are bought and sold or optioned in anticipation of variable abilities to reach emissions reduction targets. Some companies have enacted internal, competitive emissions reduction goals and trading schemes as a way to involve all corporate divisions in a search for efficiency, cleaner production methods, and identification of other opportunities for reducing their contribution to climate change.

In parallel to tracking GHG emissions, "clean tech" and "clean commerce" have become increasingly prevalent as concepts describing technologies such as wind energy and processes such as more efficient "smart" electrical grids that do not generate as much or any pollution. New investments in sustainable energy increased between 2002 and 2008, when total investments in sustainable energy projects and companies reached $155 billion, with wind power representing the largest share at $51.8 billion.[20] Also in 2008, sustainability-focused companies as identified by the Dow Jones Sustainability Index and Goldman Sachs SUSTAIN list were reported to have outperformed their industries by 15% over a six-month period.[21]

Conclusion

Our climate may always be changing, but humans have influenced it dramatically in a short time with potentially dire consequences. Greenhouse gases emitted from human activities have increased the global temperature and will continue to increase it, even if we ceased all emissions today. International policymakers have built consensus for the need to curb global climate change but have struggled to take specific, significant actions. In contrast, at a smaller scale, local governments and corporations have attempted to mitigate and adapt to an altered future. Taking a proactive stance on climate change can make good business sense. At a minimum, strategic planning should be informed by climate change concerns and the inherent liabilities and opportunities therein. Opportunities for product design and process improvements that both reduce climate change impact and increase efficiency and consumer loyalty make sense. Companies that chart a course around the most likely set of future conditions with an eye to competitive advantage, good corporate citizenship, and stewardship of natural resources are likely to maximize their flexibility—and hence their strategic edge—in the future.

---

20   UNEP, *Global Trends in Sustainable Energy Investment 2009*, 10.
21   A.T. Kearney, Inc., "Green Winners," 2009, 2.

Exhibit 10.1. Climate Change

Increase in Global Fossil-Fuel CO_2 Emissions, 1750–2006

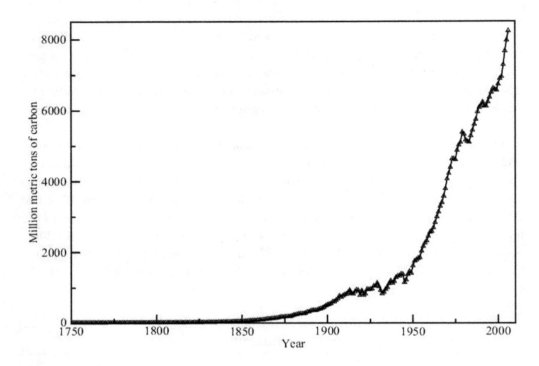

Note: Units of carbon are often used instead of carbon dioxide, which can be confusing. One ton of carbon = 3.67 tons of carbon dioxide. Hence, emissions of carbon dioxide in 2006 were roughly 8,000 million (8 billion) tons of carbon, or 29 billion tons of carbon dioxide.

Source: Oak Ridge National Laboratory, Carbon Dioxide Information Analysis Center.

Exhibit 10.2. Climate Change

Increases in the Concentration of Atmospheric CO_2 at Mauna Loa Observatory 1958–2009

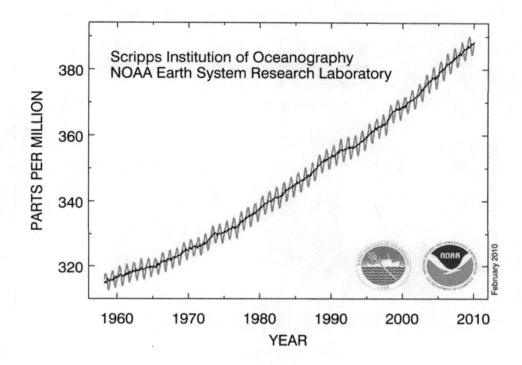

Sources: Scripps Institution of Oceanography and National Oceanic and Atmospheric Administration Earth System Research Laboratory.

Exhibit 10.3. Climate Change

Hemispheric Temperature Change, 1880–2009

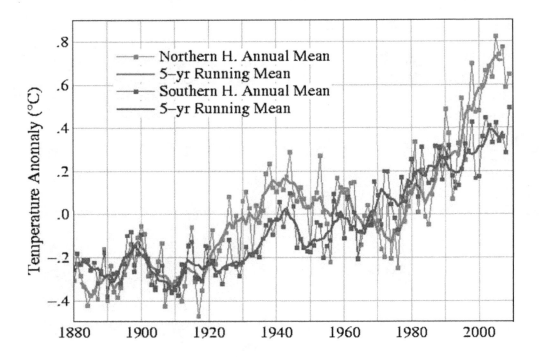

Data source: NASA, Goddard Institute for Space Studies.

Exhibit 10.4. Climate Change

Decrease in Arctic Sea Ice, 1979–2009

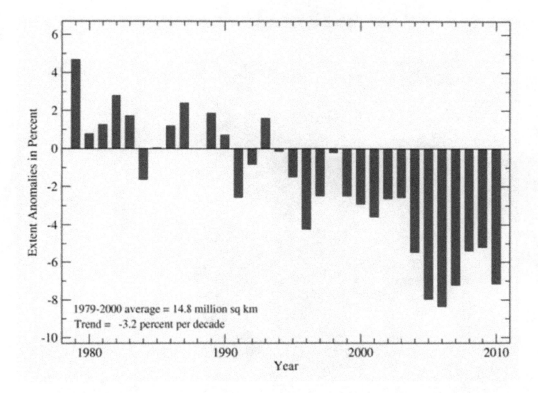

Source: National Oceanic and Atmospheric Administration National Snow and Ice Data Center.

Exhibit 10.5. Climate Change

Sources and Types of Greenhouse Gas Emissions, 2000

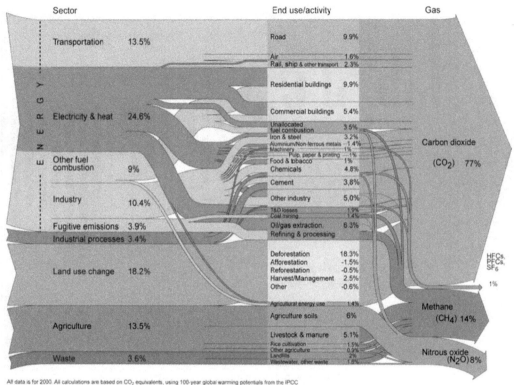

Source: Riccardo Pravettoni, UNEP/GRID-Arenda and World Resources Institute, Climate Analysis Indicator Tool (CAIT), Navigating the Numbers: Greenhouse Gas Data and International Climate Policy, December 2005; Intergovernmental Panel on Climate Change, 1996 (data for 2000), http://maps.grida.no/go/graphic/world-greenhouse-gas-emissions-by-sector2.

Exhibit 10.6. Climate Change

Increase in U.S. Energy Consumption, Total and Renewable, in Billions of BTU, 1949–2008

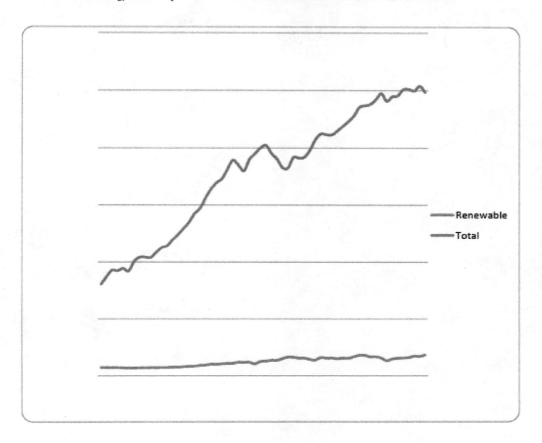

Data source: EIA Annual Energy Review, 2008.

Discussion Questions

1. Explain the basic concept and the science behind climate change.

2. What are the effects of climate change?

3. How have the U.S. and the international community responded to climate change?

4. How has the corporate community responded to climate change?

5. What is the nature of the corporate opportunity when dealing with climate change?

In-Class Assignment: Objective Questions

1. Humans have altered their environment since the beginning of history.
    a. True
    b. False
2. Certain industrial air pollutants can significantly increase rainfall downwind of their source.
    a. True
    b. False
3. Based on Paleo climatological data, the last 20 years have been the warmest of the previous 1,000 years.
    a. True
    b. False
4. Chlorofluorocarbons occur naturally.
    a. True
    b. False
5. In the absence of human activity during the years 1906–2005, natural solar and volcanic activity would have resulted in which of the following?
    a. Increased temperatures 0.74C ± 0.18C.
    b. Global temperatures would have remained unchanged.
    c. Decreased global temperatures.
    d. None of the above.

Freshwater Availability

Jill Boberg

Earth is called the water planet, and 71 percent of its surface is covered with water. In addition to the water on the surface, there is water in underground aquifers, glaciers, icecaps, and the atmosphere. But very little of this water is available for human use, even including that amount needed for ecosystems to function properly. This chapter describes the earth's water cycle, freshwater ecosystems, and the processes that naturally recycle freshwater on the earth. It also discusses water availability and the natural and human causes for changes in the water supply.

Quantity of Water on Earth

Although the total amount of water on earth is massive, estimated at some 1.4 billion cubic kilometers (km^3), the majority (97.5 percent) of it is saltwater (Figure 11.1). Of the 2.5 percent that is freshwater, an estimated stock of approximately 35 million km^3, most of it is tied up in ice and permanent snow cover. Overall, about 30 percent of the earth's freshwater is found as groundwater, while only about one percent, or about 200,000 km^3, of it is easily accessible for human use in lakes, rivers, and shallow aquifers (Shiklomanov, 1993, 2000). Humans' primary source of freshwater comes from the water that runs off after precipitation. This runoff plus groundwater recharge equals about 40,000 to 47,000 km^3 per year (Gleick, 2000b). However, the use of fossil groundwater and other water sources that are relatively slow to be replenished is common and growing.

Jill Boberg, "Freshwater Availability," *Liquid Assets: How Demographic Changes and Water Management Policies Affect Freshwater Resources*, pp. 15-27. Copyright © 2001 by RAND Corporation. Reprinted with permission.

Figure 11.1. Earth's Supply of Water

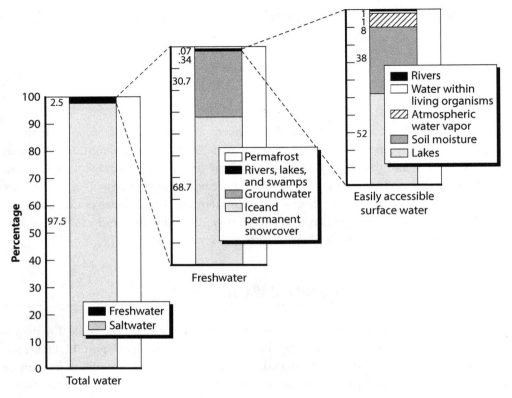

Source: Hinrichsen, Krchnak, and Mogelgaard (2002).

RAND MG358-2.1

The Hydrological Cycle and Sustainable Water Use

The hydrological cycle is the repeated process of the evaporation and redistribution of water in various forms around the earth. The complete cycle and the approximate amounts of water involved in each stage are shown in Figure 11.2. The annual precipitation on earth is more than 30 times the atmosphere's total capacity to hold water, which means that water is recycled relatively rapidly between the earth's surface and the atmosphere.

The energy of the sun evaporates water into the atmosphere from oceans and land surfaces. Evaporation is the change of liquid water to a vapor. Sunlight aids this process as it raises the temperature of liquid water in oceans and lakes. As the liquid heats, molecules are released and change into a gas. Warm air rises up into the atmosphere and becomes the vapor available for condensation. Some of the earth's moisture transport is visible as clouds, which themselves consist of ice crystals or tiny water droplets. The jet stream, surface-based circulations like land and sea breezes, or other mechanisms propel clouds and vapor from one place to another until they are condensed back to a liquid phase. Water then returns to the surface of the earth in the form of either liquid (rain) or solid (e.g., snow, sleet) precipitation. Some water on the ground, in streams, or in lakes, returns to the atmosphere as vapor through evaporation, and water used by plants may return to the atmosphere as vapor through transpiration, which occurs when water passes through the leaves of plants. Collectively known as evapotranspiration, both evaporation and transpiration occur at their highest rates during periods of high temperatures, wind, dry air, and sunshine.

Although the total amount of water on earth is fixed, the physical state of the water, on a time scale of seconds to thousands of years, is continuously changing between the three phases (ice, liquid, and water vapor), circulating through the different environmental compartments (ocean, atmosphere, glaciers, rivers, lakes, soil moisture, and groundwater), and renewing the resources. Average replenishment rates show considerable range, from thousands of years for the ocean and polar ice down to biweekly or even daily replenishment of water in rivers and in the atmosphere (EEA, 1995).

Average annual global evaporation from the ocean is six times higher than evaporation from land, while precipitation over the ocean is 3.5 times higher than over land. Based on current estimates, this results worldwide in approximately 40,000 to 45,000 km^3 of water each year transported from the ocean via the atmosphere to renew the freshwater resources (EEA, 1995). The balance of water that remains on the earth's surface is runoff, which empties into lakes, rivers, and streams and is carried back to the oceans, as well as recharging groundwater, and is potentially available for consumption each year (Institute of Water Research, 1997). This is the renewable supply of water, that part of the hydrological cycle that can be utilized each year without leading to the depletion of freshwater resources; it does not include groundwater resources that are no longer replenished.

This renewable resource is referred to as "blue water," while "green water" refers to the rainfall that is stored in the soil and evaporates from it. Green water is the primary source of water for rain-fed agriculture and for freshwater ecosystems, and is based on the level and flow of water in soil, which depends on soil texture and structure as well as climatic factors. It is significantly affected by land use and changes in land use, which are in turn affected by demographics (FAO, 1999). About 60 percent of the world's staple food is produced from rain-fed fields, along with meat production from grazing, and the production of wood from forestry (FAO, 1999). In sub-Saharan Africa, almost the entire food production, along with major industrial products such as cotton, tobacco, and wood, is produced from green water in rain-fed fields (Savenije, 1999). Water policy and planning focus almost exclusively on blue water, perhaps because green water can be managed only indirectly (FAO, 1999). This marginalization of rain-fed agriculture affects the accounting of water availability, since green water is not accounted for as water available for human use. Rain-fed agriculture in temperate zones tends to be highly mechanized and economically efficient, if energy demanding. In the tropics, small-scale farming is by far more prevalent. Small farms can also be run very efficiently, and in the wet tropics in countries such as Indonesia, Bangladesh, and Taiwan, there is often

Figure 11.2. The Hydrological Cycle

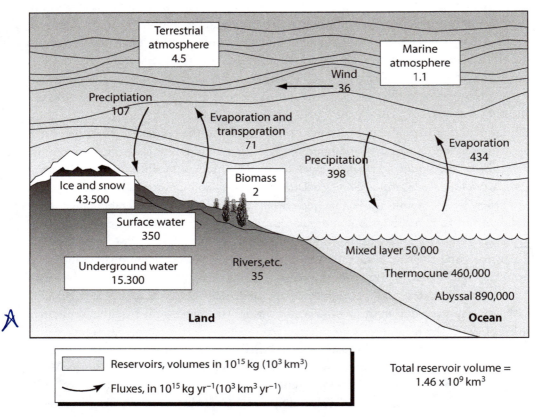

Source: NRC.

enough water to allow a second or third harvest through the use of irrigation (Savenjie, 1999). In the semi-arid tropics, especially, however, most agriculture is subsistence farming with very low efficiency. Africa, for example, gets at least 90 percent of its food from such farms. Food production could be increased in the area where it will be consumed by improving the efficiency of the use of green water for agriculture in these areas. In the semi-arid tropics, supplementary irrigation from groundwater can be used to optimize the use of green water for crop production. Relatively small quantities of blue water can safeguard crop production and, by doing so, produce high yields per cubic meter of blue water, much higher than can be attained from full-scale (dry season) irrigation (Savenjie, 1999).

Groundwater Supplies

Approximately one-third of the world's population depends on groundwater supplies, some exclusively (WRI, 2000). Almost 20 percent of global water withdrawals are taken from groundwater supplies, mostly shallow aquifers (WRI, 2000). Runoff from precipitation recharges shallow aquifers, just as runoff feeds rivers, lakes, and streams. Deep aquifers tend to contain fossil water that is hydrologically disconnected from the surface hydrologic cycle (Revenga, Brunner, et al., 2000). There are problems associated with each of these categories of sources. Fossil groundwater is finite—it recharges only over extremely long periods, if at all, so that when it is withdrawn it is mined, and, once used, cannot easily be returned to the deep aquifer. Shallow aquifers, while rechargeable, are susceptible to pollution from agricultural and industrial chemicals, among other sources. They are intimately tied to freshwater ecosystems, so pollution or overdrafting (unsustainable withdrawals) of shallow aquifers also affects adjacent freshwater ecosystems, depriving them of a significant portion of their flow and polluting their waters.

Because many developing countries depend on groundwater for irrigation, and on irrigation for food supply, the uncontrolled depletion of groundwater aquifers is seen as a serious threat to food security in many areas, in particular the North China plains and western and peninsular India (IWMI, 2001). About two-thirds of unsustainable groundwater withdrawals occur in India, in areas that are responsible for over 25 percent of India's harvest (Shah, 2000). The consequences include saline ingress into coastal aquifers, arsenic contamination, fluoride contamination, as well as declining well yields and increased pumping costs (Shah, 2000). Other areas that are also in danger of depleting their groundwater resources are urban areas across Asia, parts of Mexico, the High Plains (or Oglalla) aquifer in the Midwestern United States, Yemen (where groundwater abstraction exceeds recharge for the country as a whole by 70 percent (Briscoe, 1999), Saudi Arabia (where 90 percent of water supplies are achieved through groundwater mining) (PAI, 1993), and other Middle and Near Eastern countries. It is difficult to know the true extent of groundwater depletion, since reliable data on resource extent and use are lacking to a greater degree even than that for surface water.

The Role of Freshwater Ecosystems

Natural freshwater ecosystems play an important role in determining water quality and quantity and the sustainability of water resources. They provide a range of critical, life-supporting functions, including the cleaning and recycling of the water itself. A freshwater ecosystem is a group of interacting organisms dependent on one another and their water environments for nutrients (e.g., nitrogen and phosphorus) and shelter. It includes, for example, the plants and animals living in a pond, the pond water, and all the substances dissolved or suspended in that water, together with the rocks, mud, and decaying matter at the bottom of the pond. Freshwater ecosystems can be running waters (such as rivers and streams), lakes, floodplains and wetlands, or freshwater or brackish coastal habitats. Freshwater ecosystems provide water supply for human consumption and agricultural cultivation. They provide a host of critical goods and services for humans and other living beings including water, food fish, energy, biodiversity, waste removal, and recreation. These ecosystems, in turn, are important components of the hydrological cycle.

In theory, natural ecosystems are self-sustaining. However, major changes to an ecosystem, such as climate change, land-use changes, or the removal of a species, may threaten the system's sustainability and result in its eventual destruction (Scientific American, 1999). Destruction of or damage to a freshwater ecosystem removes it from or impairs it in its role in the hydrological cycle and diminishes the amount of freshwater the earth's environment can provide. Once a freshwater ecosystem is destroyed, it is difficult or impossible to restore it to its natural state.

Natural Influences on Water Availability

Geographic Location

Water is not distributed evenly across land masses, and much of it is far from population centers. A large percentage of global water resources is available in places such as the Amazon basin, Canada, and Alaska (Cosgrove and Rijsberman, 2000). About three-quarters of annual rainfall comes down in areas where less than one-third of the world's population lives (Gleick, 1996). Because precipitation is also temporally uneven, many people are unable to make use of the majority of the hydrologically available freshwater supply. Rainfall and river runoffs occur in very large amounts in very short time periods and are available for human use only if stored in aquifers, reservoirs, or tanks (Cosgrove and Rijsberman, 2000).

Figure 11.3 shows the large spatial variation in blue water availability, as well as the influence of population on overall availability. The two extreme cases are Asia and Australia/Oceania. Asia has the highest total water availability of the continents, but the least per-capita availability, while the opposite holds true for Australia/Oceania due to lower total population. Figure 11.4 shows the per-capita renewable water supply as calculated by river basin. The areas that have less aggregate per-capita water supply are more apparent, but additional spatial information gives a slightly clearer picture. For example, looking again at Australia, it is clear from Figure 11.4 that some areas, the most populated areas, as it turns out, have large per-capita supplies of water, while there are other parts of the continent, where very few people live, that have very small per-capita supplies of water. In Asia, it is apparent from this map that, although the continent as a whole may have relatively low per-capita water availability, some areas are quite well endowed with water. Again, the level of aggregation of water supply data makes a great deal of difference in the interpretation of the data.

Figure 11.3. Continental Total and Per-Capita Blue Water Availability

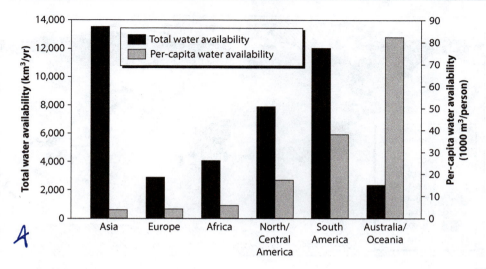

Source: Shiklomanov (1998), in Gleick (2000a).

Climate Change

Most researchers agree that climate change will result in higher variability of precipitation in many areas. Higher global and regional temperatures will mean increased evapotranspiration, changes in snowfall and rainfall patterns, and changes in the intensity, severity, and timing of major storms, among others (IUCN/WWF, 1998). This will result in changes in soil moisture, water runoff, and regional hydrology (El-Ashry, 1995). Arid and semi-arid regions will be most affected by these alterations (Frederick and Gleick, 1999). Because of these changes, the supply of and demand for water will change, as will its quality.

Changes will differ at various latitudes and will, like all weather and water resources, be locally specific. Some watersheds will experience increased runoff, more extreme storm activity, decreased soil moisture, increased groundwater recharge, or increased incidence of drought, or the opposite of these (IUCN/WWF, 1998). In some locations, the changes will be beneficial in terms of water management; in others they will be detrimental. Ecosystems are similarly variably affected. The severity of the effects will depend on the type of change, the ecosystem itself, and the nature of any human interventions. Some of these effects may include changes in the mix of plant species in a region, lake and stream temperatures, thermocline depth and productivity, lake levels, mixing regimes, water residence times, water clarity, reduced extent of wetlands, extinction of endemic fish species, exotic species invasions, and altered food web structure (IUCN/WWF, 1998).

Figure 11.4. Annual Renewable Water Supply Per Person by River Basin, 1995

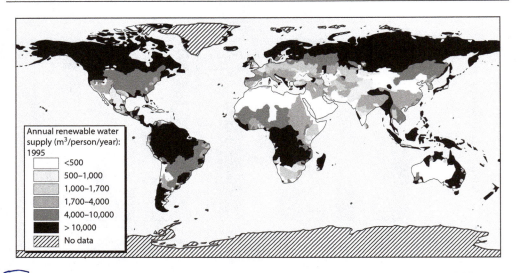

Deep aquifers will be less affected and may become even more important as a source of fresh water (Rivera, Allen, and Maathuis, 2003). Shallow groundwater may be affected by climate change—not only the shallow aquifers that may be depleted or the cause of waterlogging by changing recharge patterns, but also coastal aquifers and wells experiencing saltwater contamination through rising sea level.

Another potential repercussion of climate change is the warming of the earth's lakes. Lake temperatures may rise as much as seven degrees Celsius under carbon-dioxide doubling, decreasing the lakes' ability to naturally cleanse themselves of pollution (Poff, Brinson, and Day, 2002, pp. 13–15). Additional concern comes from the fact that these changes will evolve in a relatively short time frame. This will leave little room for adaptation or coping by the ecosystems, increasing the likelihood that they will be negatively impacted.

Climate change will have indirect effects on water resources as well, such as hydroelectric generation, human health, navigation and shipping, agriculture, and water quality (IUCN/WWF, 1998). For example, changed temperatures, flows, runoff rates and timing, and the ability of freshwater ecosystems to assimilate wastes may all affect water quality (IUCN/WWF, 1998). Many of these effects may be mitigated or worsened by human actions, so it is important to plan and take into account such changes to the extent that they can be recognized and predicted. Climate change may exacerbate changes in patterns of demand, population distribution, and poverty and food insecurity in marginalized communities. For example, arid and semiarid areas that receive less rain or rain in fewer, more concentrated time periods may experience flooding, increased crop failures, failure of wells, or other consequences that lead to out-migration or the need for water to be imported.

Calculating Water Availability

Hydrological water availability refers simply to the amount of water as liquid, gas, or solid, but not as molecularly associated with minerals and living resources in the earth's environment. However, the quantity of water practically available for humans is different from the hydrological measure of availability. The amount of water available to a population is defined by factors such as quality of water demanded, cost, environmental effects, political and legal agreements, individual wealth, and the technical ability to move water from place to place (Simmons, 1991; Gleick, 2000b). It also depends on the use to which the water will be put, because drinking water, for example, demands a higher quality of water than do industrial or recreational uses (Swan-son, Doble, and Olsen, 1999).

Although we can report continental average and per-capita water availability with some confidence, as in Figure 11.3, the actual amount available to any person or group for a particular use is less certain. Though this would clearly be a more useful and meaningful measure to have, it is very difficult to calculate, especially in areas where water is not metered, or is not closely monitored. It is also dependent not only on physical availability, but also management and investment levels that result in the development of reservoirs, dams, and other technologies to capture runoff and groundwater (Bernstein, 2002).

As will be described in more detail later, water resources are best measured in units of watersheds and aquifers. When water availability is calculated for populations in individual river basins (not including fossil water sources), 41 percent of the world's population (approximately 2.3 billion people) was found to live in river basins where per-capita water availability was less than 1,700 m³ per year. This is the level used as a cutoff by many water experts to indicate water-stressed populations (WRI, 2000). About 1.6 billion people, or 19 percent of the world's population, are found in this analysis to live in conditions of water scarcity, defined as water availability of less than 1,000 m³ per person per year (WRI, 2000). A different conclusion can be drawn when national boundaries are used as units. Earlier estimates using national water balances found 11 percent of the global population to be water stressed. As touched on earlier, the definitions of stress or scarcity used here are problematic under any calculations of water availability. The extreme differences in water availability that are created by the two different methods of calculation point out the dangers of putting too much credence in any such calculations. As discussed earlier, even the more sophisticated estimates based on watersheds are deceiving. They hide variability of access, temporal and spatial availability, and legal and political restrictions, and are therefore far from the ideal of individual water availability measures. Most importantly, they may over- or underemphasize existing or potential signs of stress, and hence poorly constrain the choices faced by policymakers, water managers, and others.

Discussion Questions

1. What is the hydrological cycle? How does it work?

2. Distinguish between green water and blue water.

3. How is water availability affected by different variables?

4. How is water availability calculated?

In-Class Assignment: Objective Questions

1. Green water is defined as
    a. Non renewable supply of freshwater.
    b. Salt water.
    c. Distilled water.
    d. Rainfall that is stored in the soil and evaporates from it.
2. Green water is the primary source of water for rain-fed agriculture and for freshwater ecosystems.
    a. True
    b. False
3. Rain fed fields produce what percentage of the world's staple food?
    a. 80%
    b. 40%
    c. 60%
    d. 0%

12 Approaches to Sustainable Water Management

Jill Boberg

As the previous chapters have made clear, various demographic factors are working together to place stress on water resources by increasing demand for water, while decreasing supplies through pollution and destruction of freshwater ecosystems. At the same time, polluted waters are adversely impacting human populations through disease and mortality. Whether or not a water crisis is imminent, measures need to be taken to reduce the pressures on water resources well in advance of their collapse. Measures aimed at ameliorating this seemingly intractable situation can approach the challenge in different ways. One is to attack the problem by influencing the demographic factors themselves—working to reduce population growth, change population composition trends, and so forth. There has been progress on this front, and the recognition of the factors that are most likely to impact or be impacted by water resources will allow planners to track these factors and, perhaps, influence them in a way that will increase water supplies or lessen water demand. This is a subject that has been covered in great detail elsewhere, and is outside the scope of this monograph.

Another tactic is to reduce the impact of these demographic factors. This can best be done through management practices. The management of water supply and demand can make large differences in water withdrawals, and can influence the quality of water and its impact on human health. Two locations with similar demographic and ecological profiles may have very different water demand and supply profiles due to any combination of the following:

- differences in management approaches
- factors both within and outside of planners' purviews such as institutional arrangements, government subsidies, land tenure, government corruption, and political factors
- technological innovation, market forces, globalization of production and trade, poverty, and armed conflict.

Jill Boberg, "Approaches to Sustainable Water Management," *Liquid Assets: How Demographic Changes and Water Management Policies Affect Freshwater Resources*, pp. 63-99. Copyright © 2001 by RAND Corporation. Reprinted with permission.

Supply management involves the location, development, and exploitation of new sources of water. Demand management involves the reduction of water use through incentives and mechanisms to promote conservation and efficiency. Because the distinction between the two is not always clear, supply management can be defined as actions and policies that affect the quantity and quality of water as it enters the distribution system, while demand management governs actions and policies affecting the use or wastage of water after that point (Rosegrant, 1997). In order to meet the needs of water users and the environment into the future, both supply and demand management approaches will need to be utilized effectively. The level of each will vary with the amount of actual and perceived water scarcity and level of development in an economy (Rosegrant, 1997). Demand management increases significantly in importance as populations and economies grow and the competition for and value of water increase. The benefits of efficient allocation of water become ever more important as these factors increase, generally as a result of a maturing economy, but also as the result of physical water scarcity.

Supply Management

Options for increasing water supplies include building new dams and water-control structures (temporal reallocation), watershed rehabilitation, interbasin transfers (spatial reallocation), desalination, water harvesting, water reclamation and reuse, and pollution control (Meinzen-Dick and Appasamy, 2002; Winpenny, undated).

Dams and Water-Control Structures

The costs of developing new water sources, and particularly of large-scale supply solutions, have become prohibitively expensive, and have reached their financial, legal, and environmental limits in most industrialized and some developing countries (Frederick, Hanson, and VandenBerg, 1996). Lending by international donors for irrigation projects has declined significantly since its peak in the late 1970s, as have total public expenditures for irrigation in many countries, especially in the developing world. The growth of irrigated land area has declined in recent years, after nearly doubling in the first half of the 20th century, before more than doubling again between 1950 and 1990 (PAI, 1993). This is due partly to investment increasingly having to be made in rehabilitating existing systems, but also because development costs are so high (Pretty, 1995). The decline has been greater in developing countries, where investment in irrigation has been lower for a longer period.

Dams and water-control structures are often extremely controversial. Currently, there are more than 41,000 large dams (defined as dams over 15 meters high) in the world, impounding 14 percent of the world's annual runoff (L'vovich and White, 1991). The dams are used for energy production, flood control, and water storage, as a way to reallocate water temporally. Most of the best dam sites in the developed world are already taken, and dam construction there has all but stopped. In the developing countries, however, demand

and potential is still high (WRI, 2000). As of 2003, there were 1,500 dams over 60 meters high under construction, with the largest number in Turkey, China, Japan, Iraq, Iran, Greece, Romania, and Spain, as well as the countries of the Parana basin in South America (Brazil, Argentina, Paraguay, Bolivia, and Uruguay) (WWF, 2004). Huge benefits may accrue from large dams, such as electric power production, irrigation, and domestic water provision. However, dams have equally large environmental and social consequences, such as the submergence of forests and wildlife, the contribution to global climate change of greenhouse gases from the decaying of this submerged vegetation, the impact of changed flow in the dammed river, the relocation of thousands of families (as discussed in the section on migration) and the destruction of villages and historic and cultural sites, and the risk of catastrophic failure, among others. Additionally, there is a significant consumptive use of water associated with dams, as the evaporative losses can be large, depending on climate and the surface area of the reservoir. In Egypt, for example, a hot and arid climate results in annual evaporative losses estimated at nearly three meters (or 11 percent of reservoir capacity) at the Aswan High Dam. This compares to evaporative losses ranging from 0.5 meters to 1.0 meter in various parts of the United States (Rosegrant, 1997).

The planned construction of large dams in India, China, Zimbabwe, and Brazil has stirred intense debate and protests within those countries and internationally. One reason for the public reluctance toward large dam projects is that the social and economic costs of construction are borne by a relatively small group, while the benefits are widely distributed. It is possible that the resistance to such projects would be lessened, though not eliminated, if those bearing the costs of the project were adequately identified and equitably compensated. Additionally, the costs of not proceeding with a project are often not analyzed. A careful, impartial, and comprehensive cost-benefit assessment may help to clarify the alternative futures associated with the construction of a large-scale irrigation or other dam project, as well as the alternative solutions to the problems being addressed.

One dam, the Salto Caxias in Brazil, took an approach to dam building that created less controversy. The builders of the dam engaged the people who lived in the area that would be inundated and worked with them to establish resettlement priorities, and then added environmental priorities such as new farming methods, tree planting, and the establishment of a national park on the land surrounding the new dam. The cooperation with displaced families eased many of the social problems associated with dam building, but concern remains over the environmental implications of the dam, despite the efforts made to relieve them (Sutherland, 2003).

Watershed Rehabilitation

As discussed in the previous chapter, one of the best and most cost-efficient ways to secure water supplies around cities is conserving forests (Dudley and Stolton, 2003). As discussed, when watersheds are deforested, rainfall turns into runoff, which causes erosion, and sediment transfers and leaves no possibility of water storage, unless a dam is used. In areas of heavy seasonal rainfall, as in many tropical areas, especially in many developing countries, this effect is exacerbated. In developed countries, watersheds are often protected, maintaining vegetative cover and limiting or eliminating human, livestock, and agricultural use of the land. In most developing countries, governance of watershed use is weak, and the long-term benefits of nonuse or rehabilitation fare poorly against short-term political or human goals (McIntosh, 2003). This is especially true in densely populated poorer countries, where competition for land and resources is fierce. It is argued that reforestation of degraded and unproductive land is cost effective in the long term if all the ensuing benefits (e.g., water supply, agricultural revenue, pollution reduction, and fuel and timber benefits) are taken into consideration ("Dry Future," 2000). The high initial cost combined with the incompatibility of long-term planning with short-term political thinking, however, make watershed rehabilitation a difficult decision in many places.

Small-Scale Irrigation Systems

Given the problems surrounding large-scale water projects, small-scale irrigation systems may provide an opportunity to expand irrigation with fewer constraints. Studies in sub-Saharan Africa have shown that small-scale irrigation systems are not inherently more or less successful than large-scale irrigation projects. Instead, their success depends on institutional, physical, and technical factors. Successful small-scale irrigation systems there had the following characteristics:

- simple, low-cost technology (usually small pumps)
- private and individual arrangements for operating the system
- supporting infrastructure adequate to allow the sale of surplus production
- high and timely cash returns generated for farmers
- farmers who are active and committed participants in the project design and implementation (Brown and Nooter, 1992).

Because each watershed is different, the large-versus-small distinction is only one of many variables that must be taken into account when projects are being considered. Both large and small projects must involve broad accounting of the project costs and benefits, including environmental and social externalities and benefits in all sectors, and fair compensation for negative impacts. Such decisions should be taken in consultation with those who will be impacted by the project, both positively and negatively.

Groundwater

The careful exploitation of groundwater resources is another potential option for increasing freshwater availability in many countries. Several large aquifers lie under the arid Middle East and North Africa regions, for instance. These aquifers store volumes of water equal to several years of the annual average renewable supply of the region, but recharge rates are often 2.5 percent of total volume stored per year or lower, which means that sustainable use of the resource can tap only that rechargeable quantity per year. Much of Asia and parts of Latin America have untapped groundwater potential. In much of sub-Saharan Africa, aquifers are small and discontinuous, with slow recharge rates, so use will be limited to local and regional areas.

Groundwater is generally a more efficient source of irrigation water than open canals. This is because the water can be accessed near to where it will be applied, reducing transportation losses; in addition, farmers can control the amount and timing of the water, giving an advantage over irrigation from surface water. In India, crop yields from farms irrigated by groundwater were found to be 1.2 to three times greater than farms irrigated with surface water (Shah, 2000).

Groundwater utilization has been seen as a means to alleviate poverty, since anyone who can afford to install a pump has access to free water (Moench, 2002). The problem is that, as with any common resource, the incentive exists to waste and overuse the resource. This has led to overuse of many aquifers and the falling of many water tables. Also, since groundwater development is undertaken, for the most part, on an individual basis, there is much less oversight and a much smaller body of law governing the allocation, management, or monitoring of groundwater (FAO, 2003b). This and the fact that groundwater, unlike surface water, cannot be readily seen, make conservation measures and improved efficiency of use of groundwater extremely difficult. Also, in developing countries, groundwater storage and recharge rates are generally poorly understood. An increase in the knowledge of aquifers and their characteristics would help to maintain sustainable use of this resource.

Interbasin Transfers and Exports

Interbasin transfers are another way to provide water to areas with dense populations, such as urban areas, or other water-short areas. Canals and pipes are the usual means of transporting the water, though new technologies such as large water bags attached to ships are another option that has been used for freshwater deliveries to emergency situations and for coastal areas with a need and desire for expensive water (Gleick, 2001b). Such large infrastructure projects are expensive and potentially environmentally harmful, both to the freshwater ecosystem from which the water is extracted and the lands over which the pipes or canals must flow. Additionally, inhabitants of the land lying in the path of the pipes or canals are demanding equivalent access to the water, adding to the demand from the source (Meinzen-Dick and Appasamy, 2002).

Water exports, where water is sold across state or national borders, are a controversial subject. Although the spatial reallocation of water makes theoretical sense, the selling of water, more than the sales of other resources, seems to strike a nerve in the public. Two recent plans to sell water from water surplus areas have met with resistance. The controversy in both cases was the idea of corporate profits taking precedence over environmental concerns, and the issue of water as a common resource held in public trust. In Michigan, a private company wanted to bottle spring water for sale within the United States, and in Canada, a private firm wanted to transport water in tanks to sell to drought-stricken

U.S. states, (Schneider, 2002; Edmonds, 1998). Residents in both cases felt that the water should not be given to a private firm to sell for private gain with no benefits, and perhaps some costs, such as environmental repercussions and resource depletion, falling to residents. There is also concern surrounding the ways in which private companies get access to water supplies. Though further discussion is beyond the scope of this paper, it is important to note that this sort of public opposition to the export of water resources comes up repeatedly. Barriers of this kind will pose a challenge to the use of this option for the augmentation of water supply.

Water Reallocation

The competition between cities, industry, agriculture, and environment for water resources is an issue that is expected to become more contentious over the next several years. Even though total domestic water use is far less than agricultural use, the rapid growth of urban areas will result in the growth of water demand in some locations. One of the ways of supplying increased demand in both the urban domestic and industrial sectors will almost certainly be a reallocation of water to these sectors from the agricultural sector.

This is already happening in many places. In the western United States, agricultural interests that had nearly unlimited access to cheap water are finding their access restricted and their payments increased. In the developing world, reallocation is also occurring, though often in a more informal manner. In India, for example, well owners and irrigators pump water from rivers into tankers, who then sell the water to households in the urban areas (McKenzie and Ray, 2004). This informal market, though illegal, is possible because of the differential value of water across sectors and because of lack of infrastructure and access to water in the domestic sector. Wealthier households connected to the public water system pay a subsidized rate more than ten times less for water than lower-income households buying from informal water sellers, and farmers can grow crops with low water requirements, sell their remaining allotment of water, and make more than their colleagues who grow traditional crops utilizing their entire water allocation.

Water reallocation can be done via supply management, by reallocating water from the top down, or via demand management, using incentives to move water between sectors. Because water in developing countries is predominantly consumed by the agriculture sector, reallocation of even small percentages of water consumption to the domestic sector can fulfill domestic needs.

Although there are many examples of farmers willingly sharing their water supplies when compensation is offered, most notably in the western United States, Mexico, and Chile (ECSP, 2002), there is a limit to the reallocation of water away from agriculture, as food production has a strong political component because of rural jobs, food as a basic need, and food security. A reduction in economic activity associated with the reallocation of water away from the agricultural sector could limit future economic development in a region and even induce out-migration (Rosegrant, 1997). However, in Chile and California, the effects have been small to this point, with agriculture revenues dropping only two to three percent in California counties where water was sold (Dixon, Moore, and Schechter, 1993). In Chile, farmers sold only small portions of their water allocations and practiced very efficient irrigation with their remaining water, resulting in only small revenue reductions (Rosegrant, 1997). This highlights a positive outcome of water reallocation: an increase in agricultural conservation practices, which become more economically appealing with high water prices. That is not to say that economic effects might not be large if participation were greater and farmers began selling larger portions of their allocations.

Desalination

Desalination is the process of turning saltwater into fresh drinking water through the extraction of salts. Desalination through evaporation (or distillation), in its simplest form by the sun, has been used for many years, but large-scale desalination by evaporation is an energy-intensive process that is extremely expensive. Additional costs are incurred when the water is moved from the sea to its destination. Currently, mostly wealthy, energy-rich, water-scarce countries, such as Kuwait and Saudi Arabia, and Israel use desalination on a large scale, although Tampa, Florida, and Houston, Texas, among other cities in the United States, are constructing desalination plants to reduce pressure on groundwater in those areas. Island nations such as Cyprus, Trinidad, and the peninsular city-state Singapore have also invested in desalination plants. Desalination accounts for only 0.2 percent of water withdrawals worldwide (Martindale, 2001). Currently, desalinated water is delivered to users at a cost of $1–4 per cubic meter, comparable to the costs of new water supplies in the most arid areas of the world, but very high compared with alternative sources in much of the rest of the world. For example, desalination costs are well above the rates paid by urban users in the United States of about $0.30 to $0.80 per cubic meter (Gleick, 2000a). Transportation costs are high for water, so attempts to use desalinated water in inland areas incur a significant cost penalty. A desalination method called membrane desalination uses reverse osmosis to produce freshwater with less energy, and therefore more cheaply, than evaporative methods. This is the method being used in the Tampa and Houston plants, and it is possible that this method will prove cost effective over a wider range of applications.

There are other issues associated with desalination, especially the disposal of the brine stream produced during the process. If desalination occurs near an ocean, with care the stream can be disposed of there, though there are reports of marine pollution from the dumping of high-temperature brine (Barlow and Clarke, 2002). Also, the energy used to fuel the desalination process is itself generally polluting, and even may use water as one of its inputs. Although solutions such as the use of renewable energy to fuel desalination plants may solve this problem, they would add unacceptable costs to the process.

Desalination research programs continue to make advances that will bring the price of desalination down. Although it is unlikely that desalination will make a large dent in the supply of world water in the near future, it is important in certain situations, such as for domestic and industrial purposes in coastal areas of water-scarce, relatively wealthy countries. Technology and costs may change, and desalination may become more important in time. At the moment, however, the expense, both economic and in terms of the environmental impacts of energy use and brine disposal, seems high, and this supply option, like many technological solutions, may be better left as a last resort option.

Water Harvesting

Water harvesting can be as simple as a water barrel under a rooftop gutter, or as complex as a sub-watershed scale storage and groundwater-recharging scheme. The capture and diversion of rain or floodwater to fields for irrigation is an old tradition in many agricultural areas, and the traditional technology can be more widely used and improved upon to increase production and income on farms in some areas. Traditional techniques include stone or earthen berms constructed across fields to either contain or redistribute rain and flood-waters, and vegetative barriers that slow runoff. Often these techniques also reduce irrigation in fields. In the urban areas in the domestic sector, water harvesting, now required in new buildings in some Indian cities, can provide positive environmental benefits by reducing the polluted runoff from urban areas into surface waters, along with reducing water use from other sources.

The storage of runoff during the rainy season is particularly important in climates where much of the available rainfall falls within a short period. The water collected can be used either directly or to recharge the groundwater, and its generally small-scale and environmentally friendly nature makes water harvesting appealing (Meinzen-Dick and Appasamy, 2002). The water that is harvested would have otherwise run off into, for example, seasonal rivers or oceans; water harvesting, by changing the temporal distribution of water, may modify the natural cycle in either positive or negative ways, particularly in rural areas. Water harvesting is a good example of more efficient use of green water, and can be an important way to increase efficiency, productivity, and soil fertility on a local and regional basis.

Water Reclamation and Reuse

More important in the industrial and domestic sectors might be reuse and recycling of water. Pure water is not needed for many tasks such as irrigation, and some types of washing and cooling tasks, as well as environmental and ecosystem restoration and the matching of water quality and water requirements can be an important part of better water management. In developed countries, most notably Japan and the United States, industrial recycling of water has been an important component of water-use reduction in the industrial sector, and in developing countries there is a movement toward seeing wastewater as a resource, rather than a problem. It is a cost-effective and reliable water resource despite significant social and economic barriers to waste-water reuse in some regions (IUCN, 2000). The advantages of wastewater reuse include water-pollution abatement, not discharging into receiving waters, and providing long-term water-supply reliability within the community by substituting for freshwater.

Water reclamation and reuse occurs when water is used, recaptured, and cleaned for reuse. Nonpotable reuse of water for irrigation of agricultural fields and landscapes, industrial cooling, and some commercial indoor uses are the most commonly encountered. The direct potable reuse of wastewater is not popular with the public, and its use is limited to date to only Windhoek, Namibia, and Denver, Colorado, in the United States (Shaw, 1999). In the future, there may be more scope for direct reuse, but the relatively high cost of treatment and community attitudes make it likely that other uses will be expanded first.

Only a small portion of industrial water, for example, is actually consumed. In most cases, water is withdrawn and returned to the freshwater system at a lower water quality. Instead, the remaining water can be treated to any level of quality, at different costs, or may be reused within a factory without treatment. This results in the reduction of both water requirements and effluents, improving the output per cubic meter of water. In developed countries, water recycling in industry has become commonplace, and has resulted in a reduction in the total industrial water use as well as the water use per dollar of output in many countries, especially Japan and the United States. Between 1973 and 1989 in Japan, for example, total industrial water use declined by a quarter, while industrial output per cubic meter of water used declined from $77 to $21 over that same period (Postel, 1997), and has remained stable since that time (Maita, 2003). In 1998, the ratio of recycled water to industrial water use was 80 to 90 percent in the steel, chemical, and allied product industries, compared to 20 to 40 percent in the paper and allied product, plastic product, and textile industries, which may be used as a goal for similar industries in other countries (Maita, 2003).

Because wastewater is discharged year-round, treated wastewater is a good match for industrial use. In Madras, India, several large industries have been paired with the local water agency to use partly treated sewage (Meizen-Dick and Appasamy, 2002). The industries purchase and treat the sewage, then use it for cooling and manufacturing processes. Both the industries and the water agency gain from the arrangement—the water agency gains revenue, reduces pollution, and saves treatment costs, while the industries have a reliable source of water, something that was not assured previously.

Wastewater and gray water, which are generally nutrient rich, can also be reused for food production and other tasks that do not require pure water. Grey water from showers and sinks, for example, can be used directly in gardens or stored in soak pits for groundwater recharge. Reclaimed water is also increasingly being used to augment natural groundwater recharge and for environmental and ecosystem restoration (Gleick, 2000a). Wastewater reuse measures are already in place in some arid areas such as Israel, Portugal, Tunisia, Namibia, and California, as well as in Japan. Countries such as Morocco, Jordan, Egypt, Malta, Cyprus, Greece, and Spain, as well as France and Italy are considering or actively pursuing wastewater reuse schemes. In Israel, for example, treated wastewater provides 18 percent of the country's total water supply, and nearly 40 percent of the agricultural water supply. The most successful wastewater-reuse irrigation projects have been located near cities, which can provide both wastewater and a market for agricultural products. The rate of expansion of wastewater reuse is dependent on both the final quality of the wastewater and the willingness of the public to embrace the use of wastewater supplies (Rosegrant, 1997).

The leading problems associated with wastewater reuse are high salinity levels and the accumulation of heavy metals in the water, which can make the water unusable in most applications, and can increase the salinity level in agricultural soils when used as irrigation water. There are also concerns about risks to workers and consumers from bacterial contamination, but this problem can be easily controlled with adequate treatment of water, if necessary, before reuse.

Pollution Control

A final supply option is to maintain the quality of the current supply of water. If water sources are protected from contamination from agricultural residues, soil erosion, runoff from urban areas, industrial effluent, chemicals, excess nutrients, algae, and other pollutants, the current supply of water can be retained, the cost of developing alternative supplies can be saved, and previous sources can be opened up after they are cleaned (Winpenny, undated).

Controlling pollution reduces the impact of water on populations by maintaining public health, maintains freshwater ecosystems and ensures their continued provision of goods and services, and provides recreational opportunities. Although most countries have pollution-control laws, including effluent standards that are required to be met by all polluting industries and municipalities, many, especially in developing countries, have not had the political will or financial resources to enforce them. Pollution control laws in developed countries have helped to clean up rivers, lakes, and streams and have promoted conservation and efficient use of water, but they have also promoted the more efficient use and conservation of water, since the most cost-effective way to meet pollution limits and water quality standards is often to reuse and recycle water, especially in industrial processes.

In many cases, municipalities may not be able to build sewage treatment plants with enough capacity to handle large and growing loads of effluents without financial assistance from higher levels of government. Even with assistance, industries will generally have to either pretreat their wastes or invest in individual or common treatment plants. However, because industries in developing countries have been reluctant or unable to do this, incentive measures are needed.

As in higher-income countries, effluent charges based on the quality and quantity of effluent discharged can encourage industries to conserve water and reduce pollution (Meinzen-Dick and Appasamy, 2002). Of course, this must be coupled with stricter regulations and enforcement, and strengthened by public disclosure and education.

Some unexploited water sources can still be harnessed, but few are left that would not have some impact on environmental systems, or that are readily accessible and economical. Therefore, the most economic option for water planners may increasingly be to meet additional requirements by reducing demand by any number of means, and by looking at alternatives such as water reclamation and water harvesting. Not only are these options often the most appealing from a strictly monetary standpoint, they are also often the most appealing from an environmental and public acceptance standpoint. Water users seek the services that water provides, and if the same services can be provided reliably in a way that causes less damage to the natural environment or at lower cost, water resources will be further protected for future use.

Demand Management

Although supply-management measures such as those discussed previously will help meet demand for water, much of the water to meet new demand will have to come through demand management, including conservation and comprehensive water policy reform.

Without appropriate management, no technical or engineering fixes will solve all of the problems of water supply and demand and the ecological damage that is associated with its misuse. Sustainable water development and management require the integration of social and economic concerns with environmental ones. Management approaches such as integrated water resources management and demand management offer effective means of providing water for human use while easing the stress on freshwater ecosystems and the ecological goods and services that they provide (CSD, 2001).

Overall, *poor governance* is at the core of many water problems, especially in developing countries. For example, relative water scarcity may be partly caused by too much water being allocated to too few activities with too high a subsidy (CSD, 2001). Poor governance results, in most cases, from corruption and outside interference, and is exacerbated by low tariffs, which lead to a host of problems such as low water supply and sanitation service, overexploitation of water resources, conflict between rural and urban users, and poor maintenance of existing water supply lines (McIntosh, 2003). Further mismanagement arises because the most efficient and effective management occurs at the scale of the watershed. However, the technical, economic, and institutional constraints on performing water management on that scale are formidable.

Recall that there are three types of water scarcity: absolute, economic, and induced. Usually scarcity in a particular location is caused by a combination of two or three of these scarcity types. *Demand management involves saving water from existing uses by reducing losses.* Losses can also be of any of three types: absolute, economic, or induced. Absolute losses occur, for example, when water vapor escapes to the atmosphere through evaporation or evapotranspiration, or when water flows to salt sinks such as oceans or saline aquifers. Economic losses occur when water flows to freshwater sinks, but is economically unavailable due to its location, when water is polluted to the extent that it is no longer economical to treat it to a usable quality, or when drainage water is too expensive to reuse due to factors such as physical characteristics of aquifers, deep percolation, slopes, and lifts (Rosegrant, 1997). Induced losses occur, for example, when the effective cost per unit of lightly polluted water is higher because crop yields per unit of water are lower (Rosegrant, 1997), or when low tariffs or high subsidies misallocate water. If the effective price of water and the true scarcity value of water are greatly different, water will be lost to the system. The effective price of water consists of more than just the tariff and subsidies; it also involves physical considerations and other factors.

The point of demand management is to reduce the losses in all of these categories. In order to do so, comprehensive policy reform may be required in many places. Because of entrenched interests, cultural and religious considerations, and tradition, policy reform, like all reforms, may be very difficult and slow. However, the reforms are generally needed if water scarcity is to be averted, since, as discussed earlier, supply augmentation alone will not provide a sustainable water supply indefinitely. Several categories of policy instruments can be used to enact demand management and improve governance of water systems. They include the following (Bhatia, Cestti, and Winpenny, 1995, in Rosegrant, 1997):

- *Enabling conditions:* actions that change the institutional and legal atmosphere in which water is supplied and used. This may include the broad improvement of governance by making policies transparent, the involvement of civil society, and regulatory bodies, as well as specific reforms such as water rights
- *Market-based incentives:* actions that use economic incentives to directly modify the behavior of consumers. Policies include tariff reform, which may raise or lower prices for all or certain users, adjust subsidies and taxes, and include charges for pollution or effluent flows
- *Nonmarket instruments:* policies that use direct, nonincentive-based laws and regulations to regulate water use. These include quotas, licenses, pollution controls, and restrictions
- *Direct interventions:* technological and other interventions to conserve water. Examples include leak detection and repair programs, investment in improved infrastructure, and conservation programs.

The exact nature and combination of policies will vary from country to country, and even within countries. The choices made will depend on the existing quality of governance, institutional capacity, level of economic development, relative water scarcity, and other factors. Some examples of demand management policies that can make a difference in many countries follow.

Enabling Conditions

Water quality matching. As noted above, high-quality water is needed for human consumption and certain industrial processes, but many other uses can be met with lower-quality water such as storm water, gray (lightly used) water, or reclaimed wastewater that can be used for landscape irrigation and some industrial applications. In order to facilitate the reuse of water, single-pipe distribution systems can be replaced with integrated systems that can deliver different qualities of water to and from the places it is needed. These systems are cost effective and practical, especially in new construction and areas without existing piped supplies.

Management of water must meet the needs of people, rather than merely supply water. In other words, the emphasis should be on providing goods and services, not water per se. As a result, efficiency is part and parcel of the management approach, helping to reduce the impact of demographic factors such as population growth, decreases in household size, and rising standards of living, by ensuring that the least amount of water is used to provide the most services. Also, as discussed previously, it is important to distinguish between water needs and water wants, and reducing nonessential wants and wasteful practices when necessary. Inefficient water-resource utilization occurs when water is allocated to activities that are not in line with the socioeconomic and environmental objectives of a country or region, especially if other activities that would better serve development objectives are deprived of water. Most societies give direct human consumption of water highest priority, an objective that is codified in national water policies, and many also give priority to agricultural use (Meinzen-Dick and Appasamy, 2002). Conflict among users results, made worse when water distribution is too inflexible to adapt to changing supply and demand conditions (Frederick, Hanson, and VandenBerg, 1996). The institutional structure of water management must be revised in many countries to meet the needs of new economic and social imperatives, and to incorporate new information on ecological and human needs.

Decentralized supplies and involvement of users. Although not appropriate in every case, decentralized water systems are cost-effective once the very cheapest centralized solutions are exhausted, and reliable when users are involved in the planning, execution, operation, and maintenance of water systems (Wolff and Gleick, 2002). In peri-urban and suburban communities, decentralized water systems may be more economical than extending the centralized system and those communities may also prefer to manage their own systems (Meinzen-Dick and Appasamy, 2002).

In addition, poor communities in urban areas may not receive wastewater and water services at the same level as do richer communities. This is especially acute where formal and informal settlements exist in an urban area, as is the case in many cities in developing countries, but exists as well in different forms in urban areas of the developed world. Low-income communities may wish to have their own community-managed systems if they will improve performance or lower costs, as might be the case in an area without infrastructure where water is brought in by tanker trucks or kiosks (Meinzen-Dick and Appasamy, 2002).

A good example of the decentralized approach to groundwater management is in Southern California, where the governance structure is agreed upon and managed by water users, is responsive to local conditions, adapts to the dynamic environment, and uses existing data. It is, in many respects, very successful and efficient, and demonstrates the importance of water managers engaging communities and individuals in water management, and interacting closely with them to aid them in decision making for water systems.

Privatization. Some see the development of water markets and the privatization of water management as ways to allocate water more efficiently, although both methods are extremely controversial. Water markets can encourage conservation by allowing the trading or selling of water rights, for example. Private-sector participation in water-resources management and financing ranges from building to operating and owning municipal water systems. Because private firms operate under profit motivations, concerns about equitable access to water for the poor, the integration of environmental concerns, and the sharing of risks must be addressed before privatization becomes a viable option (UNEP, 2002).

There have been both successes and failures in the privatization of water systems. In general, however, it is understood that some private-sector investment is required in many places, especially in developing countries, if water systems are to be successful in the future. However, the lessons learned by privatization occurring in the last several years should be taken into account, such as these (McIntosh, 2003; WSTB, 2002):

- *There are many forms of privatization of water services, and the form chosen must be adapted to the culture, political structure, and legal and regulatory framework of a particular location.* In many developing countries, there is resistance to the idea of international companies owning water systems. In these cases, a domestic owner or partnership may work better, as may a public-private partnership. An even playing field for all possible owners and operators should be a priority
- *Improved performance by public water utilities has resulted in some cases due to pressure from large national and global water companies.* Support should be given to governments to help them meet water needs without private-sector participation, if viable, as a first step. Often, tariff reform and regulatory changes are needed elements of water sector reform whether the owners and operators of the water system are public or private, and true reform will be impossible without these elements. For example, in the United States, public urban utilities can now enter into contracts of up to 20 years for the operation of private participation in the operation of plants that were funded by public bonds or grants. This was made possible by the loosening of federal tax laws. Additionally, the Safe Drinking Water Act of 1974 has pressured small- to medium-sized utilities to improve their operations or seek assistance in the private sector
- *Not all privatization efforts are successful.* In many cases, especially in developing countries, promises of efficiency and total income from water are unfulfilled. For this reason, the municipality must retain the ability to monitor performance and assume operations in case private operations fail. Regulatory arrangements must exist before privatization of assets, and must give the power to enforce such regulations to independent entities. Openness and transparency are equally important.
- *Good relations are necessary between governments and private operators.* Social, ecological, and cultural matters must be considered and agreed upon in consultation with governmental and local input

- *Tariff revision, structures, and mechanisms for tariff setting must be agreed upon.* The public has shown its willingness to pay for reliable, high-quality water. However, many governments are reluctant to raise tariffs. This is essential to the success of water systems, and must be agreed upon in advance. The poorest sector of the urban population must be adequately served, often a major constraint in developing countries
- *The water services industry faces a great need for maintenance and replacement.* This is true in both high and low-income countries, as demographic expansion has outpaced municipal investments. However, using privatization as a means of financing investment is unlikely to be the cheapest way. Host countries bear almost all of the risk of price demand or exchange rate, and investors assume almost none. Additionally, unlike debt, such a transaction does not attract relief measures (Lobina and Hall, 2003)
- *Some types of contracts have proven successful; others have proven risky to the host country.* For example, the design-build-operate contractual arrangement and its variants have been successful in some large water-service systems in the United States. However, take-or-pay agreements often oblige public authorities in developing countries to buy bulk quantities of water irrespective of future demand. In this case, public funds guarantee multinationals' profits at taxpayer and consumer expense (Lobina and Hall, 2003)
- *Good and reliable water sources are essential for reliability in the long term, and therefore will be critical to water services privatization.* This includes issues pertaining to the development of watershed lands and the protection of environmental quality around reservoirs. This is essential whether privatization of water services takes place, and is something upon which the local or national government involved must act
- *Employee rights should be protected, including offering retraining or transfer.* Labor force issues are important both as a possible source of cost savings and a focal point for public concern
- *Most water utilities will continue to be publicly owned and operated.* However, in the most troubled cases, private help will be needed, either permanently or temporarily. It is important that all contracts set up to transfer ownership or operation of water services to a private entity are carefully, independently, and transparently negotiated so that all parties wind up better off.

Economies of scope

Economies of scope can reduce the total cost of water systems by combining the decisionmaking process of several authorities, integrating thinking about land-use patterns, flood control, and water demands. Integrated water-resource management involves the management of water resources at watershed level. It recognizes the need for institutions and policies to consider the watershed basin as a whole when management decisions are made, to assure that, for example, interventions made upstream do not negatively affect users located downstream.

Several requirements must be met in order to realize the potential of integrated water-resource management. Adequate funding, human and institutional capacity-building, and a better grasp of the extent of the freshwater resource and supporting ecosystem need to be combined with education and training and the transfer of appropriate technological solutions. This is particularly important in areas in developing countries that are already water-scarce (UN, 2002a). As mentioned earlier, a critical step in improving water resources management is the recognition of the primacy of water catchment areas as the appropriate scale for effective water and ecosystem management. Management decisions taken at the level of the watershed allow all impacts to be accounted for, as opposed to decisions at the level of an individual water source. This is even more imperative in arid and semi-arid climates where changes in land use and vegetation have clear implications for land and water (SIWI, 2001). This will often require cooperation and coordination between countries or administrative units, which is extraordinarily complicated due to political prerogatives, but is nonetheless the best way for water resources and freshwater ecosystems to be managed competently and resources equitably distributed.

Market-Based Incentives

Water pricing. The recognition of the social and economic value of water and of freshwater ecosystems allows the resources to be compared with other social and economic goods and reinforces their status as scarce and essential resources. Water supply has in many cases not been treated as a commercial enterprise. In most developing countries, farmers and households either do not pay for water per unit used, or pay a low price (Meinzen-Dick and Appasamy, 2002). Governments, businesses, farmers, and consumers must treat water not as a free good, as they often do now, but rather as a scarce resource that comes at a price. (Rosegrant, 1997). If water is seen as a free or nearly free good, the incentive for conserving it is absent. Limited access to water adds costs to water the same way that higher economic prices do. Water pricing can help to assure adequate supplies of water, as well as encourage environmentally appropriate usage. In the water-scarce arid and semi-arid western United States, farmers traditionally have paid nothing for water and only a small amount for its transport to their farms. They then have applied it liberally to low-value crops, imparting a marginal value of well under $50 an acre-foot—the amount of water needed to cover an acre of land with a foot of water (Frederick, 1998). The water would have more value left in stream to provide hydropower, fish and wildlife habitat, and recreation than it does when diverted to irrigation in this way. The value of water might also rise by selling it to urban areas that are spending ten times as much to augment supply in other ways (Frederick, 1998).

Keeping water prices low can be presented as controlling inflation or making service affordable to the poor. However, in reality, it ensures service deficiencies such that the poor never receive their supply and have to pay very high rates to vendors. Raising prices, on the other hand, can result in civil unrest, especially if it has to be done before service can be improved (Rogers, Kalbermatten, and Middleton, 1999). This may be prevented by raising prices gradually, and by involving the community in the implementation, especially if the stated goal of such price increases is to improve reliability and to invest in connecting the urban poor to piped water.

There are many benefits connected with increased prices, such as the following (McIntosh, 2003):

- Demand is reduced, conservation is made viable, substitutes are relatively cheaper, and consumption preferences change
- Supply is increased, as marginal projects become viable and efficiency measures are economically attractive
- Reallocation between sectors, especially from agriculture to domestic and industrial, is facilitated, as are reallocations between off-stream and in-stream uses
- Increased revenues at utilities allow for improved managerial efficiency, partly by making modern monitoring and management techniques and staff training affordable
- The per-unit cost of water to the poor is reduced if piped supply is extended, since many (most, in some places) poor people rely on expensive water from vendors.
- Environmental sustainability is more easily attained due to reduced pollution loads (especially due to recycled industrial water) and reduced demands on water resources, leaving more water available for ecosystems (Rogers, 2001).

A common problem in developing countries is the flawed nature of price structures. This is especially true when lifeline tariffs are set too high, as is frequently the case in developing countries. Often such tariffs benefit only those who have a house connection, who are more likely to be urban rich than urban poor or rural dwellers of any income bracket. Prices should increase with increasing consumption. Low-income users need to be protected by lifeline rates, but the higher levels of consumption should be charged at the marginal cost of developing new supplies (which is what excessive use will eventually force). This is likely to be two or three times higher than the cost of current supplies, and should act as a deterrent to frivolous use (Rogers, Kalbermatten, and Middleton, 1999). Many studies have documented willingness to pay for water in developing countries with unreliable or nonexistent access to public water supplies. Many times, consumers are purchasing water from private vendors at a level (as much as 25 times the unit rate paid by the rich) that matches what would be required to finance the development of a water infrastructure, or are willing to pay large amounts relative to their income to secure a continuous supply of water (see, e.g., Whittington, Lauria, and Mu, 1991; Altaf et al., 1993; Brookshire and Whittington, 1993; McIntosh, 2003).

Nonmarket Instruments

Natural infrastructure. Natural infrastructures and their need for water must be included in management of water resources. Human users of water demand water-based services such as fishing, swimming, tourism, and the delivery of clean water to downstream users. Water that is not being withdrawn by humans goes to these and other productive uses. Reducing the amount and quality of water available for use by ignoring the importance of natural systems ignores the impact of water resources, especially their degradation, on health, mortality and other demographic factors. This has been well recognized in most developed countries, but has yet to become a factor of importance in most lower-income countries. Many demand-management instruments promote environmental sustainability and water quality, and the goals of use efficiency, and conservation, economic efficiency, and environmental sustainability are usually, when carefully examined, complementary (Rosegrant, 1997). California again provides an example of this: between 1960 and 1990, urban water use, water use in irrigated agriculture, and legally mandated runoff for environmental purposes all increased, despite competition among these sectors.

The challenge of water management is that, in order to guarantee adequate water for all users, there is a menu of trade-offs from which choices can be made. Some of these choices in allocation are between sectors and demands, in upstream-downstream water sharing, and in allocation of water between societal uses and ecosystems (SIWI, 2001). In the case of freshwater ecosystems, a management decision such as the building of a dam or the pumping of groundwater may do damage that is irreversible or widespread. Much of the damage that will be done may be unaccounted for by those managers making the management decisions. When humans manipulate soil, vegetation, and water systems, they produce benefits, but they also produce environmental feedback due to intricate interdependencies and interactions in the ecosystem (Falkenmark, 1994, p. 99). Unfortunately, the combination or degree of abuse that will bring about a system collapse is generally unknown, and the stress of a single small change may seem harmless. This makes it difficult to know how to alter management policies to ensure that ecosystems can be maintained at a level at which they are providing the full complement of goods and services that they originally furnished. All of this is complicated by the fact that the decisions made in the past by water planners were made under the assumption that future climatic conditions would be the same as past conditions. Because of the effects of climate change on climate and water resources, a reliance on the historical record may lead to incorrect and potentially disastrous decisions (Gleick, 2001b).

Education measures and civil society involvement. In order to maintain reductions in water consumption over time due to dramatic increases in water prices, measures such as education campaigns should be implemented concurrently (UNCHS, 1996). This is true because the conservation of water relies, ultimately, on behavioral changes. These changes are more likely to be achieved if the public is aware of and understands the reasons behind such changes. Education can take place through the media, schools, and user groups, and can inform the public about water conservation methods, sanitation, water scarcity, the need for water-tariff increases, if any, government policies and plans, and other topics (Meinzen-Dick and Appasamy, 2002). Civil society, which includes all stakeholders with interests in the water sector (e.g., consumers, nongovernmental organizations [NGOs], academics, journalists, and utilities), can become involved by putting pressure on the government and other water managers, as well as on the general public, to better manage the water resources, and ensure equitable distribution of resources. This can take the form of education or advocacy. Education can help communities understand the link between water, sanitation, health, and productivity, and help industries learn about the efficient use of water, efficient treatment and discharge, and the need for higher water prices, in some cases (McIntosh, 2003). It can also help to move toward transparency in the management of water systems, as educated consumers demand accountability from water managers and others.

Direct Intervention

Conservation. Demand management can include laws and regulations aimed toward conservation, such as restrictions on certain types of water use or efficiency standards for plumbing fixtures. It can also include changing management practices at the scale of individual enterprises. This includes changing such wasteful practices as irrigating during the day and using potable water for irrigation purposes, as well as shifting to more water-efficient crops and shifting industrial processes away from water-intensive production.

Demand Management by Sector

The different categories of policy instruments can be assessed in more detail if considered sectorally. Each sector—agricultural, industrial, and domestic—has a range of specific demand-management options open to it, ranging from technological changes to changes in management practices.

Agricultural Sector

Improving agricultural water productivity. Changing patterns of demand, increased populations, and poverty and food insecurity in marginalized communities, exacerbated by changes in rain patterns (e.g., higher variability, increased or decreased total rainfall) due to climate change, are likely to require further productivity gains from agriculture and from the water that feeds it. This will take place as a result of agricultural water management practices implemented in a way that will improve equitable access and the conservation of the water resource base (FAO, 2003b). Water productivity in agriculture, defined as the water consumption per kilogram of output, is estimated to have doubled since the 1960s (FAO, 2003b). This is in part due to improved water management and conservation in both rain-fed and irrigated agriculture, and partly because of the intensification attributable to fertilizer application and the use of high-yield varieties (FAO, 2003b). By one calculation, a 10-percent increase in water productivity would save the same amount of water as current domestic water consumption, making investing in water management in agriculture an attractive option for water managers looking to free water for use in other sectors (FAO, 2003b).

Water productivity can be improved by doing four things: increasing the marketable yield of a crop for each unit of water transpired by it; changing the variety of crops grown; reducing drainage, seepage, percolation, and evaporative outflows; and increasing the effective use of alternative water resources such as rain, stored water, and water of marginal quality (FAO, 2003b). These actions can be taken at different scales, from plant to basin-wide levels. One example is the reuse of urban wastewater as irrigation water. Depending on its first use and its intended reuse, the water may be treated or used directly. The water might be discharged into wetlands for treatment, and then drawn for use as irrigation water.

The agricultural sector has water requirements not only for crops, but also for livestock and fish. Because of the relatively small volume of water required for these uses compared to crops, they are less discussed. However, with growing demand for animal products due to changes in eating habits with rising living standards, as well as increases due to population growth, this part of the agricultural sector may become more important. Aquaculture and livestock can produce a very high value of output for the water they use (Bakker, et al., 1999; Meinzen-Dick and Appasamy, 2002), and, much like potential shifts to industry from agriculture in the face of reduced water supplies, there may also be a shift to these kinds of agricultural products. There is, however, a risk of overgrazing, which can severely traumatize land and water resources, or, if grain-fed, the virtual water that is embedded in the food fed to the livestock. There may be a local reduction in water resources used, but perhaps not an overall reduction. This may be, however, a way to shift resources from areas with ample supplies to those without.

Irrigation efficiency improvements. There are several avenues by which the irrigation of crops may be modified to reduce the agricultural sector's demand for water: irrigation efficiency improvements, the reuse of urban wastewater in agriculture, and the reduction or elimination of subsidies for irrigation water.

Irrigation efficiency is a measure of the amount of water required to irrigate a field, farm, or watershed. An efficient irrigation system can irrigate a crop with a minimum of waste or losses. In agriculture, irrigation efficiency is low. In the western United States, efficiencies are at about 55 percent; in the Indus region of Pakistan, they are less than 40 percent; and in many other countries they are also less than 50 percent. Although there is much room for improvement, the figures can be somewhat misleading. The water that one farm does not use may be used by the next farm downstream, raising the apparent system efficiency (Gleick, 2000a). Nonetheless, there is much room for more efficient irrigation technologies.

Irrigation water is lost in several ways. As Figure 12.1 shows, 55 percent of irrigation water is not used effectively by crops. In general, about 45 percent of the water losses occur when the water is applied to fields, while the rest is lost in equal amounts to transmission to and during use on the farm. The most successful efficient technology has been drip irrigation, which replaces the flooding of fields with the precise application of water at the roots of plants, resulting in large reductions in evaporative losses as well as productivity gains, sometimes as large as 200 percent or more (Gleick, 2000b). Other technological improvements include laser leveling of fields and advanced drainage systems that reduce salinization. These technologies can raise irrigation efficiencies from 60 percent to 95 percent (Wolff and Gleick, 2002).

These technologies are relatively expensive, and poorer farmers do not generally have the resources to pay for them. In these cases, traditional techniques can also save water and improve productivity.

Figure 12.1. Water Losses in Irrigation

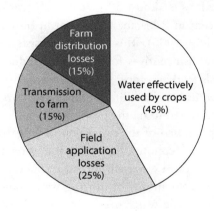

Source: FAO (2003a).

These technologies include water-harvesting techniques, using rocks to slow runoff on terraced land, and using small reservoirs to store rainwater from the rainy season to be used for crops in the dry season (Gleick, 2001b).

It is important, however, to investigate fully the implications of all such technologies. Increases in water productivity do not necessarily result in greater economic or social benefits (FAO, 2003b). Water has many uses in rural parts of developing countries, such as the production of timber firewood and fiber, raising fish and livestock, and domestic and environmental uses. The water is both a public and a social good, and the reduction of irrigation seepage or runoff may impair these other uses. For example, in Sri Lanka, a leaky irrigation project was found to be the source of sustenance for agroforestry that was important socially and economically for the local community (FAO, 2003b). In arid parts of Mali and Burkina Faso, similar gains are realized from agroforestry components of irrigation projects.

These examples show the importance of understanding the multiple roles of water in a community before proceeding with any projects.

Water prices. Releasing water from the agricultural sector frees it for other sectors, or for further irrigation projects, thereby reducing conflicts and avoiding development of new supplies (El-Ashry, 1995). Agricultural productivity would improve at the same time.

However, the current management of water in most places provides far more disincentives than incentives for conservation in agriculture. Low water prices are one of the most visible disincentives to investment in conservation technology and management expertise (El-Ashry, 1995). Huge publicly funded price subsidies are common in most countries, including the United States and Mexico, where users pay about 11 percent of the full cost of water (El-Ashry, 1995). This encourages producers to grow crops that are both low value and water intensive, and removes any incentive to use water efficiently (Gleick, 2000a).

Salinization. Poor management of irrigation systems has resulted in inefficiencies, poorer than expected crop yields, and salinization problems that have affected productivity on as much as 20 percent of irrigated lands, or about three percent of total cultivated land worldwide (Ghassemi, Jakeman, and Nix, 1995). About 22 percent of irrigated land in the United States is affected, as is almost 39 percent in Argentina, 33 percent in Egypt, 30 percent in Iran, 26 percent in Pakistan, 17 percent in India, and 14 percent in China (Ghassemi, Jakeman, and Nix, 1995). This problem persists anywhere large-scale irrigation is practiced. There are disincentives for farmers to reduce the pollution caused by the application of chemicals to their crops. Drainage water from many agricultural areas contains toxic levels of chemicals for wildlife. Because irrigators do not feel the impact of this runoff, they are not motivated to fix the problem.

Food imports and storage. In lieu of the expansion of irrigated agriculture or importation of water, it may be more economically viable for water-short countries to import water-intensive food products from water-surplus areas. Food can be thought of as virtual water, and is a relatively efficient means of moving water, since the product is perhaps a thousandth of the weight of the water needed to grow it (Zaba and Madulu, 1998). Along with the water savings enjoyed by the country importing the food, there may be global water savings due to the greater water productivity in the producing versus the importing country. For example, corn grown in Europe uses approximately 0.6 m^3 water per kilogram of corn, versus 1.2 m^3 per kilogram in Egypt, a savings of 0.6 m^3 of water for every kilogram of corn imported by Egypt instead of grown there (FAO, 2003b). Additionally, the water that is used in the growing of the imported crop may be renewable in the exporting country, but would be nonrenewable in the importing country. For example, wheat grown in Saudi Arabia uses fossil groundwater, unlike that grown in parts of Europe (Smith, 2003).

Imports of food assume a surplus in other parts of the world and that the importing country has the income required for imports. In many countries where water and food are in short supply, the poorly developed economy provides low purchasing power. The situation is worsened by the fact that logistical problems, trade barriers, and political and social instabilities encumber international trade or food-relief programs. Also, an emphasis on national food security for political, cultural, or other reasons impairs the rational trade of food, and hence a more rational use of water. A reliable global food security system would allow water-scarce countries to target their water for domestic and industrial use while importing food (WWC, 2000).

Some plants and animals provide relatively little nutritional value per amount of water required. In fact, another method of conserving water is to choose different foods to eat, since, for example, a pound of beef takes 15 to 30 times more water to produce than a pound of corn, and three to 12 times more water than a pound of chicken.

Another water-saving tactic in arid and semi-arid countries is food storage. If surplus production in a wet year is stored, it can be used during dry years where yield is reduced. Because water productivity is higher in the surplus year, real water savings can be realized. This tactic has been practiced since biblical times. More recently, in Syria in the years 1988 and 1989, some 2.8 billion m^3 of water was saved by storage of food (FAO, 2003b).

Domestic Sector

Many municipalities have implemented successful conservation programs, achieving reductions in water demand ranging from 10 to 50 percent (Wolff and Gleick, 2003; Postel, 1997). They have achieved these successes using integrated water-resource management, including such tools as monitoring and evaluation of water use; indoor and outdoor efficiency standards; technical assistance programs; education programs such as demonstration gardens, landscape seminars, and school education programs; economic incentives such as conservation rate structures, rebates for decentralized investments; and full economic integration of efficiency improvements when developing new water supplies (Wolff and Gleick, 2003).

Reducing losses from nonrevenue water. A large proportion of the water wasted in the municipal and domestic sector is in the delivery of water to and within towns. Much like losses associated with irrigation systems, cities lose from three percent to 70 percent of their water as nonrevenue water (NRW) or unaccounted-for water (UFW) (Table 12.1). NRW is the difference between the quantity of water supplied to a city's network and the metered quantity of water used by customers. Rates of NRW are far higher in developing countries, as is intimated in Table 12.1. Throughout Latin America, losses of 40 to 70 percent are common (WHO, 1992). In India, 35 urban centers average 26 percent of water lost, while Asian cities have losses ranging from four to 65 percent (McIntosh, 2003). This can be compared with Geneva, Switzerland, where losses have been reduced to 13 percent, or Singapore, where losses are six percent (WHO, 1992; McIntosh, 2003). NRW tends to be higher in areas where piped water coverage is lower.

NRW has three components: (a) real losses (physical losses due to leakage and overflow from pipes and reservoirs), (b) apparent losses (administrative losses due to illegal connections and inaccuracies of water meters), and (c) unbilled authorized consumption (water for firefighting, main flushing, and process water for waste-treatment plants, for example). While every case is different, often real losses and apparent losses contribute roughly equally to UFW, while unbilled authorized consumption is a minor component of losses (Saghir, Schiffler, and Woldu, 2000). There are three well-known and technically uncomplicated steps toward solving this problem:

Table 12.1. Nonrevenue Water, by Region

Region	Percent of Supply
Africa	39
Asia	30
Latin America and the Caribbean	42
North America	15

Sources: WHO/UNICEF (2000); McIntosh (2003).

- Reduce physical losses to the lowest economically feasible level
- Meter at least all major consumers (universal metering may have to be a longer-term project, since meters are a major foreign-currency expenditure in developing countries)
- Bill everyone for water supplied, and enforce payment (Rogers, Kalbermatten, and Middleton, 1999).

The benefits of reducing NRW, especially in developing countries, include the following:

- allowing investments in new works to be deferred or at least reduced in scope, with significant savings
- greatly increased revenue to pay for water treatment and distribution as well as operation and maintenance when the meter-reading and billing system and the detection and billing of illegal connections are improved. For example, in urban areas in Thailand in the 1980s, each 10 percent of unaccounted-for water saved was estimated to generate immediately an additional $8 million per year from the 3.5 million people served (WRI, 1997)
- reduction in wasteful consumption, which has been shown to be connected to metering and adequate water rates (Mcintosh, 2003). This will likely lead to decreases in total consumption
- improved understanding of consumption patterns, allowing the optimization of distribution systems and improved demand projections (McIntosh, 2003)
- reduced wastewater and pollution.

These benefits depend not only on improved management of water losses, but also on reasonable pricing of water and water services. If water customers pay low tariffs, there is little incentive for utilities to reduce apparent losses and for customers to deal with leaks and wasteful use.

Improved household and municipal water efficiency can also be attained, especially by using water-saving devices. In developed countries, toilet flushing is the largest indoor use of water in single-family homes, and one flush can use as much water as an entire day's worth of water in a rural village in a low-income country (Gleick, 2000b). Many municipalities have revised their building codes to require low-flow toilets, faucets, and showers. These devices can save large amounts of water at very low cost. Some developing countries are following suit. In Mexico City, for example, 350,000 six-gallon-per-flush toilets were replaced by 1.5-gallon-per-flush toilets, saving enough water to supply 250,000 additional residents (Gleick, 2001b).

Toilet flushing, via new technology, can even be eliminated. New toilets manage human waste without the use of water in a convenient, safe, and odor-free way. Although they are still expensive and unfamiliar to most consumers, their existence points to new ways of conserving water.

Pricing. In the domestic sector, it is argued that wasteful household practices such as watering thirsty lawns and maintaining swimming pools can be reduced through pricing. However, the demand for water is inelastic for some uses over a wide range of quantities (Brookshire and Whittington, 1993). In the domestic sector of developing countries, even poor households show willingness to pay for a reliable supply of water, as discussed earlier. However, it is thought that demand is only price inelastic in the short term, and that water users will change their behavior if the final, water-based service they desire is maintained (Wolff and Gleick, 2002).

Industrial Sector

In industry, the redesign of production processes to require less water per unit of production and the reuse and recycling of water in current production processes have resulted in large efficiency improvements in some industries. In steel production, new technology uses less than six cubic meters of water per ton of steel produced, as compared to the 60 to 100 m^3 used by the old technology. Aluminum can be made with 1.5 m^3 per ton produced, so even greater water savings can be realized by substituting aluminum for steel in the production process. Potential and actual savings in the United States have been estimated to range from 16 to 34 percent (Wolff and Gleick, 2002). Often the payback period for these conservation measures has been found to be short, as little as a year or less.

In the developing world there is even more scope for improvement. In China, steel making still uses 23 to 56 m^3 of water per ton of steel, and paper manufacturers use 900 m^3 of water per ton compared to the 450 m^3 per ton used in Europe. In all parts of the world, the valuing of water at market prices improves water-saving and efficiency measures undertaken by industry. Inexpensive water subsidizes inefficiency (de Villiers, 2000).

Demand management and supply management are both necessary for the efficient use of water resources. As seen here, they are multifaceted categories of management tools. As previously discussed, the combination of tools selected for a particular location will depend on its social, political, economic, and physical climate. Tools of different types are also applicable to different levels of management. In other words, a water system's manager and a small city's government will not have influence on food imports, but they can work toward improving local irrigation systems' efficiency or reallocating irrigation water to industry, for example. Other tools may be facilitated by policies at the state or national level, even if the choice of tools will be locally based. Whatever the case, these choices, and the mitigating factors illustrated in the framework at the beginning of this study, will determine the ultimate sustainability of water resources.

Discussion Questions

1. What is the role played by dams and water-control structures in managing the supply of water?

2. How does groundwater affect the supply of water?

3. What is desalination? How does water desalination work?

4. Explain in detail the concept of water reclamation and reuse.

5. Explain in detail the four main variables affecting demand-management of water.

6. How do market-based incentives affect water demand?

7. What is the nature of non-market instruments that affect water demand?

8. How can agricultural water productivity be improved?

9. What is non-revenue water (NRW)?

10. What are the benefits of reducing NRW?

In-Class Assignment: Objective Questions

1. What percentage of the world's population depends on groundwater supplies?
    a. About 10%
    b. About half
    c. About one third
    d. None of the above
2. What percentage of annual rainfall comes down in areas where less than one third of the world's population lives?
    a. About half
    b. About one third
    c. About three fourths
    d. About 10%
3. Of all the continents, what continent has the largest total water availability?
    a. Europe
    b. Africa
    c. North/Central America
    d. Asia

13 Product Modularity and the Design of Closed-Loop Supply Chains

Harold Krikke, Ieke le Blanc, and Steef van de Velde

Business Logistics Management has gone through dramatic changes over the last few decades. Customers have become more and more demanding, and increased transparency—through information and communication technology—has shifted the balance of power in their favor, giving them the opportunity to actively configure the final product. Firms now must focus more on creating value through personalized and individualized offerings to customers. Moreover, businesses are increasingly expanding into international markets, which requires the ability to manage manufacturing and distribution functions on a global basis. Supply Chain Management concepts—such as Efficient Consumer Response (ECR), Continuous Replenishment, Collaborative Planning, Vendor Managed Inventory, Virtual Integration, Mass Customization, Postponement, and so on—are necessary to meet these increasing demands. Companies have spent much effort on the design, analysis, and management of their forward supply chain. Soon, they may need to pay the same effort on their *reverse* supply chain.[1]

Most people associate "reverse logistics" with the environment. Indeed, over the last decade, new environmental laws in Europe and Asia have introduced the principle of extended producer responsibility. This "polluter pays" principle makes original equipment manufacturers (OEMs) and other "forward" supply chain actors responsible for take-back and recovery of their products once discarded by their last users. Moreover, increased recycled content for new products manufacturing is required.[2] Commercial returns are also increasing due to trends such as product leasing, catalogue/internet sales, shorter product replacement cycles, and increased warranty claims. This type of return, related to the sales or service process, has nothing to do with environmental legislation. Moreover, companies are increasingly willing to buy back returns actively for economic gain.

Reverse supply chains add complexity to closed-loop supply chain management due to new coordination issues. Examples include cross-border waste transportation, more complex trade-offs in supply chain objectives, (perceived) conflicts of interest amongst actors, micro internalization of macro externalities, and so on. The intrinsic complexity may be a reason for companies not to spend much effort on designing their closed-loop

---

Harold Krikke, Ieke le Blanc, and Steef van de Velde, "Product Modularity and the Design of Closed-Loop Supply Chains," *California Management Review*, vol. 46, no. 2, pp. 23-39. Copyright © 2004 by California Management Review. Reprinted with permission.

supply chains. The lack of managerial attention very often leads to "quick and dirty" solutions, resulting in inefficient, non-responsive, and sometimes even environmentally unsafe reverse chains. Most companies definitely do not see reverse chains as a means by which to thrive in today's marketplace. However, there is sufficient evidence that closed-loop concepts can strengthen a company's competitiveness. Companies such as Kodak, Océ Technologies, Mercedes-Benz, Xerox, ReCellullar, Philips, Volkswagen, and IBM have already discovered this. These pioneering firms have learned that making returns profitable relies on good design of reverse chain business processes—including the possible integration with the forward chain. Moreover, they have learned that product design is crucial.

Designing Closed-Loop Supply Chains

Some Definitions to Start With

Closed-loop supply chains consist of a *forward* supply chain and a *reverse* supply chain. Loops can be closed by several options: reusing the product as a whole, reusing the components, or reusing the materials. Most closed-loop supply chains will involve a mix of reuse options, where the various returns are processed through the most profitable alternative.

There are five key business processes involved in the reverse chain.[3] The importance and sequence of these processes may differ from chain to chain:

- *Product Acquisition*—This concerns retrieving the product from the market (sometimes by active buy-back) as well as physically collecting it. The timing of quality, quantity, and composition needs to be managed in close cooperation with the supply chain parties close to the final customer.
- *Reverse Logistics*—This involves the transportation to the location of recovery. An intermediate step for testing and inspection may be needed.

Table 13.1. Outline of Recovery Options

Options	Operations	Resulting Output	Applied In:
Direct Reuse	Check on damage and clean	As is, e.g., for refill	Original or similar markets
Repair	Restore product to working order, some component repaired or replaced	Original product	Original or similar markets
Refurbishing	Inspect and upgrade critical modules, some modules repaired or replaced by upgrades	Original product in upgraded version	Original or similar market
Remanufacturing	Manufacture new products partly from old components	New product	Original or similar markets
Cannibalization	Selective retrieval of components	Some parts and modules reused, others scrapped	Both original and alternative markets
Scrap	Shred, sort, recycle, and dispose of	Materials and residual waste	Alternative markets

Adapted from Martijn Thiery, Marc Salomon, Jo van Nunen, and Luk van Wassenhove, "Strategic Issues in Product Recovery Management" *California Management Review*, 37/2 (Winter 1995): 114–135.

- *Sorting and Disposition*—Returns need to be sorted on quality and composition in order to determine the remaining route in the reverse chain. The sorting may depend on the outcome of the testing and inspection process. However, the disposition decision not only depends on product characteristics, but also on market demand. Table 13.1 presents six recovery options on a conceptual level: direct reuse, repair, refurbishment, remanufacturing, cannibalization, and scrap.
- *Recovery*—This is the process of retrieving, reconditioning, and regaining products, components, and materials. In principle, all recovery options may be applied either in the original supply chain or in some alternative supply chain. As a rule of thumb, the high-level options are mostly applied in the original supply chain and the lower-level options in alternative supply chains. In some areas, the reuse in alternative supply chains is referred to as "open-loop" applications.
- *Re-Distribution and Sales*—This process largely coincides with the distribution and sales processed in the forward chain. Additional marketing efforts may be needed to convince the customer of the quality of the product. In alternative chains, separate channels need to be set-up and new markets may need to be developed.

There are four main types of returns: end-of-life returns, end-of-use returns, commercial returns, and reusable items.[4] Figure 13.1 shows how the recovery options fit in closed-loop supply chains as well as which basic category of returns is generally processed by which recovery options.

- *End-of-Life Returns*—These are returns that are taken back from the market to avoid environmental or commercial damage (so-called "negative externalities"). EU legislation based on Extended Producer Responsibility and corresponding EU directives on material collection targets drove Dutch consumer electronics OEMs/importers to implement a collective recycling system.[5] The large volumes of returns and the fees charged to the end-consumer at the time of purchase let the system run at a breakeven level. Similar collective recycling systems are in use for end-of-life vehicles, packaging, batteries, tires, and construction waste throughout the European Union and in some Asian countries. Foremost, these recycling systems aim at "playing it safe," to make sure that all discarded items are collected and processed according to formal prescriptions. Very often, these reverse chains are alternative supply chain applications.
- *End-of-Use Returns*—These are products and components returned after some period of operations due to the end of the lease, trade-in, or product replacement. These goods may be remanufactured into new products (e.g., copiers at Xerox) or traded (e.g., eBay or similar alternative markets). A specific category concerns suspect components from field exchange activities, which can often be repaired and in many cases represent a high economic value. Similar to forward chains, the time-based economic value of end-of-use returns necessitates a responsive closed-loop supply chain.
- *Commercial Returns*—These are returns that are linked to the sales process. Increasingly, customers return their products shortly after sales. Developments in mail order and e-commerce have strongly increased the volume of these returns, sometimes up to 25% of turnover. Estèe Lauder redesigned the reverse logistics processes for cosmetics products returned by the sales process, cutting the volume of returned products that are destroyed in half.[6] Overstocks of non-sold products (e.g., due to seasonal effects) are also part of this category. Other examples include products under warranty or a product recall. The time-based value of these returns may require a responsive closed-loop supply chain.[7]
- *Re-Usable Items*—These returns are related to consumption, use, or distribution of the main product. This type concerns many different items, e.g., reusable containers and pallets, refillable cartridges, bottles, and "disposable" cameras. The common characteristic is that they are not part of the product itself, but contain and/or carry the actual product. Many examples can be found in society, for example, the reusable trays for food in the residence homes, restaurants, and so on. These returns present a stable but relatively low value. Collection is relatively simple and reuse markets are internal, hence easy to find. For example, Kodak's "disposable" cameras are almost

always returned, since the customer needs the films developed and the reuse markets coincide with the primary market. The main driver is cost; and, similar to forward chains, this requires efficient closed-loop supply chains.[8]

Table 13.2. Returns and Closed-Loop Supply Chains

	Control	Responsive	Efficient
End-Of-Life Returns (Negative Externalities)	FIT		
Commercial Returns		FIT	
End-Of-Use Returns		FIT	
Reusable Items			FIT

Adapted from Harold Krikke, Jo van Nunen, Rob Zuidwijk and Roelof Kuik "E-Business and Circular Supply Chains, IT-Based Integration of Reverse Logistics and Installed Base Management," manuscript.

Fisher developed a simple but compelling framework for designing the forward chain.[9] Following his example, we present a similar framework for closed-loop supply chains in which the design depends on the type of return (see Table 13.2). The idea to apply Fisher's framework to closed-loop supply chains was developed by Blackburn et al.[10] To Fisher's framework, we have added the "control type." This type is mostly to be found in Europe and Asia due to the new environmental legislation on extended producer responsibility and to a far lesser extent in Northern America.

Honeywell Industrial Automation and Control: Matching Type of Return with Type of Closed-Loop Chain

Honeywell Industrial Automation and Control (IAC) is a global player in industrial automation.[11] It produces, supplies, and maintains distributed control systems, that is, hardware and software to monitor and control production processes at large industrial installations. The high capital value of the installations and their dependence on control systems makes fast and reliable service is an absolute necessity.

Printed Wiring Assemblies (PWAs) are critical and valuable components, serviced by well-trained field service engineers. Service agreements oblige Honeywell to dispatch a field service engineer to the customer location with 24 or 48 hours in response to a customer request. Figure 13.2 represents the closed-loop service supply chain of PWAs of the TDC-3000 system in Europe, including return flows for repair. The national affiliates deal directly with the customer, whereas logistics are organized on a Pan-European level.

Once a PWA has been replaced and returned, the inventories are replenished, i.e., the affiliate replenishes the engineer's car, the central warehouse replenishes the affiliate depot, and the production factories replenish the central warehouse, and so on. Replenishment

Figure 13.1. The Full Closed-Loop Supply Chain

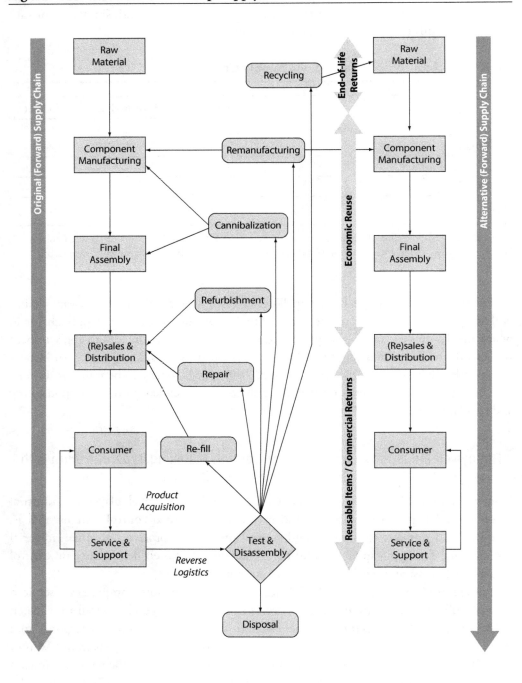

rush orders—carried out by DHL—can skip one or more echelons, depending on the cause and location of demand.

Figure 13.2. Honeywell European Closed-Loop Supply Chain for Service of TDC 3000-PWAs

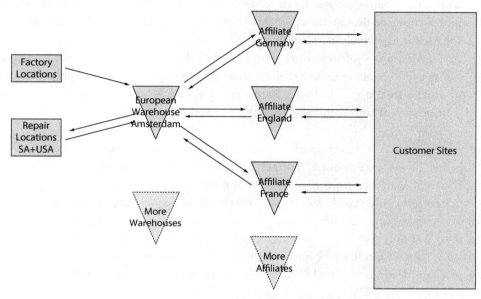

Inventories are kept at the production and repair sites, the central warehouse, affiliate locations, and in the service engineer's car. About 3,000 units are kept in stock.

At the time of our research, the reverse chain situation was as follows:

- *Product Acquisition*—The engineer replaced the bad or suspicious PWA and brought it back in his car to the affiliate depot. Often, returned PWAs were not handled correctly, because there was no clear procedure and responsibility structure for return shipping. Also, the service engineers often packaged the replaced PWAs in the wrong way, as a result of which good or repairable PWAs were sometimes broken during transport.
- *Testing and Inspection*—At the affiliate depot, the PWA was visually inspected. Bad parts would be scrapped and good parts were restocked, either at the depot or the engineer's car. The testing proved to be inaccurate, leading to the scrapping of good or repairable parts. Similarly, too many broken or obsolete PWAs were returned, leading to unnecessary costs. Product information was often lost on the way, thus complicating the testing and inspection process.
- *Reverse Logistics*—Malfunctioning parts were returned for repair after authorization from the central logistics department of the Industrial Service Logistics Center (ISLC), which controls all logistics and tactical support for European customers. PWAs are returned by truck to the central warehouse for Europe in Amsterdam,

operated by Van Ommeren Intexo (VOI) and controlled by ISLC. Returned PWAs are consolidated at this central warehouse and from there are transported by plane in large batches to the repair sites in Phoenix (USA) and Johannesburg (SA). Long lead times were costly due to the high capital value of PWAs and obsolescence.
- *Recovery*—In Phoenix and Johannesburg, returned PWAs were again inspected and, if feasible, repaired or upgraded. Johannesburg got the majority of returns, because it was more dedicated to repair. However, regular production also took place in Johannesburg, which increased repair lead-times.
- *Re-Distribution*—Johannesburg sent repaired PWAs to the central European warehouse and Phoenix restocked them in their own facility for possible supply to Europe or other warehouses worldwide. Return of repaired PWAs from Johannesburg to the European warehouse ran independently from the replenishment procedure.

About 500 parts were returned each year, of which two thirds were repaired. Total reverse chain costs were about € 360 per item, whereas the reuse value was estimated € 700. The total lead-time from the moment of return at the customer site through repair and back to serviceable stock was approximately 34 weeks, leading to high obsolescence rates of repaired parts. In comparison, the lead-time from production to stock for new spare parts was only 3–4 weeks. Moreover, as a result of the long and stochastic lead-time, returned PWAs were not taken into account in the availability planning of ISLC for the forward chain, leading to double orders. Hence, repair was not sufficiently reliable as an internal supplier of PWAs.

The existing reverse chain was based on using the forward supply chain as much as possible and minimizing out-of-pocket costs per stage in the reverse chain. To realize supply chain improvements, issues such as lead-time reduction, better inspection at the source, and better product information had to be taken into account. Improvements were found in simplifying and speeding up the reverse chain. The following solution was implemented: an improved return shipping procedure for service engineers and clients made sure the right PWAs were returned; and direct shipping by DHL from affiliates to the repair center in Johannesburg reduced lead-time and prevented the loss of product information. The forward system and the shipping from Johannesburg back to stock in the central European warehouse remained the same. This closed-loop supply chain improved performance. Although unit out-of-pocket cost *increased* from €360 to € 373 per PWA, total shipping costs were reduced because useless returns were avoided by the improved shipping procedure. Obsolescence rates, lead-times, and double ordering were drastically reduced, thereby increasing the overall profitability and reliability of the repair activities. In short, the time-based value of the returns has led to a more responsive and efficient (rather than "out-of-pocket") closed-loop supply chain.

The lesson from this case is clear. Returns need to be processed by their matching closed-loop chain as indicated in Table 13.2. If the main take-back does not match the supply chain focus, management needs to re-design the closed-loop supply chain.

The PWAs are critical spare parts in Honeywell's installed base. Their cross-product and cross-generation applicability gives them a high reuse value, although obsolescence risk is a major issue. This raises the question whether modular product design can help further optimize the closed-loop supply chain.

Is Product Modularity the Answer?

Mass customization requires modular product designs for cross compatibility of components. The overall functionality of a product is divided into sub-functions, provided by individual components. Flexible product designs allow component substitution without adjusting other components. For example, in most PCs it is easy to replace an older hard disk with a new one, because the interfaces between the hard disk and the rest of the computer have been well defined.[12]

Reuse on a modular level has the greatest economic and ecological potential, because much of the added value of the forward chain is regained while quality upgrades are possible to assure customer value. Krikke et al. have shown economic and environmental benefits of product modularity in a quantitative study on refrigerators.[13]

Océ Technologies: Product Modularity and Value Separation

Océ Technologies is an international document-processing firm headquartered in Venlo, The Netherlands. It develops, produces, and sells high-quality copying, printing, and plotting systems as well as consumables and imaging supplies for these systems. The main markets are the office and the drawing table markets. The company aims at a prominent position in these markets by producing advanced products of excellent quality, reliability, durability, and environment friendliness.

Responsive closed-loop supply chains play a key role in Océ's strategy:[14]

- *Product Acquisition*—Océ machines are returned from the market due to the ending of lease contracts or active buy-back because of market demand for recovered machines. The business model for Océ is thus favorable to a closed-loop supply chain, since product acquisition is relatively simple.
- *Reverse Logistics*—Customers return a machine to the local operating company. The operating company is allowed to refurbish the machine and put it back on the market. If operating companies themselves are not interested in refurbishing, they return the machine to an Océ recovery location, for which they receive a fee. A large stock of returned machines is built up in Venlo. The machines are actually "pulled in" on market demand.
- *Testing and Inspection*—Machines returned to Venlo are classified in four quality classes: A (excellent); B (reasonable); C1 (poor); and C2 (extremely poor). The

inspection and classification is done visually before disassembly, using limited data from maintenance records.
- *Disassembly and Recovery*—Depending on its quality class, a returned machines is revised, remanufactured (made as good as new), or cannibalized/scrapped. In the revision strategy, a returned machine is disassembled to a limited level and cleaned. New or repaired parts, that are the (sometimes improved) equivalent of the released parts, are built in. The new machine is an upgraded version of the returned machine is leased at a lower than new price. This strategy is applied to good quality machines of somewhat older generations. Remanufactured machines are converted into a new model machine that contains all the features and functions of the old model and adds new ones. The return process is basically the same as for revision, except that not only released parts are replaced, but also entirely new parts and units are added to provide additional functionality. Moreover, in case there are insufficient returns to fulfill market demand, these machines can also be built brand new, although at a higher cost than the remanufactured ones. Thus, Océ can choose between new production and remanufacturing of the same machine, a phenomenon known as dual sourcing. The resulting machine has a new quality status and is leased as such. Note that the identity of the machine, expressed by the type and even serial number is changed in the remanufacturing process. The re-assembly process is integrated with the assembly process in regular production.

  The lowest-quality machines are cannibalized and scrapped. The returned machine is disassembled to the part level. Some parts are cleaned and repaired for re-use. Other parts are recycled or disposed of.
- *Re-Distribution and Sales*—This process is completely integrated with the forward chain.

Océ's product-leasing business model has served as a catalyst to create demand for secondary machines. By integrating the reverse and forward chain, Océ was able to make the difference between new and secondary machines invisible to the customer. In addition, mass customization has been applied too. It has been often suggested that both these elements of the business model are the reason for the success of copier remanufacturing. However, this is only partly true.

In optimizing the closed-loop supply, Océ spends a notably high percentage of its turnover to "Design for X" (DfX), where X implies that products should be reusable at large. In other words, X stands for disassembly, repair, recycling standardization, source reduction, and so on. Modularity is important at Océ at several levels—not just the copiers themselves, but also the refillable toner cartridges and reusable packaging. It enables the application of components along multiple product lines and across product generations.

Much attention has been paid in the literature to DfX in general. Little attention, though, has been paid to an important principle with respect to the modularity of product, which is essential at Océ, namely, the principle of *value separation*. This means that for capital-intensive parts, a distinction must be made between parts that contain stable technology and those that do not. By containing valuable, technologically stable components in separate modules, value can be separated easily from the rest in the disassembly process.

There are three important lessons from the Océ case. First, value recovery from returns is best accomplished in closed-loop supply chains based on modular reuse. Second, the difference between new and secondary products should be made invisible to the customer. Finally, applying modular value separation creates optimal reuse opportunities in original forward supply chains. Interestingly, physical characteristics of the product (modularity) in fact help to create market demand. By making the difference between new and remanufactured machines invisible to the customer, the "market loop" is also closed.

Auto Recycling Nederland: Product Life Cycle Information

Anticipating the European legislation on Extended Producer Responsibility, all car importers in The Netherlands (including Mercedes-Benz, Ford, Opel, Volkswagen, and the Japanese trademarks) jointly set up Auto Recycling Nederland (ARN). ARN is a collective organization for coordinating the recycling of end-of-life vehicles (ELVs). Existing ELV dismantlers, collection companies, and recycling companies carry out the work for ARN on a commercial basis, recycling over 275,000 vehicles yearly. The closed-loop supply chains managed by ARN are setup as follows.[15]

- *Product Acquisition*—Consumers can hand in their end-of-life automobile for free, regardless of the brand, at one of the 266 ELV dismantlers affiliated with ARN. The consumer receives all the necessary documents and the warranty that their automobile will be appropriately recycled.
- *Testing and Disassembly*—The ELV dismantler determines, based on the condition of the vehicle and the market opportunities, whether marketable parts can be separated and stored for trade. Parts are made available to the forward supply chain by offering a choice to the customer: repair with old ("green") or new parts. The remainder of the ELV is recycled on a material level, including hazardous elements (such as fuels, oils, and batteries) and easily separable materials (such as bumpers and tires). The remaining carcass is shredded to regain the ferrous materials.
- *Reverse Logistics*—The repair shops pick up the parts at warehouse of the ELV dismantler at the time needed. The other dismantled materials are stored in dedicated storage equipment provided by ARN. Once the ELV dismantler has a minimal number of filled storage units, the collection company is contacted for collecting the materials. After collection, the materials are consolidated in containers for transport and transferred to the selected recycler for the material in consideration. The remaining carcasses are transported to the shredder for further processing.
- *Recovery*—Currently, about 86% of the weight of the car is recycled, exceeding the current EU target. The trade is left to the ELV dismantlers as much as possible. Parts and materials for which reuse is difficult or impossible are collected in the ARN

system and brought to a recycler. These recyclers are selected by ARN on recycling quality to minimize environmental impact. For example, used oil is recycled into lubricating oil, while tires are recycled to granulated rubber which is used in the production of sport floors and tiles for playgrounds. The ferrous materials retrieved from the shredders are used again as raw materials in all sorts of applications. The new EU directive and the Dutch decree on ELV recycling has set strict recycling targets up to 95% of the average vehicle weight.
- *Re-Distribution*—The ARN contractors are responsible for re-sale and distribution of the recycled materials. The matching of supply and demand of repairable parts is assured by online databases.[16]

The ARN system represents a typical closed-loop supply chain with a "control" focus. Compulsory registration and reporting of mass balances—but also the determination of recovery options and the development of new recycling technology—requires detailed product information. ARN often faces a lack of good product information. The EU has acknowledged this problem and therefore the new EU-directive (2000/53/EC) on car dismantling prescribes the following:

- The development of component and material coding standards for vehicle manufacturers and their suppliers.
- Within 6 months after a product's market introduction, the producer is obliged to provide dismantling information on the vehicle (a recycling passport).

In the future, authorized vehicle dismantling sites will be able to more easily check the material composition of certain parts and determine the appropriate recovery option. The automotive industry in general needs to consider the value of information over the total product life cycle. This information is also particularly valuable in case of product recalls and for design platforms.

Product recalls occur more and more often in all kinds of industries. Many product design and functionality errors reveal themselves only during actual use, frequently resulting in product recalls for modifications.[17] The costs of these recall operations in the automotive industry are huge, and therefore the industry should not only correct them, but also learn from them.

Advanced CAD/CAM systems enable the *reuse of design information* and automobile manufacturers have increasingly adopted platform strategies. Different cars of the same size in different market segments are built on the same basis. For example, Fiat and Saab have a significant number of parts in common.[18] Developing new products based on selectively upgraded and/or jointly developed components reduces time-to-market, generates tremendous savings on development costs, and reduces design flaws.

Despite the design platforms, there are a few component/module-based closed-loops in the industry. Exceptions include the spare parts business (where insurance companies introduced so-called "green" insurance policies) and the reuse of engines at Mercedes-Benz.[19] There appear to be at least two major reasons why modular reuse remains limited.

The first reason is the reduced economic profitability of reuse due to low-cost forward chains. For example, the recapping of passenger car tires was profitable business for a long time. However, competition from Eastern Europe reduced the prices of new tires considerably and the market for recapped tires is under severe pressure. In general, the automotive (component) industry is, like many other industries, increasingly becoming a low-cost production-distribution sector, thus reducing reuse revenues.

The second reason lies in the overall Life Cycle Analysis (LCA). Although manufacturers are considering easy disassembly techniques for maintenance and end-of-life purposes, LCAs have been primarily focused on weight and energy reduction rather than recyclability. The material composition of cars is changing due to requirements for low-fuel consumption and passenger safety. The increased use of aluminum, composite fibers, and electronics reduces possible recycling yields. Moreover, new post-shredder technology (PST) and "smart," self-retrieving materials (materials that, once exposed to the correct temperature range, will be triggered to automatic disassembly) increase the economic feasibility of alternative applications, such material recycling or even incineration options combined with energy generation.

The ARN case shows that product life cycle information is an essential for optimizing closed-loop supply chains, both in the development, re-engineering, and disassembly/recovery of products. Due to low-cost forward supply chains and life cycle analysis, physical reuse in alternative supply chain applications may sometimes be preferred to original supply chain applications. Still, the reuse of information is focused on the original application, since design feedback is given to the original supply chain.

Product modularity cannot stand on its own. It needs to be supported by product life cycle management.

Product Life Cycle Management: Beyond Modularity

As the ARN and Océ cases demonstrate, product design issues need to be put in the broader context of product life cycle management, which is the process of optimizing service, cost, and environmental performance of a product over its full life cycle. Key issues include product design for recovery, re-engineering, product data management, installed base support, and evaluating end-of-life scenarios. Product life cycle management is supported by such methodologies as life cycle analysis and life cycle costing. Figure 13.3 shows the relationship between product life cycle management and closed-loop supply chains.

Figure 13.3. Product Life Cycle Management and (Closed-Loop) Supply Chain Management

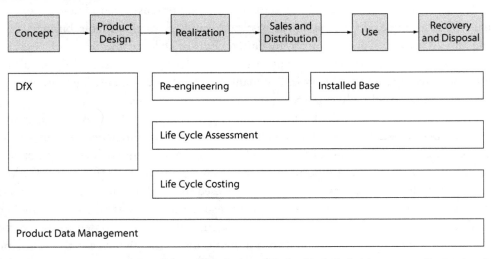

Source: E. West Kaemper and David Osten, "Methods and Tools of Life Cycle Management Optimizing the Performance of Manufacturing Systems," in J. Jeswiet, TN. Moore, and PH. Oosthuizen, eds., *Proceedings of the CIRP Conference,* Kingston, Ontario, Canada, June 21–23,1 999, pp. 401–409.

The reverse chain represents only one phase of the full product life cycle. A firm must assess the relative impact of its reverse chain, both economically and environmentally. Life cycle analysis aims measures the environmental impact of a product (including its energy use, waste volumes, and toxicity) next to its economic cost (life cycle costing).[20]

The installed base is the total number of placed units of a particular product in the entire primary market or product segment. Installed base management concerns the care of products during operations. Customers have installed bases that generally have longer life cycles than the replacement cycles at the OEM. The installed base is an important market, for example, in overhaul, upgrades, or repair. Traditionally viewed as a technical matter, many organizations fail to recognize that installed base support offers many commercial opportunities. For example, early take-back and re-customization by retrofit or product exchange might be an appropriate strategy for a customer with a hardly used high-volume copier.

Advances in information technology—such as POS registration, 2D barcoding, electronic marketplaces, and RFID—offer similar opportunities. Tracking and tracing for logistics purposes has great potential for easing collection. Information technology makes the monitoring of a large number of parameters possible, generating lots of useful data.

Figure 13.4. Summary of Enforcing and Enabling Factors in Closed-Loop Supply Chains

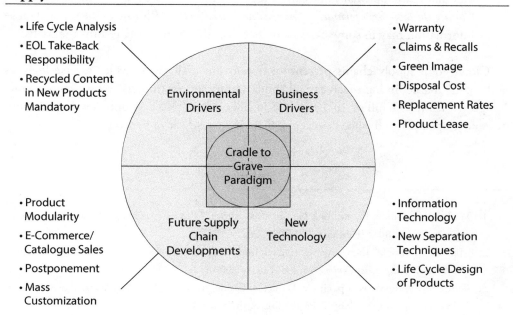

Product Data Management (PDM) is then needed to tie all these systems and data together. PDM serves to maintain accurate data on complex products (many parts, variants, and alternatives), record maintenance changes on a product during its lifecycle, and disseminate product data at an intra-organizational or inter-organizational level. PDM improves the quality of data and reduces labor intensity, since it reduces the amount of manual data transfer of information in the chain. Bosch, for example, implemented a "green port" chip in its power tools for end-of-life optimization.[21]

Lessons Learned and Outlook

New environmental laws and commercial trends will force companies to adopt a cradle-to-grave paradigm. At the same time, new forward supply chain developments—in particular, product modularity and new recycling technology—offer new possibilities for establishing efficient, responsive, and environmentally sustainable closed-loop supply chains. Figure 13.4 summarizes both the enforcing and enabling factors.

The intrinsic complexity of closed-loop supply chain concepts may hamper their adoption. Presently, companies tend to focus on the physical aspect of reuse, if not ignoring the subject at all. Optimal closed-loop supply chain management requires:

- *Closing the loop for goods flows, as evidenced by the Honeywell case*—The type of return needs to be matched with the appropriate closed-loop supply chain.

- *Closing the loop for markets, as evidenced by the Océ Technologies case*—Through modular reuse, optimal value can be regained in closed-loop applications.
- *Closing the loop for information flows, as evidenced by the ARN case*—The value of reuse information may in some cases be higher than the value of the returns themselves.

Closed-loop supply chains present an opportunity. The worldwide "reverse logistics and repair" market in high-tech capital goods, consumer electronics, and packaging alone has grown from $17 billion in 1994 to $34 billion in 2000.[22] As only few organization have developed truly advanced concepts, there is a world yet to be won.

Notes

1. V. Daniel R. Guide, Jr., and Luk N. Van Wassenhove, "The Reverse Supply Chain," *Harvard Business Review*, 80/2 (February 2002): 25–26.
2. Michael W. Toffel, "The Growing Strategic Importance of End-Of-Life Product Management," *California Management Review*, 45/3 (February 2003): 102–129.
3. Guide and Wassenhove, op. cit.; V. Daniel R. Guide, Jr., and Luk N. Van Wassenhove, "Closed-Loop Supply Chains," Proceeding of the International Working Group on Distribution Logistics, *Lecture Notes in Economics and Mathematical Systems* (Berlin: Springer Verlag, 2002), pp. 47–60; Moritz Fleischmann, Harold Krikke, Rommert Dekker, and Simme Douwe Flapper, "A Characterization of Logistics Networks for Product Recovery," *Omega, The International Journal of Management Science*, 28/6 (2000): 653–666.
4. Harold Krikke, Jo van Nunen, Rob Zuidwijk, and Roelof Kuik, "E-Business and Circular Supply Chains, IT-Based Integration of Reverse Logistics and Installed Base Management," manuscript.
5. Rene de Koster, Simme Douwe Flapper, Harold Krikke, and Sander Vermeulen, *Consumer Electronics Recycling in The Netherlands: Corporate Implementation of Producer Responsibility for Large White Goods* (forthcoming, Revlog).
6. Bruce Caldwell, "Reverse Logistics," *Information Week online*, April 1999, <www.information-week.com/729/logistics.htm>, accessed November 2002.
7. For a full discussion, see Joseph D. Blackburn, V. Daniel R. Guide, Jr., Gilvan C. Souza, and Luk N. Van Wassenhove, "Reverse Supply Chains for Commercial Returns," *California ManagementReview*, 46/2 (Winter 2004).
8. Ibid.
9. Marshall L. Fisher, "What Is the Right Supply Chain for Your Product?" *Harvard Business Review*, 75/2 (March/April 1997): 105–116.
10. Blackburn, op. cit.
11. Harold Krikke, Jacqueline Bloemhof-Ruwaard, Costas Pappis, and Giannis Tsoulfas, "Design Principles for Green Supply Chains, Optimizing Economic, Logistic and Environmental Performance," Proceedings of the International Working Group on Distribution Logistics,

Lecture Notes in Economics and Mathematical Systems (Berlin: Springer Verlag, 2002), pp. 61–74.

12. Prof. Ron Sanchez of IMD Lausanne, interviewed by Alan McCluskey, "Modularity: Upgrading to the Next Generation Design Architecture," <www.connected.org/media/modular.html>, accessed November 2002.

13. Harold Krikke, Jacqueline Bloemhof, and Luk Van Wassenhove, "Concurrent Design of Closed-Loop Supply Chains: A Production and Return Network for Refrigerators," *International Journal of Production Research,* 41/16 (2003): 3689–3719.

14. Harold Krikke, Aart van Harten, and Peter Schuur, "Business Case Océ: Reverse Logistic Network Re-Design for Copiers," *OR Spektrum,* 21/3 (1999): 381–409.

15. Ieke Le Blanc, Hein Fleuren, and Harold Krikke, "Network Design for LPG Tanks in The Netherlands," (forthcoming, *OR Spektrum*).

16. Onderdelenlijn, <www.onderdelenlijn.nl>, accessed October 13, 2003.

17. Recalldata, <www.recalldata.org>, accessed October 13, 2003.

18. A. Camuffo and G. Volpato, "Partnering in the Global Auto Industry: The Fiat-GM Strategic Alliance," *International Journal of Automotive Technology and Management,* 2/3 (2002): 335–352.

19. H.M. Driesch, S.D.P. Flapper, and J.E. van Oyen, "Logistieke besturing van motorenhergebruik bij Daimler-Benz MTR," *Reverse Logistics* (1998), p. C2410.

20. Krikke, Bloemhof, and Van Wassenhove, op. cit.; Jacqueline Bloemhof, "Integration of OR and Environmental Management," Ph.D. thesis, Wageningen University, 1996.

21. Markus Klausner, Wolfgang Grimm, and Chris Henderson, "Reuse of Electric Motors in Consumer Products, Design and Analysis of an Electronic Data Log," *Journal of Industrial Ecology,* 2/2 (1998): 89–102.

22. Donald F. Blumberg, "Strategic Examination of Reverse Logistics and Repair Service Requirements, Needs, Market Size and Opportunities," *Journal of Business Logistics,* 20/2 (1999): 141–159.

Discussion Questions

1. Explain the concept of reverse logistics.

2. What are the different product-recovery options?

3. Explain the nature of the closed-loop chain at Honeywell Industrial Automation.

4. What is product modularity?

5. What is life-cycle assessment (LCA)?

6. Explain the various enforcing and enabling factors in closed-loop supply chains.

In-Class Assignment: Objective Questions

1. Direct Reuse is used in original or similar markets.
   a. True
   b. False
2. Refurbishing is used in alternative markets.
   a. True
   b. False
3. Mass customization requires modular product designs for cross compatibility of components.
   a. True
   b. False
4. Cannibalization involves which of the following?
   a. Shred, sort, recycle, and dispose of
   b. Check on damage and clean
   c. Restore product to working order
   d. Selective retrieval of components
   e. None of the above.

14 EDP Renewables North America:
Tax Equity Financing and Asset Rotation

Pedro Matos and Griffin Humphreys

In November 2015, João Manso Neto, the CEO of EDP Renováveis, S.A. (EDPR), was arriving at the Houston, Texas, headquarters of EDP Renewables North America (EDPR-NA), its platform for the United States, Canada, and Mexico. He was going to be briefed by Gabriel Alonso (CEO of EDPR-NA), Bernardo Goarmon (EVP Finance of EDPR-NA), and the finance team on a new 200 MW wind farm investment located in Florida.[1] Right before entering the meeting room, Manso Neto passed the state-of-the-art dispatch center that oversaw all of the 34 wind farms located across 12 U.S. states and one Canadian province. Over 2,300 wind turbines, totaling more than 4.2 GW of installed capacity, were monitored in real time for load factors and energy output. The new Rising Star wind farm under construction in Florida consisted of 90 turbines, and its start of operations was scheduled to occur at the end of 2015.

Seven years after its IPO on the Euronext Lisbon (Portugal) stock exchange, the company had reached almost 10 GW of renewable energy capacity and was operating in 10 countries across Europe and the Americas. Its renewable-energy business comprised the development, construction, and operation of wind farms and solar plants to generate and deliver clean electricity (**Exhibit 14.2**). It was the world's third-largest wind energy producer and a strong believer in the case for wind energy onshore (**Exhibits 14.3 and 14.4**). EDPR was also actively working on offshore wind projects in three different countries in Europe. The company's stock price had significantly outperformed both its local index and the utilities sector since the start of the year, and its market capitalization exceeded EUR5.7 billion (**Exhibit 14.3**).[2] Its leadership in environmental, social, and governance (ESG) practices was decisive in leading EDP (incumbent utility in Iberia and

---

[1] The term "wind farm" (or wind power plant) is used to reference the land, wind turbine generators, electrical equipment, and transmission lines for the purpose of generating wind energy. The basics of wind power, as well as technical terms such as megawatts (MW), are defined in the glossary in **Exhibit 14.1**.

[2] As of March 29, 2016, the exchange rate was EUR 1 = USD 1.12. EUR = euros; USD = U.S. dollars.

Pedro Matos and Griffin Humphreys, "EDP Renewables North America: Tax Equity Financing and Asset Rotation," pp. 1-28. Copyright © 2016 by Darden Business Publishing. Reprinted with permission.

EDPR parent) to be ranked as the number-one utility company worldwide in the FTSE4Good Index.[3]

In a few weeks' time, the 2015 United Nations' Climate Change Conference (COP21) was being held in Paris in an attempt to limit greenhouse gas emissions. Manso Neto reflected on how wind energy was no longer a European oddity. Wind energy was already becoming a sizable source for new power capacity in the United States, and this was expected to continue as the aging conventional fleet was replaced (**Exhibit 14.5**). The evolution of taller towers and larger generators increased production and decreased the number of turbines per wind farm. With the improvements in technology, wind farms were economical even in low-wind areas in the United States (**Exhibit 14.6**). EDPR was well positioned to take advantage of this opportunity (**Exhibit 14.7**). In fact, Manso Neto had stated that "the U.S. market is a priority," and more than 60% of EDPR's investment was going to be in the United States until 2017.[4]

The long meeting room was filled with the logos of all 34 wind farms, named for each unique location, such as Lone Star (Texas), Rising Tree (California), Pioneer Prairie (Minnesota), and Top Crop (Illinois). As the team hung a picture with the logo of the new Rising Star wind farm, Alonso half-joked that "selecting the right name is hard, the financing is easy!" In fact, the project financing would be EDPR's 14th tax equity and 5th U.S.-based cash equity asset-rotation transaction. In the room were Pedro Almeida (director of finance), Sandhya Ganapathy (director of investments and M&A), and Leslie Freiman (general counsel) who had been at EDPR-NA since its origin as Horizon Wind more than a decade before. Goarmon, and the team were going to present a simple valuation model that explained the key drivers of returns to EDPR shareholders with this particular financing strategy.

The Renewables Industry in the United States

U.S. electricity had traditionally been provided by coal, natural gas, and nuclear power plants and, in the first half of 2015, these energy sources combined to generate more than 85% of all electrical output. During the same period, renewable sources, such as wind and solar, generated approximately 6.5% of total electricity production.[5] Albeit still relatively small, 6.5% of total generation of the United States was already remarkable for such a new industry; moreover, the U.S. electrical mix was expected to change considerably in the coming years. The U.S. Energy Information Administration projected an annual

---

3   In the ESG evaluation process, more than 2,300 companies were considered and fewer than 50% were included in the index by FTSE. The process took into consideration the impact/risk that the company had on each one of six ESG criteria.
4   André Cabrita-Mendes, "United States Will Be the 'Priority' of EDP Renováveis in the Next Two Years," *Jornal de Negócios*, November 13, 2015.
5   U.S. Energy Information Administration, *Electrical Power Monthly*, October 2015.

electricity demand growth of only 0.3% per year until 2040, but approximately 42 GW of coal capacity was scheduled to retire by 2020.[6]

Wind was becoming an increasingly attractive alternative to traditional power generation. In 2014, wind was the third-largest source of newly installed generating capacity, adding 4.9 GW or 24% of total new capacity (**Exhibit 14.5**).[7] Due to improvements in technology, such as increased hub height (the height from the ground to the center of the rotors), larger-diameter rotors, and increased name plate capacity, wind had become cost competitive with traditional power sources (**Exhibit 14.6**). Levelized cost of energy (LCOE) was a comparison metric for energy sources equating the average cost of producing a megawatt hour (MWh) over a facility's lifetime, including all capital expenditures (CapEx), input prices, maintenance, and carbon emission costs. Technological improvements had lowered wind's LCOE over the previous five years (**Exhibit 14.6**).[8]

EDPR was a strong believer in the case for wind energy onshore not only because it was green but because it was competitive (**Exhibit 14.7**). EDPR's view was that the wind energy market had a number of growth drivers in the United States as well as in Europe and emerging economies. EDPR was well positioned to take advantage of this opportunity as more than 75% of the worldwide renewable additions were in regions where EDPR already operated.

Unlike traditional power plants, where the majority of facilities were owned by utility companies responsible for bringing power directly to customers, independent power producers (IPPs) owned 82% of wind capacity.[9] Due to the predominance of IPPs, power purchase agreements (PPAs) were the industry standard. PPAs outlined the MWhs and the fixed price (typically with an escalator) in dollars per MWh that an off-taker—typically a utility, local municipality, or corporation—would purchase from the IPP. Typically, a PPA was with one off-taker for the entire facility's generated MWhs and covered a period of 20 years.

Should a wind facility not have a PPA, its power was still guaranteed to be purchased at spot prices (sometimes hedged in the short to medium term, when these financial instruments were available, in order to provide more certainty to the revenue stream). Wind benefited from a free, reusable power source, placing it early in the "generation stack" in almost all energy market regions in the United States. A generation stack was the pecking order of merit in which facilities provided power to the grid for consumption (**Exhibit 14.1**). The facilities were stacked based on their variable cost to produce, and generation was paid for at the most expensive variable cost. At the bottom of the stack was nuclear, followed by renewables (hydroelectric, solar, and wind), natural gas, and the stack ended primarily with coal (and, in rare occasions, more expensive oil "peaker" plants). Having

---

6   U.S. Energy Information Administration, *Annual Energy Outlook 2015*, April 2015; EDPR annual report, 2014.
7   U.S. Department of Energy, *2014 Wind Technologies Market Report*, August 2015.
8   Lazard, Lazard's Levelized Cost of Energy Analysis—Version 8.0, September 2014.
9   U.S. Department of Energy, *2014 Wind Technologies Market Report*, August 2015.

the lowest variable cost meant that, if the turbines were generating power, it was usually purchased at the highest price available.

Tax Incentives for Wind Energy

Most of the wind farms in the world were located in Europe and in the United States, where primarily federal programs supported wind power development as an alternative energy source to reduce dependency on fossil fuels and lower greenhouse gas emissions. In recent years, renewable portfolio standards (RPSs) had become popular and increasingly more stringent across U.S. states. RPSs were state-level regulations that set a minimum required generation from renewable sources on either a statewide or a producer level. As of 2015, 31 states had RPSs that typically required 10% to 25% of electricity to come from renewable sources by 2020 or 2025.[10] Standards increased periodically, setting higher and higher targets. This framework drove many utilities to set up auction systems to seek long-term PPAs with renewable energy generators.

In addition to state-level RPSs, the federal government encouraged investment in renewable energies through federal subsidies to lower the cost to developers and boost the industry. The subsidies were typically in the form of production tax credits (PTCs). Originally enacted in the Energy Policy Act of 1992, PTCs incentivized generation by providing the facility owner USD23 per generated MWh credit (as of 2015) on the tax bill for the first 10 years of the project's life.[11] There were concerns regarding the cost of these incentives to taxpayers. Since their inception, eligibility for PTCs had been possible for a couple of years at a time. There had been extensions (i.e., PTCs were allowed to expire and revived retroactively), but this stop-and-go approach limited the visibility on the investment horizon for wind energy companies. PTCs had expired at the end of 2014, and the U.S. Congress had yet to extend these tax credits. Analysts were confident that PTCs would be extended by mid-December 2015 as part of the 2015 Tax Extenders Bill, but there was still some uncertainty.[12]

Additionally, the federal Modified Accelerated Cost Recovery System (MACRS) allowed wind project owners to depreciate a relevant part of project capital costs on an accelerated basis. Both tax credits (PTCs and MACRS) were important despite wind energy having become cheap on a LCOE basis for new projects, because wind farms had to be built, while the fossil fuel–powered plants were already operating after massive infrastructure

---

10   EDPR annual report, 2014.

11   Wind projects also had the option to choose investment tax credits (ITCs) in lieu of the PTCs. ITCs were tax credits equal to a percentage of the project's qualified capital expenditures, not linked to production, but were more rarely used by wind developers because of their nature of not being linked to output and productivity.

12   Bank of America Merrill Lynch, "Renewable Energy—Only One Degree of Separation, Bullish on Renewables," November 27, 2015.

projects had been built (and having received public-sector support, including tax credits, low-cost loans, and outright grants from the federal government).[13]

EDPR

Headquartered in Madrid, EDPR was a subsidiary of EDP Group, the third-largest electricity generator in the Iberian Peninsula, present in several European countries and the Americas (**Exhibit 14.3**). Headquartered in Lisbon, Portugal, EDP-Energias de Portugal S.A. was a listed company in the NYSE Euronext Lisbon.

The first EDP wind farms were built in 1996, although the EDP Group did not form the holding company EDPR (encompassing EDP's renewable assets, with the exception of hydro energy) until 2007. EDPR was listed on the Euronext Lisbon stock exchange in June 2008, with EDP floating 22.5% of outstanding shares to outside investors (**Exhibit 14.3**). The EDP Group opted for an IPO to increase the visibility of a "pure player in renewables with a unique investment case." "An IPO at this time is the best way to harness the value of the company," António Mexia, the CEO of EDP, had stated at the time of the IPO.[14] Back in 2008, EDPR's IPO was the biggest of the year in Europe.

EDPR had been created as a high-growth platform focused on "clean" assets and aimed at becoming one of the top five global renewable companies. By 2015, EDPR had succeeded in being the world's third-largest wind energy producer (after Iberdrola Renovables and NextEra Energy Resources), having grown its portfolio to about 10 GW of installed capacity, and concentrated in North America and Europe (**Exhibit 14.4**). EDPR had mostly focused on onshore wind energy but also had solar photovoltaic (PV) farms and was developing several GWs of pipeline in offshore wind (United Kingdom, France, and the Netherlands). In addition, EDPR had an experimental wind float offshore project in the north of Portugal, particularly targeted at deep waters (see **Exhibit 14.8**).

The majority of its portfolio was developed, constructed, and operated internally. The installed asset base was young, with an average age of less than six years, and achieved a 30% load factor, indicating the quality of the sites (**Exhibit 14.4**). In 2014, EDPR revenues had totaled close to EUR1,277 million and an EBITDA of EUR903 million, with a net profit reaching EUR126 million and a robust operating cash flow of EUR707 million. The first nine months of 2015 were proving to be even stronger (**Exhibit 14.3**).

CEO Manso Neto ran the business through two operating units (**Exhibit 14.3**). The European and Brazil platform was based in Madrid and controlled the seven European and one Brazilian business units. The North America platform, headed by Gabriel Alonso, was based in Houston, Texas, and controlled the three EDPR-NA units (United States, Canada, and Mexico). After entering the U.S. market with the 2007 acquisition of Horizon Wind

---

13   Kate Gordon, "Why Renewable Energy Still Needs Subsidies," *Wall Street Journal*, September 14, 2015.
14   Peter Wise, "EDP Prices €1.8bn Renewables IPO," *Financial Times*, June 2, 2008.

LLC from Goldman Sachs, EDPR-NA more than doubled its wind-power production.[15] The United States had become EDPR's largest market by installed capacity and production (**Exhibit 14.3**). EDPR-NA had 34 wind farms and one solar park, and maintained 10 offices across the United States. With approximately 400 employees, it operated more than 2,300 turbines with a cumulative installed capacity of more than 4 GW across 12 U.S. states and one Canadian province (**Exhibit 14.9**). Of the existing portfolio at the end of 2015, 85% of this capacity was under existing PPAs, while the remaining was merchant power, free to be sold at market prices (**Exhibit 14.3**).

The 2014–2017 Business Plan and the Self-Funding Growth Model

In May 2014, at the EDP Group Investor Day held in London, Manso Neto presented to the financial community EDPR's business plan for 2014 through 2017. The plan focused on three pillars: selective growth, increased profitability, and self-funding growth. Between 2013 and 2017, this plan targeted a 9% electricity output CAGR, a 9% EBITDA CAGR, an 11% net profit CAGR, and a 25% to 35% dividend ratio.[16] To achieve this growth, EDPR planned to increase its portfolio capacity by more than 2 GW by 2017, but only on low-risk projects with established PPAs to lock in project profitability. Growth would come from PPAs already signed in the United States (60%), European markets with low risk (20%), and emerging markets with long-term PPAs (20%). Increased profitability would arise from maintaining availability above 97.5% by keeping maintenance and downtime to less than 2.5%, and increasing the assets' load factor to generate electricity 31.5% of the available hours, up from its current approximately 30% net capacity factor (so called "load factor").

EDPR's self-funding model had been one of the three supporting pillars of the company's strategy. This self-funding model excluded any increase in corporate debt, and therefore relied on the combination of cash flow from operating assets, external funds from tax equity for the United States (and other structured project finance for non-U.S. projects), as well as proceeds from asset-rotation transactions to finance the profitable growth of the business.

---

15 EDPR-NA transitioned through several stages and business models. The company began in 1998 as a "build-transfer" developer known as Zilkha Renewable Energy, followed by a period of ownership by Goldman Sachs as a "developer-owner-operator" beginning in 2005. Following its acquisition in July 2007 by EDP, the company became an integrated strategic renewable energy company. In December 2007, EDPR was incorporated to hold and operate the growing renewable energy assets and activities of EDP.

16 EDPR annual report, 2014.

Operating Cash Flow

The primary source of funds for the company was the operating cash flow generated from the existing assets, which was first used to pay for the debt service and capital distributions to equity partners, while the excess was available to pay dividends to EDPR's shareholders or to fund new investments. The company expected operating cash flow generation of about EUR3.5 billion for the period from 2014 to 2017. Cash flow available for dividends and new investments, about EUR400 million in 2014, was expected to have an annual average double-digit increase until the end of the current business plan period in 2017. EDPR had indicated a dividend payout ratio policy in the range of 25% to 35% of its annual net profit, thus allowing that most of the cash flow was available to fund growth.[17]

U.S. Tax Equity

EDP did not want to raise the level of debt at the group level. EDP procured all debt for its group to achieve the most competitive rates in the market. Secondly, EDPR's development in the United States during a period where a cash grant[18] could be collected from the U.S. Treasury to develop wind farms had generated a net operating loss (NOL) over the years, due to the MACRS depreciation of these investments, which was appropriated by the company. EDPR was still offsetting its taxable income against this legacy NOL, not being able to take advantage of the entire PTCs and accelerated fiscal depreciations (MACRS) associated with new projects.

In the U.S. market, the solution was "tax equity" investment, a hybrid between debt and equity.[19] Tax equity investors were able to more efficiently use the entire tax deductions and were willing to pay up front for the reduction of future tax bills. EDPR structured these transactions as "partnership flips," where the cash distributions and tax allocations were split differently between its members. The tax equity investor was allocated 99% of EBIT during the first project phase (typically 10 years, the length of the expected PTC credits) and, therefore, retained 99% of the PTCs. During that first phase, cash allocations would be initially allocated 100% to the developer until a cash flip date, and 100% to the investor from that cash flip date until the flip date. After the flip date, the developer would start receiving the vast majority of EBIT and cash flows (typically 95%).[20]

---

17 The dividends paid in 2014 amounted to about EUR35 million, corresponding to the low end of the range relative to the net profit of the previous year.

18 No longer available for wind development, cash grants were an incentive similar to the ITC but in the form of cash and not a tax credit.

19 In other markets, such as Mexico, EDPR relied more on nonrecourse project finance debt deals.

20 IRS rules meant that the tax incentives had to go to the legal owner of the wind project company. If ownership of the wind project company changed within the first years, the initial owner would have to repay the IRS a prorated amount of the tax incentives (called "recapture").

Tax equity investors were usually large tax-paying financial entities such as banks, insurance companies, and, in a few cases, big corporations that used these investments to reduce future tax liabilities. A tax equity investor was passive and only got involved for certain major decisions outside the normal course of business of the asset. In 2014, EDPR had established three institutional tax equity financing structures for a total amount of USD332 million, in exchange for an interest in the 200 MW Headwaters wind farm (counterparty being Bank of America Merrill Lynch), the 99 MW Rising Tree North (with MUFG Union Bank) and the 30 MW Lone Valley solar PV plant (undisclosed). These transactions had brought total tax equity financing proceeds raised by EDPR close to USD3.5 billion.

Asset Rotation

The second source of external financing consisted of the "asset rotation" or "cash equity" investment. EDP's capital structure was close to its target of debt to capital, and further debt might negatively impact its debt rating and cost of debt. The purpose was to sell a minority stake (typically 49%) in a renewable-generation asset such as a wind farm or solar plant and still maintain full management control over these assets. With the start of operations and the PPA, a wind farm project offered a low-risk profile with stable cash flows that were attractive to institutional investors, such as infrastructure private equity funds, pension funds, and strategic partners, from whom EDPR could source financing at competitive costs. The yields these institutional investors demanded (5% to 8%) were lower than those of EDPR equity investors.[21] With a successful asset rotation, EDPR would crystallize the value of the asset up front and reinvest the proceeds to fund investment in additional accretive projects (**Exhibit 14.10**).

The one disadvantage with asset rotation was its short-term EPS dilution. In the 18-to-24 months following a rotation, the earnings typically decreased because they were split over two parties. However, the long-term accretive benefits outweighed the short-term depressed earnings, which was why EDPR pursued this strategy.[22]

EDPR entered its first asset rotation in 2012 and had since entered a total of 11 transactions in the United States (six deals), France (two), Portugal (one), Canada (one), and Brazil (one) with different investors (**Exhibit 14.10**). For its first transaction, EDPR had been approached by Borealis Infrastructure, which managed infrastructure investments on behalf of the Ontario Municipal Employees Retirement System (OMERS), one of Canada's largest pension funds. At the deal's announcement, Michael Rolland, president and CEO of Borealis, stated that "our investment in this portfolio marks a significant commitment by Borealis to the renewables sector, and is the type of large-scale infrastructure asset we look for to generate

---

21 Based on Morgan Stanley, "EDPR - The cost of capital game", May 12, 2015.
22 Morgan Stanley, "EDPR - The cost of capital game", May 12, 2015.

stable and consistent returns for the pension plan."[23] Many investors wanted to invest in green infrastructure, and the process became more competitive such that subsequent transactions had been conducted in a bidding process. In 2014, EDPR received EUR215 million of proceeds regarding the EFG Hermes, Northleaf Capital Partners, and Axpo Group (signed in October 2013) transactions. The settlement of the Fiera Axium transaction occurred in 2015. Two transactions, worth EUR339 million, had already closed in 2015.

Asset rotations allowed EDPR to sell at an increased multiple compared to precommissioned projects. EDPR used an EV/MW (enterprise value per megawatt) multiple to value most of its asset rotations. A Morgan Stanley research report estimated that asset-rotation transactions were adding value since these were at multiples above that implied by EDPR's share price. Another way to look at attractiveness of asset rotation was a peer comparison. Iberdrola, the largest wind energy producer, had sold a few noncore wind farms as part of its 2012–2014 disposal plan.[24] **Exhibit 14.11** provides details of precedent asset-rotation transactions for EDPR in North America versus M&A transactions in 2015. One caveat was that simple EV/MW valuations did not consider the load factors of each plant, or reflect the plant's age, the PPA life, the tax attributes, among other factors.

The Rising Star Project[25]

The Rising Star wind farm was a 200 MW wind farm located just south of Jupiter, Florida (**Exhibit 14.12**). It was expected to generate enough electricity to power more than 51,000 average homes for a year and prevent more than 335,000 tons of carbon emissions. Approximately 150 landowners had written long-term leases for the land where turbines, access roads, and transmission corridors were to be installed. The Rising Star wind farm would contribute significant economic benefits to the surrounding community in the form of payments to landowners, local spending, and annual community investment. The development, construction, and operation of the wind farm generated a significant number of jobs in the area.

The project was first announced when a Florida utility entered a 20-year PPA with EDPR-NA. FPLC was a public, regulated utility serving more than 1,800,000 customers in 47 counties in Florida. With regard to Rising Star and an additional project, FPLC CEO John Cruz had stated: "These investments continue our commitment to move toward a more sustainable energy future in an affordable way. These two new wind projects will nearly double the amount of clean, renewable generation in our energy portfolio. This addition will be another step in diversifying our generation mix."[26]

---

23 "Borealis Infrastructure Buys Equity Stake in Four US Wind Farms from Spain's EDP Renováveis," Borealis press release, November 6, 2012.
24 Morgan Stanley, "Utilities - Asset Rotation, The Money Multiplier," May 18, 2015.
25 The identity of the project has been disguised. Specific data for the project have been altered for confidentiality reasons and pedagogical purposes.
26 "KCP&L Will Increase Wind Power and Energy Conservation," *Kansas City Star* January 9, 2014.

As with previous wind farms, the Rising Star project was capital intensive. EDPR was using its own balance sheet during the construction phase and, as explained previously, did not have sufficient tax liability in the United States to fully benefit from the federal tax incentives. The first step was to find a third-party tax equity investor to monetize them. Earlier in 2015, a nontraditional tax equity investor announced it would invest USD240 million of institutional equity financing in the form of a tax equity agreement. These funds were to be paid as construction would close in December 2015. Banks were traditionally the tax equity investors, but this was one of the new investors joining the class and increasing the supply in the tax equity market. With its investments in renewable energy and the purchase of tax credits, the company was able not only to reduce the carbon footprint of its operations but also "green" the communities it operated in.

A Simple Model for Assessing the Tax Equity and Asset-Rotation Strategy[27]

After signing the PPA with FPLC and the USD240 million proceeds from the tax equity investor, the asset-rotation partner was the remaining piece of the project financing for EDPR. For the asset rotation, the institutional investor who was willing to purchase the minority stake (49%) in the Rising Star wind farm was looking to earn a 6% yield. This was a very attractive yield in a low-interest-rate environment at the end of 2015.

The EDPR-NA finance team sketched a diagram to represent the main participants in the project financing for Rising Star and the "waterfall" of the project cash flows (**Exhibit 14.13**). The team had also developed a financial model based on the investors' proposals (**Exhibit 14.14**). Based on the engineers' projections for the turbines' performance and wind forecasts for the project site, the finance team had put together the revenue calculations and the deductions for operating expenses for 25 years.

The next step was to calculate the tax benefits, based on the rules for both the MACRS' depreciation schedule (for the most part, five years) and the PTC benefits (10 years), which started at $23 per MWh for the electricity produced at the facility and would be adjusted for inflation.[28] Based on the "partnership flip" model, the tax equity investor received 99% of the EBIT and PTCs during the first project phase (the first 10 years, until the flip date). After the flip date, the second phase began and class-A equity owners would from there on receive the majority of tax credits and cash flows.

The cash proceeds were modeled to occur at the end of 2015, when the project reached commercial operations date (COD). At that time, both tax equity and cash equity minority investors would simultaneously invest their funding amounts into the project. For the

---

27 Specific numbers and other data in this model have been altered for confidentiality reasons and pedagogical purposes.

28 U.S. Department of Energy, "Renewable Electricity Production Tax Credit (PTC)," http://energy.gov/savings/renewable-electricity-production-tax-credit-ptc (accessed April 13, 2016).

first project years, the non–tax equity holders would be paid first (because the tax equity investor was making its return from tax benefits) until the cash reversion date. From the cash reversion date until flip date (the blackout period), all cash flows would flow to the tax equity investor until they reached the return on their investment. The residual cash flows (after the flip date) were then split pro rata based on ownership between the EDPR (the sponsor equity investor with 51%) and the asset-rotation minority investor (49%). EDPR was interested in finding its internal rate of return (IRR) as the sponsor equity investor for the project. EDPR wanted to earn a spread over its cost of equity in its projects.

The Meeting

Manso Neto asked the team to start the meeting by presenting the analysis for the Rising Star project. He followed up with a bigger question to Alonso, Goarmon, and the rest of the team. How much did the tax equity and asset-rotation financing strategy really boost the returns for EDPR shareholders? What made more sense, an attractive IRR on residual cash flows or a higher NPV for the complete project?

The meeting also gave a chance to brainstorm alternatives. Some of EDPR's competitors had completed public market transactions where a portfolio of renewable energy–generating assets were formed into corporations and sold to the public such as NRG Yield, NextEra Energy, and Abengoa Yield in U.S. markets (**Exhibit 14.15**). A "YieldCo" owned and managed operational assets with contracted revenues and aimed to maximize cash distributions to its shareholders and grow such distributions through asset acquisitions. The sponsor company could then use the proceeds from the IPO to fund future projects. EDPR had indeed been studying this possibility in the U.S. market for almost three years. Primarily, the pause was related to the growth "promise" component for the cash dividend, which for U.S. assets seemed to be too expensive in the long run.[29]

Manso Neto wondered, should EDPR slow down growth to best comply with EDP optimal capital structure targets? Or, on the contrary, should EDPR continue expanding tax equity financing and the asset-rotation program as means to fund growth? Should EDPR change its business model into one of "build and sell" (flip 100% of the asset at commercial operation date)? Some competitors seemed to favor a public YieldCo; should EDPR consider it an alternative financing solution? The questions piled up as the team prepared to go through the analysis.

---

29 In summer 2015, EDPR had indeed considered a possible YieldCo focused only on European wind-generation assets to be listed on the Spanish stock exchange. Capital markets conditions were not favorable and it decided to hold this potential listing. "EDP Renováveis Informs about Its Asset Rotation program," EDPR press release, September 14, 2015. http://www.edpr.com/investors/market-notifications/?id=44626 (accessed April 13, 2016).

Exhibit 14.1. EDP Renewables North America: Tax Equity Financing and Asset Rotation

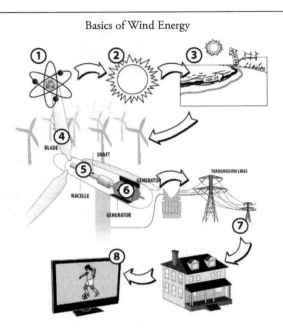

Basics of Wind Energy

Energy from moving air: People have been harnessing wind energy for thousands of years. Wind is caused by the uneven heating of the earth's surface by the sun. Because the earth's surface is made of different types of land and water, it absorbs the sun's heat at different rates. During the day, the air above the land heats up faster than the air over water. The warm air over the land expands and rises, and the heavier, cooler air rushes in to take its place, creating wind. At night, the winds are reversed because the air cools more rapidly over land than over water.

How wind turbines work: Wind turbines use blades to collect the wind's kinetic energy. Wind flows over the blades creating lift (similar to the effect on airplane wings), which causes the blades to turn. The blades are connected to a drive shaft that turns an electric generator, which produces electricity.

Wind farms: Wind farms (or wind power plants) can be located either on land (onshore) or offshore in the ocean. It is important to consider how fast and how often the wind blows at the site to position wind turbines. Wind speed typically increases with altitude and increases over open areas without windbreaks.

Wind is an emissions-free source of energy: Wind is a renewable energy source. Wind turbines do not release emissions that can pollute the air or water, and they do not require water for cooling.

Source: Created by the authors based. Image from the National Energy Education Development Project (public domain).

Glossary of Technical Terms:

Availability: Percentage of time a wind turbine is technically available (i.e., not malfunctioning or out of service) to capture the wind resource and convert it to electricity.

Blades: The large "arms" of wind turbines that extend from the hub of a generator. Most turbines have either two or three blades. Wind blowing over the blades causes the blades to "lift" and rotate.

The "generation stack:" The combination of each power plant's marginal cost in a given electrical market. As demand increases, the price increases and more power plants are used to generate electricity to serve the market. All plants receive the prevailing market price, making renewable sources with free input prices profitable in almost all markets. The curves are traditionally made with hydroelectric plants to the left, followed by other renewables, coal, natural gas, and, finally, "peaker" oil plants.

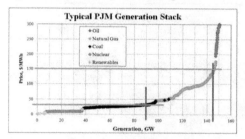

Greenhouse gases: Gases that trap the heat of the sun in the Earth's atmosphere, producing the greenhouse effect. The two major greenhouse gases are water vapor and carbon dioxide; lesser greenhouse gases include methane, ozone, chlorofluorocarbons, and nitrogen oxides.

Levelized cost of energy (LCOE): A comparison metric for energy sources equating the average cost of producing a megawatt hour (MWh) over a facility's lifetime. LCOE takes into account the installed system price and associated costs such as financing, land, insurance, transmission, operation and maintenance, and depreciation. The LCOE is a true apples-to-apples comparison of electricity costs and is the most common measure used by electric utilities or purchasers of power to evaluate the financial viability and attractiveness of a wind energy project

Load factor (or net capacity factor): A measure of how constant the device's activity is. The higher the load factor, the more regular the electricity production.

Watt (W): Unit of power. Power corresponds to an amount of energy per second. Kilowatt, megawatt, gigawatt, etc. are multiples of watt. 1 KW = 10^3 W, 1 MW = 10^6 W , 1 GW = 10^9 W

Watt hour (Wh): Unit of energy. Corresponds to the amount of energy produced/spent by a device with 1 watt of power operating during 1 hour.

Source: Created by authors based on EDPR annual report, 2014.

Exhibit 14.2. EDP Renewables North America: Tax Equity Financing and Asset Rotation

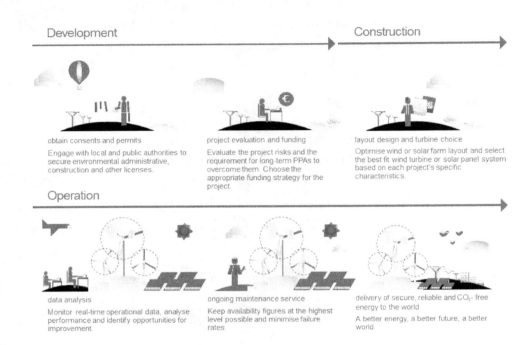

Source: Company document, used with permission.

Exhibit 14.3. EDP Renewables North America: Tax Equity Financing and Asset Rotation

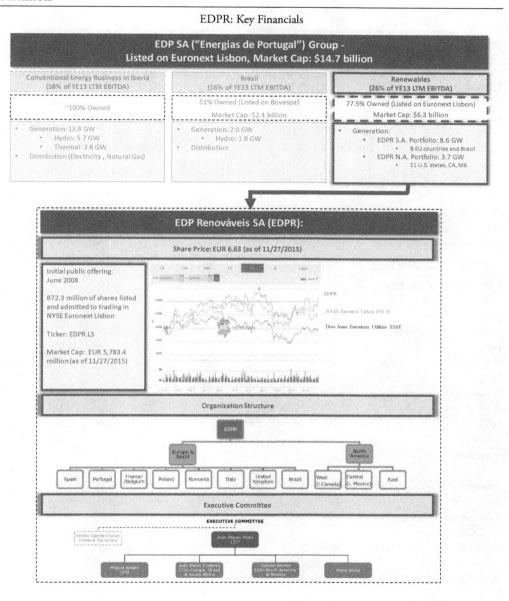

Source: EDPR investor relations website, http://www.edpr.com/investors/ (accessed April 13, 2016).

Exhibit 14.3. (continued)

EDPR - Financial Data (€m)	2008	2009	2010	2011	2012	2013	2014	9M15
Revenues	581.4	724.7	947.7	1,068.8	1,285.2	1,316.4	1,276.7	1,078.9
Operating Costs & Other Operating Income	(143.6)	(182.1)	(234.9)	(268.1)	(347.6)	(395.8)	(373.5)	(296.8)
EBITDA	437.9	542.6	712.8	800.7	937.6	920.5	903.2	782.1
EBITDA / Revenues	75%	75%	75%	75%	73%	70%	71%	72%
EBIT	231.6	230.8	289.9	347.5	450.1	473.0	422.4	374.1
Net Financial Expenses	(74.9)	(72.2)	(174.2)	(233.6)	(274.9)	(261.7)	(249.9)	(211.6)
Net Profit (Equity holders of EDPR)	104.4	114.4	80.2	88.6	126.3	135.1	126.0	99.6
Operating Cash-Flow	294	392	567	643	666	677	707	536
Capex	2,091	1,846	1,401	829	612	627	732	595
PP&E (net)	7,142	8,635	9,982	10,455	10,537	10,095	11,013	12,349
Equity	5,199	5,328	5,394	5,454	5,749	6,089	6,331	6,778
Net Debt	1,069	2,134	2,848	3,387	3,305	3,268	3,283	3,686
Institutional Partnership Liability	852	835	934	1,024	942	836	1,067	1,114

EDPR - Operating Data	2008	2009	2010	2011	2012	2013	2014	9M15
Installed Capacity (EBITDA MW + Eq. Consolidated)	4,400	5,576	6,676	7,483	7,987	8,565	9,036	9,231
Europe	2,477	2,938	3,439	3,977	4,266	4,796	4,938	5,034
North America	1,923	2,624	3,224	3,422	3,637	3,685	4,014	4,113
Brazil		14	14	84	84	84	84	84
Electricity Generated (GWh)	7,807	10,907	14,352	16,800	18,445	19,187	19,763	14,994
Europe	3,900	4,975	6,632	7,301	8,277	9,187	9,323	7,201
North America	3,907	5,905	7,689	9,330	9,937	9,769	10,204	7,638
Brazil		26	31	170	231	230	236	156
Load Factor (%)	30%	29%	29%	29%	29%	30%	30%	28%
Europe	26%	26%	27%	25%	26%	28%	27%	26%
North America	34%	32%	32%	33%	33%	32%	33%	30%
Brazil		22%	26%	35%	31%	31%	32%	28%
Average Selling Price (€/MWh)	65.9	58.8	58.4	57.7	63.5	62.6	58.9	65.0
Europe (€/MWh)	98.0	87.2	84.2	88.0	94.2	89.3	80.3	83.5
North America ($/MWh)	33.2	34.7	34.3	32.8	47.1	48.4	50.8	52.1
Brazil (R$/MWh)	0.0	0.0	109.4	119.7	286.4	309.2	346.4	370.6
Employees	630	721	822	796	861	890	919	1,009
Europe	324	365	398	393	393	467	434	448
North America	276	303	332	260	251	298	316	367
Brazil		8	17	16	21	23	26	32
Holding	30	45	75	127	196	102	143	162

(continued)

Exhibit 14.3. (continued)

EDPR - North American Data							
Installed Capacity (MW)	2008	2009	2010	2011	2012	2013	2014
EBITDA MW	**1,923**	**2,624**	**3,224**	**3,422**	**3,637**	**3,506**	**3,835**
US PPA/Hedge	1,596	1,888	2,459	2,659	2,876	2,907	3,251
US Merchant	327	735	764	763	761	569	554
Canada						30	30
Avg. Load Factors (%)							
US	34%	32%	32%	33%	33%	32%	33%
Canada						-	27%
North America	**34%**	**32%**	**32%**	**33%**	**33%**	**32%**	**33%**
Electricity Output (GWh)							
Total GWh	**3,907**	**5,905**	**7,689**	**9,330**	**9,937**	**9,769**	**10,204**
US PPA/Hedge	3,708	4,798	5,367	6,716	7,409	7,795	8,384
US Merchant	198	1,107	2,323	2,614	2,528	1,974	1,761
Canada						-	59
U.S. Tax Incentives							
MW under PTC	1,925	2,024	2,024	2,123	2,123	1,962	2,291
MW under cash grant flip		202	401	500	500	500	500
MW under cash grant		398	799	799	1,014	1,014	1,014
Average Selling Price ($/MWh)							
US PPA/Hedge	48.3	52.4	53.9	50.8	51.7	52.6	52.3
US Merchant	60.5	29.8	31.1	30.1	31.2	31.9	41.4
Canada	-	-	-	-	-	-	132.0
North America	**49.0**	**48.2**	**47.7**	**45.7**	**47.1**	**48.4**	**50.8**
Income Statement (US$m)							
Electricity sales and other	193.9	282.7	364.5	421.6	456.8	461.9	507.6
Income from Institutional Partnerships	90.4	114.9	141.9	155.4	163.6	166.2	164.2
Revenues	**284.3**	**397.6**	**506.4**	**577.0**	**620.4**	**628.0**	**671.8**
Other operating income	34.5	45.4	61.0	24.7	25.4	39.9	22.6
Operating costs	(114.4)	(145.0)	(185.2)	(225.6)	(237.7)	(230.3)	(217.1)
EBITDA	**204.5**	**298.0**	**382.3**	**376.1**	**408.1**	**437.6**	**477.4**
EBITDA / Revenues	72%	75%	75%	65%	66%	70%	71%
Provisions	-	-	-	-	-	(1.55)	-
Depreciation and amortization	(129.5)	(220.9)	(294.7)	(291.8)	(299.9)	(287.9)	(292.1)
Amortization of deferred income (government grants)	-	2.2	13.1	19.1	18.1	23.1	23.1
EBIT	**75.0**	**79.3**	**100.7**	**103.3**	**126.3**	**171.2**	**208.4**
USD/EUR exchange at the end of the period	1.39	1.44	1.34	1.29	1.32	1.38	1.21
Income Statement (€m)							
Electricity sales and other	131.8	204.7	276.5	306.4	355.5	347.8	382.0
Income from Institutional Partnerships	61.2	82.7	107.0	111.6	127.4	125.1	123.6
Revenues	**192.6**	**286.1**	**382.0**	**414.5**	**482.9**	**472.9**	**505.6**
Other operating income	23.4	32.7	46.0	17.7	19.8	30.0	17.0
Operating costs	(77.4)	(104.3)	(139.7)	(162.0)	(185.0)	(173.4)	(163.4)
Supplies and services	(45.4)	(65.4)	(93.0)	(101.3)	(116.5)	(108.0)	(108.8)
Personnel costs	(18.0)	(21.0)	(24.3)	(25.9)	(29.0)	(28.8)	(27.8)
Other operating costs	(14.0)	(17.9)	(22.3)	(34.8)	(39.5)	(36.6)	(26.8)

(continued)

Income Statement (€m)							
EBITDA	138.5	214.4	288.3	270.2	317.7	329.5	359.3
EBITDA / Revenues	72%	75%	75%	65%	66%	70%	71%
Depreciation and amortization	(87.7)	(159.0)	(222.3)	(209.7)	(233.5)	(216.8)	(219.8)
Amortization of deferred income (government grants)	-	1.6	9.9	13.7	14.1	17.4	17.4
EBIT	50.8	57.1	75.9	74.2	98.3	128.9	156.8

Source: EDPR 9M2015 Key Data, http://www.edpr.com/investors/key-data/ (accessed Nov. 27, 2015).

Exhibit 14.4. EDP Renewables North America: Tax Equity Financing and Asset Rotation

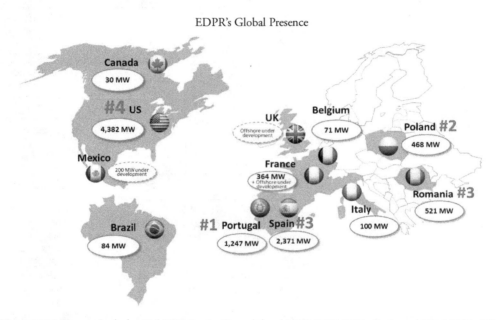

Notes: 2015 Figures; Includes 356 MW Equity Consolidated MW: 177 MW in Spain and 179 MW in US

Exhibit 14.4. EDPR: a market leader with a top quality portfolio... (continued)

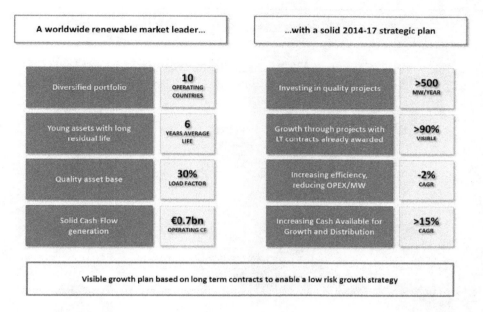

Source: EDPR Investor Presentation, September 2015.

Exhibit 14.5. EDP Renewables North America: Tax Equity Financing and Asset Rotation

U.S. Wind Energy Market

New Capacity Additions in the United States by Year

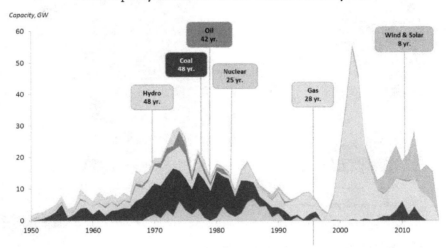

U.S. Wind Energy Share of Electricity Generation, by State

Wind as Percent of Total Energy Generated – by State (Year-End 2014)

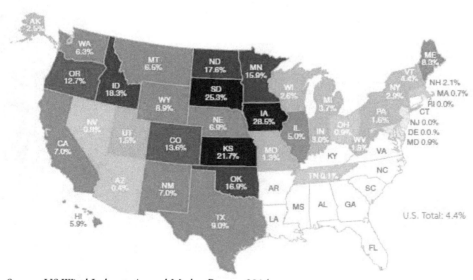

Source: US Wind Industry Annual Market Report, 2014

< 1% 1% to <5% 5% to <10% 10% to <15% 15% to <20% 20% and higher

Source: EDPR.

Exhibit 14.6. EDP Renewables North America: Tax Equity Financing and Asset Rotation

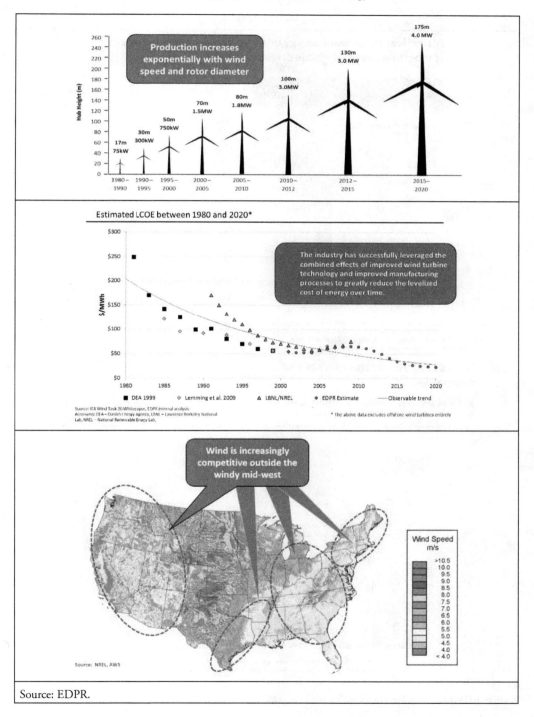

Source: EDPR.

Exhibit 14.7. EDP Renewables North America: Tax Equity Financing and Asset Rotation

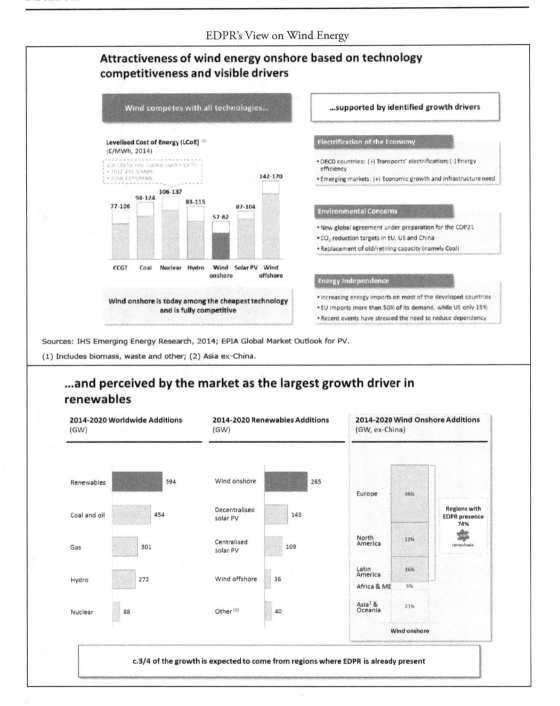

Source: EDPR Investor Presentation, September 2015.

Exhibit 14.8. EDP Renewables North America: Tax Equity Financing and Asset Rotation

EDPR's Windfloat project

Source: EDPR.

Exhibit 14.9. EDP Renewables North America: Tax Equity Financing and Asset Rotation

Map of EDPR-NA's Existing Capacity, Development, and the Rising Star Wind Farm

Source: EDPR.

Exhibit 14.10. EDP Renewables North America: Tax Equity Financing and Asset Rotation

EDPR's Self-Funding Growth Plan

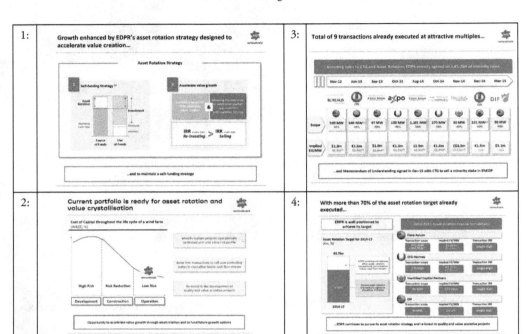

Notes: Slide 1: Illustrative and non-exhaustive. Slide 3: (1) 615 MW in operation + 29 MW under development; (2) 84 MW in operation + 237 MW under development; (3) including all cash flows generated by the projects since inception. Slide 4: (1) Total of 1,101 MW considers 801 MW in operation plus 300 MW under construction; (2) including all cash flows generated by the project since inception.

Source: EDPR Investor Presentation, September 2015.

Exhibit 14.11. EDP Renewables North America: Tax Equity Financing and Asset Rotation

List of Precedent Transactions

EDPR's North American Asset-Rotation Transactions

Date	Minority Investor	Asset Type	Minority Stake	Project Location	Project Size (MW)	Implied EV/MW ($mm)
Mar. '15	DIF Infrastructure	Solar	49%	United States	30	3.1
Nov. '14	Northleaf Capital Partners	Wind	49%	Canada	30	3.3
Aug. '14	Fiera Axium	Wind	36%	United States	1,101	1.5
Sept. '13	Fiera Axium	Wind	49%	United States	97	1.0
Nov. '12	Borealis Infrastructure	Wind	49%	United States	599	1.3

2015 Market Transactions

Date	Asset Name	Asset Type	Acquirer Ticker	MW	Equity ($mm)	EV ($mm)	Expected CAFD ($mm)	EV/MW ($mm)	Yield
11/3/2015	12 Asset Portfolio	Wind	NYLD	611	210	520	30.0	0.90	5.8%
10/5/2015	Jericho Wind Farm	Wind	NEP	149	210	523	22.5	3.51	4.3%
9/30/2015	Solaben 1/6	Solar	ABY	100	290	290	0.50		
8/4/2015	Desert Sunlight	Solar	NYLD	137.5	285	572	22.0	4.16	3.8%
7/21/2015	Solacor 1/2	Solar	ABY	100	370	370	31.5	3.70	8.5%
	ATN2	Transmission	ABY						
7/21/2015	Gulf Wind	Wind	PEGI	170	86	178	NR	1.05	
7/20/2015	Vivint Solar	Solar	TERP	523	922	922	81.0	1.76	8.8%
7/6/2015	Invenergy	Wind	TERP	930	1,100	1,918	141.0	2.06	7.4%
6/25/2015	Duke Energy Solar	Solar	TERP	9	45	55	5.0	6.11	9.1%
6/22/2015	K2 Wind	Wind	PEGI	89.1	128	350	NR	3.93	
6/4/2015	Integrys Portfolio	Solar	TERP	23	45	55	5.0	2.39	9.1%
5/18/2015	Wind Capital	Wind	PEGI	351	242	344	NR	0.98	
5/12/2015	Ashtabula Wind III	Wind	NEP	62.4					
	Baldwin	Wind	NEP	102.4	424	693	30.0	1.04	4.3%
	Mammoth Plains	Wind	NEP	198.9					
	Stateline	Wind	NEP	300					
5/11/2015	Spanish Solar Port.	Solar	ABY	320	669	669	63.0	0.50	9.4%
	KaXu	Solar	ABY	51					
5/4/2015	Amazon Wind	Wind	PEGI	116	127	127	12.1	1.09	9.5%
4/1/2015	Atlantic Power	Wind	TERP	521	350	460	44.0	0.88	9.6%
1/29/2015	First Wind	Solar/Wind	TERP	521	862	862	73.0	1.65	8.5%
1/7/2015	Solar Drop Downs	Solar	TERP	26	47	$71	5.00	2.73	7.0%
1/5/2015	Walnut Creek	Natural Gas	NYLD	500					
	Tapestry	Wind	NYLD	204	480	1,217	35.0	1.55	2.9%
	Laredo Ridge	Wind	NYLD	81					

Source: Created by authors based on YieldCo press releases.

Exhibit 14.12. EDP Renewables North America: Tax Equity Financing and Asset Rotation

Rising Star Wind Farm

edp renewables

FLORIDA:
Rising Star Wind Farm

Location
Rising Star Wind Farm will be located nearby Jupiter, Florida.

Energy Output
Rising Star Wind Farm will have an installed capacity of 200 megawatts (MW) – enough to power approximately 51,306 average homes with clean energy each year.[1]

Benefits to the Community
Rising Star Wind Farm will contribute significant economic benefits to the surrounding community in the form of payments to land owners, local spending, and annual community investment. The development, construction, and operation of the wind farm will also generate a significant number of jobs in the area.

Environmental Benefits
Rising Star Wind Farm will avoid adding more than 331,349 tons of carbon dioxide to the atmosphere per year, the equivalent of removing 58,439 cars from the road.[2]

Landowners
Approximately 150 supportive landowners participate under long-term lease and easement options that cover turbines, access roads and transmission corridors.

Rising Star Wind Farm
Jupiter, FL 66871
P: 785.243.4446
F: 785.243.9742

Source: Created by authors. Please note that the identity of the "Rising Star" project has been disguised. Specific data for the project have been altered for confidentiality reasons and pedagogical purposes.

Exhibit 14.13. EDP Renewables North America: Tax Equity Financing and Asset Rotation

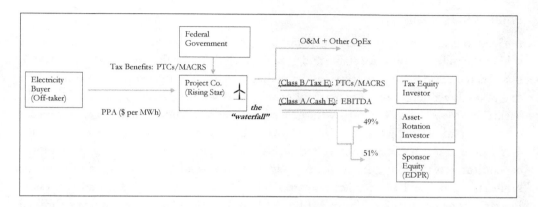

Rising Star Wind Farm—Simplified Diagram of the Project Financing

Source: Created by authors. Please note that the identity of the "Rising Star" project has been disguised. Specific data for the project have been altered for confidentiality reasons and pedagogical purposes.

Exhibit 14.14. EDP Renewables North America: Tax Equity Financing and Asset Rotation

Rising Star Wind Farm Simplified Valuation Model

PROJECT CASH FLOWS

Capacity (MW)	200	Wind Turbine Net Capacity	41%	
Hours/year	8,766	Generation (MWh)	718,812	

	(2-10y)	(10-20y)	(20-25y)
Turbine Operation & Maintenance (O&M)	35	40	50
Other Operational Expenses (Opex)	15	inflation	2%

TAX BENEFITS

Tax Rate	35%

TAX EQUITY (TE)

Year	Capex	Energy price ($/MWh)	Cash Revenues	Turbine O&M	Other Opex	EBITDA	MACRS Depreciation	Tax Benefit (Tax Liability)	PTC Price ($/MWh)	PTC Revenues	Tax Allocation to TE	Cash Allocation to TE
2015	-387,090,729	36.0					-5,970,971	2,089,840			99%	0%
2016		36.7	26,394,777		-3,000,000	23,394,777	-141,589,345	41,368,099	23	16,532,676	99%	0%
2017		37.5	26,922,672		-3,060,000	23,862,672	-85,836,379	21,690,797	24	17,251,488	99%	0%
2018		38.2	27,461,126	-7,000,000	-3,121,200	17,339,926	-52,325,660	12,245,007	24	17,251,488	99%	0%
2019		39.0	28,010,348	-7,000,000	-3,183,624	17,826,724	-40,464,197	7,923,115	25	17,970,300	99%	0%
2020		39.7	28,570,555	-7,000,000	-3,247,296	18,323,259	-38,725,202	7,140,680	25	17,970,300	99%	0%
2021		40.5	29,141,966	-7,000,000	-3,312,242	18,829,724	-1,898,403	-5,925,962	26	18,689,112	99%	60%
2022		41.4	29,724,806	-7,000,000	-3,378,487	19,346,318	-1,898,403	-6,106,770	26	18,689,112	99%	60%
2023		42.2	30,319,302	-7,000,000	-3,446,057	19,873,245	-1,898,403	-6,291,195	27	19,407,924	99%	60%
2024		43.0	30,925,688	-7,000,000	-3,514,978	20,410,710	-1,898,403	-6,479,307	27	19,407,924	99%	60%
2025		43.9	31,544,201	-7,000,000	-3,585,278	20,958,924	-1,898,403	-6,671,182	28	20,126,736	99%	60%
2026		44.8	32,175,085	-8,000,000	-3,656,983	20,518,102	-1,898,403	-6,516,895			5%	5%
2027		45.7	32,818,587	-8,000,000	-3,730,123	21,088,464	-1,898,403	-6,716,521			5%	5%
2028		46.6	33,474,959	-8,000,000	-3,804,725	21,670,234	-1,898,403	-6,920,141			5%	5%
2029		47.5	34,144,458	-8,000,000	-3,880,820	22,263,638	-1,898,403	-7,127,832			5%	5%
2030		48.5	34,827,347	-8,000,000	-3,958,436	22,868,911	-1,847,952	-7,357,336			5%	5%
2031		49.4	35,523,894	-8,000,000	-4,037,605	23,486,289	-45,428	-8,204,302			5%	5%
2032		50.4	36,234,372	-8,000,000	-4,118,357	24,116,015	-45,428	-8,424,706			5%	5%
2033		51.4	36,959,060	-8,000,000	-4,200,724	24,758,335	-45,428	-8,649,518			5%	5%
2034		52.4	37,698,241	-8,000,000	-4,284,739	25,413,502	-45,428	-8,878,826			5%	5%
2035		53.5	38,452,206	-8,000,000	-4,370,434	26,081,772	-45,428	-9,112,721			5%	5%
2036		54.6	39,221,250	-10,000,000	-4,457,842	24,763,407	-45,428	-8,651,293			5%	5%
2037		55.7	40,005,675	-10,000,000	-4,546,999	25,458,676	-45,428	-8,894,637			5%	5%
2038		56.8	40,805,788	-10,000,000	-4,637,939	26,167,849	-45,428	-9,142,848			5%	5%
2039		57.9	41,621,904	-10,000,000	-4,730,698	26,891,206	-45,428	-9,396,222			5%	5%
2040		59.1	42,454,342	-10,000,000	-4,825,312	27,629,030	-45,428	-9,654,261			5%	5%

Source: Created by authors. Please note that the identity of the "Rising Star" project has been disguised. Specific data for the project have been altered for confidentiality reasons and pedagogical purposes.

Exhibit 14.15. EDP Renewables North America: Tax Equity Financing and Asset Rotation

A Short Primer on YieldCos

What is a YieldCo?

A YieldCo is a portfolio of energy-generating assets that is formed into a corporation and sold to the public to produce stable and growing cash flows for the investors. The portfolio of assets is created by a sponsoring company that can either be an energy development firm (e.g., NextEra Energy) or a financial sponsor (e.g., Riverstone Holdings LLC). Typically, YieldCos are set up as C Corps with general partner (GP) units owned by the sponsor company, and limited partner (LP) units owned by the sponsor company and the public. A YieldCo is a vehicle that allows renewable energy developers to recoup their capital expenditure costs more quickly than a traditional buy-and-hold strategy and is in the same family of yield companies as master limited partnerships (MLPs) and real estate investment trusts (REITs). Unlike MLPs and REITs, a YieldCo is not a tax pass-through entity. The Internal Revenue Service determines which assets can be included in MLPs and determines the earnings payout ratio for REITs to enable them to pay no taxes. YieldCos pay little tax based on net operating losses and tax credits.

List of YieldCos (as of November 2015)

YieldCo Name	Ticker	Sponsor Company	Share Price ($)	Market Cap ($mm)	Portfolio Size (MW)	ROFO Size (MW)	Portfolio Mix	2015E CAFD ($mm)	LT DPS CAGR	Return since IPO
NextEra Energy Partners (06/2014)	NEP (NYSE)	NextEra Energy	27.05	803	2,072	1,364	Wind (84%) Solar (16%)	110	12–15%	(15%)
NRG Yield (07/2013)	NYLD (NYSE)	NRG Energy	14.98	1,459	5,745	1,585	Wind (35%) Solar (9%) Other (57%)	165		(46%)
Pattern Energy Group (09/2013)	PEGI (Nasdaq)	Riverstone	23.56	1,760	2,282	1,270	Wind (100%)	84	12–15%	(17%)
Trans-Alta Renewables (08/2013)*	RNW (TSE)	TransAlta	10.35	1,975	1,830	2,042	Wind (63%) Hydro (6%) Other (31%)	98	6–7%	(1%)
8point3 (06/2015)	CAFD (Nasdaq)	First Solar & Sun Power	13.88	278	301	1,143	Solar (100%)	15	12–15%	(20%)
TerraForm Power (07/2014)	TERP (Nasdaq)	SunEdison	18.78	1,502	1,918	NA	Wind (26%) Solar (74%)	225	21%	(69%)
Abengoa Yield (06/2014)	ABY (Nasdaq)	Abengoa	19.27	1,931	1,741	NA	Wind (6%) Solar (76%) Other (18%)	174	20%+	(53%)

*Traded in Canadian dollars (CAD).

Source: Created by authors based on company filings and Yahoo! Finance.

Exhibit 14.15. (continued)

Under a YieldCo IPO, the sponsor company combines a group of operating assets into a portfolio, which is sold to a newly created YieldCo corporation in exchange for stock and cash. Simultaneously, the YieldCo sells stock to the public market and uses the cash proceeds to pay for the portfolio from the sponsor company. The sponsor company can then use the proceeds from the IPO to fund future projects. In return for the IPO proceeds, investors receive a steady, predictable stream of distributions over the life of the YieldCo from PPA-contracted operating assets. YieldCos distribute over 70% of cash available for distribution (CAFD) as dividend payouts.

CAFD Calculation

Quarterly EBITDA
Less: Interest Paid
Less: Tax Paid (if applicable)
Less: Debt Amortization/Principal Payments
Less: Maintenance Expenditures and Growth CapEx
Less: Sponsor Company Management Fee (typically 3%–5%)
= Cash Available for Distribution

The structure matches a developer's need for capital to fund future projects with the investors' need for yield in a low-interest-rate environment. The structure is advantageous for developers because it immediately monetizes the assets rather than receiving monthly cash flow. Developers have a high cost of capital because of development risk. Once a project is operational and generating cash, it is immediately de-risked and cost of capital decreases. YieldCos have an even lower cost of capital than single-asset investors because of the diversified portfolio. A single asset might produce poor returns one year due to poor natural resources, but a portfolio of assets should continue to produce stable CAFD even if one asset has a poor year.

Following the 2008 financial crisis, investors were looking for investments that would create greater yield than they were receiving through traditional investments. YieldCos offer predictable returns by distributing more than 70% of CAFD to investors and significant distribution growth through "dropping" additional assets into the portfolio from the development company through a right of first refusal list (ROFO). The ROFO tells the market what assets will eventually be included in the portfolio and how future distributions will grow over the coming one to five years. The sponsor company is incentivized to increase the investor distribution due to incentive distribution rights (IDRs) that allow them to share in the increased cash flow. As distributions increase, IDRs allow the parent company to share in larger portions of the distributions.

YieldCos are valued based on market transactions distribution yield. YieldCos have traded on yields ranging from 2% to 10%. At IPO, the yield is smallest, meaning investors are willing to pay the most per dollar of distribution, because the ROFO assets have not been dropped and there is the largest distribution growth potential. As YieldCos age, the yield increases because much of the future growth is already realized. Some Wall Street analysts also perform discounted cash flow models and dividend discount models to project distributions using the ROFO assets.

Market Conditions in 2015

The first six months of 2015 boded well for the renewable energy industry as technology continued to reduce costs and installed capacity continued to increase. However, the summer had seen a free fall of prices among YieldCos. As oil prices dropped, energy investors, particularly MLP investors, sold their investments, including their YieldCo holdings, flooding the market and decreasing price. In addition, YieldCos issued $2 billion worth of equity, oversaturating the market. Over the summer, the Federal Reserve hinted at raising interest rates before the end of the year, and many yield-seeking investors sold their YieldCo positions. The confluence of factors began the price tailspin. YieldCos survive on getting cheap cost of capital and as prices tumbled, the yield approached the equity cost. As yield increases, YieldCos can no longer achieve immediately accretive acquisitions, halting drop-downs and stopping distribution growth, further increasing yield. The fourth quarter of 2015 was expected to be tough for YieldCos as they struggled to change their capital structure to suit the changing industry environment.

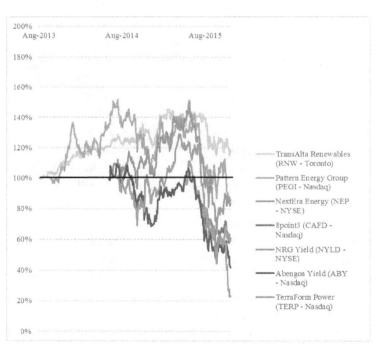

Price Performance of YieldCos (IPO Price = 100%)

Source: Created by authors.

Discussion Questions

1. Explain the nature of the renewables industry in the United States.

2. What are the different tax incentives provided for wind energy?

3. Explain the nature of EDPR's self-funding model incorporating the three pillars of sustainability.

4. Explain the technical and business characteristics of the "Rising Star" project.

5. What are the different stages of EDPR's business model?

6. What is YieldCo?

In-Class Assignment: Objective Questions

1. U.S. electricity had traditionally been provided by coal, natural gas, and nuclear power plants in the first half of 2015.
    a. True
    b. False
2. Wind has become an increasingly attractive alternative to traditional power generation.
    a. True
    b. False
3. In 2014, wind added 24% of total new electricity generation capacity.
    a. True
    b. False
4. Independent power producers own what percentage of wind capacity?
    a. 50%
    b. 75%
    c. 82%
    d. None of the above.
5. As of 2015, 31 states had RPSs that typically required what percentage of electricity to come from renewable sources by 2010 or 2025?
    a. 50%-60%
    b. 75%-90%
    c. 5%-8%
    d. 10%-25%

15

Coffee Cup Woes
Starbucks' Environmental Sustainability Challenge

Debapratim Purkayastha

"Because cups are so tangible, touchable, and really part of the ethos, our customers feel partially responsible for those cups when they choose to dispose of them. If this is their only option—an overflowing garbage can on the street—they are going to hold Starbucks partially responsible too. The consequence is that we've got to solve these issues to be able to address the other major wins that we have on our environmental and sustainability front."[1]

— **Jim Hanna, Starbucks' Director of environmental impact, Starbucks Coffee Company**

In 2013, the world's largest coffee chain, Starbucks Coffee Company (Starbucks), introduced a US$1 reusable cup in its stores in the US and Canada and later on, in the UK. The company came up with the idea of a plastic reusable cup to reduce its environmental footprint as it often faced criticism from environmental enthusiasts for generating 4 billion single-serve cups as trash each year that ended up in landfills or as litter. Starbucks also offered a 10 cent discount on the drink every time the customer used the reusable cup in its store. According to Starbucks' Director of environmental affairs, Jim Hanna (Hanna), *"Energy security and price spikes and brownouts and climate change are really now part of our everyday lives and everyday lexicon. Those who choose to spend their dollar at Starbucks, or make their careers at Starbucks, expect us to really use our size for good in that arena."*[2]

One of the most pressing challenges before the world's largest coffee chain, and more specifically Hanna, was to solve the problem caused by the branded trash it generated each year. Hanna contended that the 4 billion cups it generated as waste each year was a small percentage of the 500 billion cups generated globally, and it represented a "sliver" of Starbucks' environmental footprint.[3] *"Around 75 percent of our environmental footprint comes from the operation of our stores…. That being said, if you ask any customer what is Starbucks' greatest environmental footprint, most of them will assume it's our cups… It's essential that we at Starbucks, and that we as an industry that's using single-serve packaging, solve

Debapratim Purkayastha, "Coffee Cup Woes: Starbucks? Environmental Sustainability Challenge," Copyright © 2014 by IBS Center for Management Research. Reprinted with permission.

these end-of-life issues for our customers regardless of the contribution they make to our total footprint, so that we can shift the conversation to more pressing environmental issues,"[4] he said.

Starbucks had taken various initiatives over the years to address the issue of the cup, but these had had limited impact. In 2008, the company had taken up the target of making 100% of its cups recyclable by 2015, and serving 25% of beverages made in its company-operated stores in personal cups by 2015. Though the company had been offering the 10 cent discount to customers who brought along their own cups since the 1980s, only a small fraction of customers availed of the offer. In 2012, the company was forced to reduce the target of serving beverages in personal cups to 5% by 2015 as it was able to achieve less than 2%.[5] The company faced an uphill task in changing consumer behavior and critics felt that the new US$1 reusable cup too would have limited impact. The company also faced huge problems on the recycling front as the existing ecosystem was not "designed to take the individual Starbucks cups" due to the presence of plastic coating inside the cup.[6] The questions before Hanna were how to solve the problem of the Starbucks cup in a way that would be acceptable to all stakeholders: How could more customers be encouraged to use reusable mugs? Was the 10 cent discount incentive enough to drive behavior change of its customers? How was Starbucks to achieve the target of making 100% of its cups recyclable by 2015?

About Starbucks

As of 2013, Starbucks was the world's largest coffee house chain. It served hot and cold beverages, whole-bean coffee, microground instant coffee, full-leaf teas, pastries, snacks, packaged food items, hot and cold sandwiches, and items such as mugs and tumblers. More than the offerings, the company focused on selling a 'third place' experience, and the stores became places for relaxing, chatting with friends, reading the newspaper, holding business meetings, or browsing the Web. The 'experience' brought spectacular success for the store. Starbucks was considered an iconic brand and most of its customers were passionate about it. Of the customers, 8% were repeat customers.[7]

For the year ended September 30, 2012, the company had earned revenue of US$ 13.29 billion and net income of US$ 1.38 billion[8] **(See Exhibit 15.1 for Starbucks' financial data).** As of 2013, Starbucks operated in 62 countries, employing around 200,000 people worldwide whom it called partners.[9] Its operations were organized in three geographic regions: the Americas; Europe, the Middle East and Africa (EMEA), and China and Asia Pacific (CAP). The company had company-operated and licensed stores in these regions **(See Exhibit 15.2 for a break-up of company operated and licensed stores; Exhibit 15.3 for details on its company-operated stores and growth; and Exhibit 15.4 for the details about its licensed stores).**

Starbucks and Sustainability

In 1990, Starbucks started 'contributing positively to the community and environment' and had included this as one of the guiding principles in conducting its business. In 1992, the company also framed its Environmental Mission statement, in which it pledged that, *"Starbucks is committed to a role of environmental leadership in all facets of our business."*[10] Since then, it had integrated environmental policies and programs in all its operations, and held its partners accountable for the cause. The company fulfilled this environmental commitment by understanding the prevailing environmental issues and sharing the information with its partners and then developing innovative and flexible solutions to bring about change. It bought, sold, and used environmentally friendly products and encouraged its partners to share its mission. *"I think the most loyal and connected consumers understand this to be part of our values proposition. There's a lot of consumers out there who if they know something about the company they might expect us to have these standards, but I think there's also a very small subset of influential customers who do care about this as a purchase criteria,"*[11] said Ben Packard, Starbucks' VP for Global Responsibility, who was previously the company's Director of Environmental Affairs.

The company's Environmental Affairs group, led by Hanna, implemented Starbucks' environmental strategy, including setting and tracking the progress of environmental goals. Hanna, a BS in Environmental Sciences from Washington State University and a US Green Building Council LEED-accredited professional, had joined Starbucks in November 2005 as its Environmental Affairs manager. Before this he had served as Director of environmental affairs for Xanterra Parks & Resorts at Yellowstone National Park.[12] As the Director of environmental affairs at Starbucks, Hanna's job was to lend his expertise to the company's initiatives to minimize its environmental footprint through green building, energy conservation, international procurement, waste minimization, and collaboration with partner corporations and NGOs. An important part of his job was also to maintain the reputation of the Starbucks brand.

Climate Change

Starbucks relied on high-quality Arabic coffee beans, which grew only in certain environmental conditions. The company had encountered changing soil conditions, increasing pest infestations, soil erosion, changing weather, temperatures, and rain patterns and other threats to the Arabica coffee bean harvest. Starbucks recognized that a change in climate could have a negative impact on coffee-growing communities and the long-term sustainability of its business. *"From a business perspective, it's impacting the longevity of our company.... We need to make sure the bean still exists,"*[13] explained Hanna.

In 2004, Starbucks voluntarily took stock of its Green House Gas (GHG) emissions with support from CH2M HILL[i], to evaluate its contribution to climate change. It included areas of retail, coffee roasting, administration operations, and its distribution network for its GHG evaluation and committed itself to reducing its emissions by purchasing renewable energy, addressing the impact of its transportation operations, monitoring its roasting operations, taking leadership, creating awareness, and setting reduction targets. Increasing energy and water usage per square foot of retail space spurred the company to intensify its focus on conservation measures in its retail stores. The company also looked for ways to install company-wide metrics and data-collection systems to measure the performance and environmental footprint of each store.

In 2008, Starbucks set its environmental and global responsibility goals for 2015 **(See Exhibit 15.5 for Starbucks Global Responsibility Goals and their progress by 2012).** In June 2009, the company re-evaluated its water consumption, and in September 2009, implemented a new water saving solution (saving 570 l of water per day in each store) in company-operated stores in Canada and the US. In 2010, the company made a commitment that every company-owned store it built in the world would be LEED-certified. In 2012, Starbucks also swapped conventional bulbs for LEDs in all company-operated stores. This would help each store reduce energy requirement for lighting by around 7%.[14]

Starbucks encouraged recycling among its partners. The company was also a major participant in the Sustainable Packaging Coalition, a group of packaging professionals that adopted sustainable principles and applied them in the development and procurement of new packaging materials.

Ethical Sourcing and Fair Trade

In 1994, Starbucks partnered with Conservation International to draft plans and audit its Coffee and Farmer Equity (C.A.F.E) program. It supported sustainable production of coffee through its C.A.F.E. practices, and socially responsible coffee buying guidelines. In 2000, Starbucks introduced a line of fair trade products. In 2006, 6% of the coffee beans purchased (136,000 metric tons) by the company was fair trade certified (FTC). And since then, the company had become the largest buyer of FTC coffee in North America. By launching the FTC line of coffee products in its stores worldwide, the company gave greater visibility to FTC products and in turn made a significant contribution to the coffee farmers by increasing the coffee volume purchased from them with each passing year. The entire espresso roast sold in the UK and Ireland bore the FTC label. The company further claimed that it paid above market prices for the coffee it purchased.

In March 2013, Starbucks announced that it would invest in a new 240 hectare farm and research center on Poas Volcano in Costa Rica. The company claimed that the farm

---

i   A US-based global full-service provider of consulting, design, construction, and operations services.

would be the company's global agronomy center, which would enhance its ethical sourcing program and help it in fulfilling its commitment to source all its coffee ethically by 2015.

Ethos Water

In 2003, Starbucks acquired Ethos – a bottled water brand sold in North America. Its bottle featured a label – helping children get clean water. The brand itself was not incorporated for charity and critics claimed that the label on the bottle misled consumers into thinking that Ethos was a charitable organization. However, the company stated that the brand was intended to raise awareness about clean water issues in developing nations and to provide socially responsible consumers to support the cause. Starbucks then changed the label of the bottle, to include a reference to the fact that 5 cents from each US$1.80 bottle sold were used to fund clean water projects in under-developed countries.

Inspiring Partners

Starbucks used targeted internal communication tools – like the weekly Scoop operations bulletin and the monthly Siren's Tale newsletter – and its Partner Portal online to educate partners about its sustainability efforts and to encourage them to act in a similar way. Major environmental programs were announced through the company-wide voicemail system. The communication system at the company was a two-way channel, where the partners could also ask questions or provide feedback about the environmental programs. The company also acknowledged its partners with the Bravo Award, MUG Award, Green Apron Award, etc., for their environmental efforts.

Starbucks Cups – The Problem

Around 60 million customers walked into Starbucks stores every week, according to the company. Around 80% of Starbucks customers in the US took their beverage on the go, while only 20% of customers did the same in other markets.[15] Starbucks had been using disposable paper cups to serve its customers on the go for a long time. It claimed that it had little option other than paper cups. Starbucks discarded an estimated 3 billion paper cups and 1 billion plastic cups each year, which bought the company under intense scrutiny. Nearly 80% of the paper cups ended up in landfills or as litter.

Though the discarded cups were made of paper, the thin plastic coating used to make the cups impermeable kept them from being recycled. Most recycling processors in the US did not take paper coffee cups because they were lined with a tiny amount of plastic or wax to make them watertight. Even Waste Management, the largest recycler of North America, did not accept Starbucks' cups for recycling. Commenting on the reluctance of the recyclers to accept Starbucks' cups, Hanna said, *"The biggest roadblock to recycling is the lack of demand."*[16] Eric Lombardi, a zero-waste pioneer who ran Eco-Cycle, US's largest not-for-profit recycling company, in Boulder, Colorado, explained, *"The recycling industry is already established for paper, plastic, and glass, and sorting out cups is not economically feasible. It doesn't go with newspaper or cardboard or office paper. I have to stand there and handpick it off a conveyor belt."*[17]

Paper recyclers were not sure of getting volumes of cups cheaply enough to invest in the equipment required to recycle such content. On the processing side, the first problem was that of scale, as the processing required different equipment and the capability to make the investments worthwhile and earn profits for recyclers. The second problem that arose from the use of paper cups was that customers often walked away with the cups and threw them into trash cans at some other place along with other trash. So the possibility of such cups returning to the recycler was very low. Then, there was the possibility of the consumer throwing the cup into the wrong trash can.

Hanna, speaking about making a fully recyclable cup, said, *"We define a recyclable cup not by what the cup is made out of but by our customers actually having access to recycling services."*[18] For instance, San Francisco was equipped to handle the material and hence Starbucks' paper cups there were recyclable, whereas in another city, the cups would be diverted to the trash. Hanna indicated that until 75% of Starbucks customers had access to recycling, it would not call its cups recyclable. In 2008, the company set a goal of making 100% of its cups recyclable by 2015.

Notwithstanding the various problems, Starbucks found new ways to recycle its cups. In 2010, Starbucks stores in Chicago sent all of their used cups to Wisconsin, where they were recycled into napkins by the Georgia Pacific paper mill. The company also had set up recycling and compost bins in its 90 stores. Critics opined that while the company's efforts would help promote recycling, it would not truly solve the problem of 'unrecyclable' Starbucks cups, and that there was a major disconnect as only 5% of its stores recycled the cups where mandated by law. As of 2011, in only a few cities like Seattle, Toronto, and San Francisco did the company offer recycling; most other communities did not recycle them as there was hardly any market for – and very little incentive to come up with such recycling facilities.

However, Starbucks continued to work toward its long-term goal of having 100% reusable or recyclable cups by 2015. But not everyone was enthusiastic about the goal. Conrad MacKerron, senior program director at As You Sow[ii], said, *"Our concerns focus on how they're going to implement it and meet the goals. It's going to be a lot of work in an area in which there's no precedent…you've got hundreds of communities far from a paper mill and that's a challenge."*[19] In 2011, As You Sow had filed a shareholder resolution asking Starbucks to broaden its recycling commitment. The company was asked to set goals for glass (used for Frappucino), plastic (Ethos water), and metal containers (Doubleshot) as well as paper

---

ii A non-profit organization promoting environmental and social corporate responsibility through shareholder advocacy, coalition building, and innovative legal strategies.

cups, and to keep track of how many customers used disposable versus ceramic cups so that it would know when that rate improved.[iii] The proposal received 8.1% support from the shareholders. A similar proposal in 2010 received 11% support.[20]

Early Efforts to Solve The Problem

In the 1990s, Starbucks formed a relationship with Environmental Defense Fund to work out a solution to this problem. At that time, the company used 100 percent bleached virgin paperboard and also used double-cupping on a regular basis. The company tried to increase the use of reusable ceramic or glass cups and dishes in its stores. In comparison to paper cups, ceramic cups and glass cups were better for the environment after 70 and 36 uses respectively. Moreover, Starbucks' financial analysis showed that the company could save more than US$1 million per year if it served its in-store customers their beverages in serverware made of ceramic. These cups could also be reused up to 1,000 times and reduced solid waste by more than 80% by weight compared to paper cups.

As an alternative to double-cupping, in 1997, the company introduced a corrugated paper sleeve to serve as an insulating layer. This was made from 60% postconsumer recycled fiber.[21] In the same year, the company introduced the recycled-content cup. After almost a decade of working with its suppliers and after securing the approval of US Food and Drug Administration, in 2005, it incorporated 10% post-consumer fiber into its cups that reduced the company's use of tree fiber by five million pounds annually.[22]

Further in 2008, Starbucks replaced its polyethylene terephthalate (PET) cold cup with a polypropylene alternative in the US, Canada, and Latin America, based on research which proved that polypropylene cups used 15% less plastic than PET cups and emitted 45% less GHG during their production.[23] Apart from these initiatives, the company had also encouraged customers since 1985 to opt for reusable options like ceramic mugs for in-house consumption and rewarded customers who bought their own travel tumblers with 10 cents off on the price of their beverages. Starbucks also sold ceramic tumblers priced between US$10 and US$18 in its stores.

Cup Summits

To come up with a solution to the cup problem, Starbucks hosted a series of Cup Summits. In May 2009, Starbucks hosted the first Cup Summit with representatives from its paper and plastic cup value chain. The goal of the summit was to prioritize and address the

---

iii Starbucks kept track of customers who brought their own reusable mugs into stores and carried them out, but it did not count customers who used its in-store serverware for hot and cold beverages consumed in its stores.

obstacles and opportunities of coming up with a recyclable cup. The company through the summit strove to achieve its long-term commitment to reduce the environmental impact of its cups. The company invited 30 cup, cupstock, and coating manufacturers, recyclers, peer retail and beverage partners, local municipal governments, environmental NGOs, waste managers, and university researchers to Seattle to discuss how to improve the recyclable cups. An equal number of Starbucks professionals led by CEO Howard Schultz too participated in the discussion. Peter Senge, senior lecturer at the Massachusetts Institute of Technology (MIT) and founding chair of the Society for Organizational Learning, was the moderator of the summit. According to Ben Packard, Starbucks' VP of Global Responsibility, *"We are taking this step to drive meaningful dialogue with an unprecedented cross-section of stakeholders who have the expertise and infrastructure to create a fundamental shift in the entire packaging industry. With this initial meeting, we hope to foster collaboration and define the specific actions that will allow us to achieve our goal of 100 percent recyclable cups."*[24]

Every aspect of the cups was discussed in the summit. The participants of the summit also agreed that a new standard disposable cup needed to be developed. Joel Kendrick, Western Michigan University's director of Paper & Coating facilities, reported that the trials done in his laboratory indicated that paper coffee cups including that of Starbucks were readily repulpable and recyclable. Kristen Newman, International Paper's marketing and business strategies director, exemplified the certified compostable container at the summit. Martha Stevenson, a senior fellow of GreenBlue Institute[iv], said, *"Compass is an online software tool for packaging designers and engineers to assess the human and environmental impacts of their packaging designs. In achieving the Starbucks promise, each element of the cup's composition will need to be considered insofar as its sustainability criteria."*[25] Hanna said that the company would use Compass to help achieve its recycling goals. Stephanie Jones (Jones), vice president of the Canadian Restaurant & Foodservices Association, highlighted the importance of involving the stakeholders in developing common, realizable protocols and standards for recycling and composting. Jones also highlighted the importance of educating customers and the responsibility of the government officials in providing proper infrastructure. The summit also addressed other topics like – Starbucks cups were already recyclable but often weren't recycled; collection and sorting of recyclable cups from the unrecyclable waste; and recycling of cold cups.[26]

During 2009, Starbucks introduced front-of-store recycling in Toronto, San Francisco, and many other places. However, in April 2010, when the company tracked its progress in terms of recycling, it found that only 299 of its company-operated stores had implemented the recycling program.[27] None of the licensed stores had any recycling program.

On April 22, 2010, the company hosted the second Cup Summit at MIT. The two-day summit attracted a wide variety of participants as in the earlier summit. However, the

---

iv A nonprofit organization that had established criteria for determining costs, carbon footprint, compatibilities and end-of-life properties through its Sustainable Packaging Coalition. Martha Stevenson oversaw the creation of SPC's software 'Compass' (comparative packaging assessment).

company's goal this time was to ensure that 100% of its cups were reusable or recyclable by 2015. While the company considered its cups to be recyclable, the major challenge that it faced was the variance in local recycling capabilities across its global locations. The company became the lead sponsor for the 'betacup challenge' launched by social entrepreneur Toby Daniels and mass collaboration specialist Colaboratorie Mutopo, to solve the problem of disposable cup waste. In June 2010, the winner was chosen out of a pool of 430 designs. In addition, the company encouraged customers to opt for reusable tumblers to reduce cup waste. In 2009, the company served around 26 million beverages in reusable cups in its company-owned stores in the US, Canada, and the UK, saving 1.2 million pounds of paper from wastage in the process.[28]

Starbucks conducted its third Cup Summit on September 9, 2011, at Boston, moderated by Hanna. Among the participants were Joe Burke, who worked for Action Carting Environmental Services[v], experts from MIT, and representatives from Canadian coffee chain Tim Hortons and paper giant Georgia-Pacific. Hanna said, *"There are folks from paper mills (here) who are fierce competitors of each other and vying for business and a piece of the puzzle. But the willingness of the system and of these competitors to come together for a common initiative is something that I think has been one of the shining achievements of this process, and essentially a model for moving forward on how to solve big issues across our industry sector."*[29] In the summit, Georgia-Pacific's representative opined that though Starbucks generated 3 billion waste cups per year at that time, this would only amount to four days' worth of paper. The discussion also highlighted that the 10 cent discount the company offered for bringing customer-owned tumblers was not enough of an incentive to get people to bring their own cups.

Noting that 80% of the paper cups went out into the community, Hanna said, *"Commercial recycling is 100 percent reliant on the local market. If we can work with local governments to make sure that cups are included as part of local recycling ordinances that prompts the creation of the local market where one may not exist."*[30] The huge difference in recycling infrastructure around the world and even from city to city only compounded the problem.

The company also implemented around 175 of the 750 customer ideas that were received through its 'My Starbucks Idea' website.[31]

Low-Cost Reusable Cups

In January 2013, Starbucks rolled out a US$1 reusable plastic cup at its cafés. Customers could carry the cup while on the move and have it refilled with a drink whenever they required. Reusable mugs and ceramic cups had been available at the Starbucks stores for a long time, but the company had been unable to achieve its targets for customer adoption.

---

v A US-based company that worked with Starbucks to collect its used cups for recycling and turning them into a commodity for the secondary market.

The reusable plastic cups were roughly of the same size and appearance as the company's paper cups. These cups were made in China and had fill lines inside denoting the size of the drinks – tall, grande, and venti. The cup was made of 100% polypropylene and was lighter than the Starbucks ceramic tumblers. The cups could be used 30 times. The words "Reusable Recyclable* Re-enjoyable" were emblazoned on the cups with the asterisk denoting that they were not recyclable in all areas.[32] **(See Exhibit 15.6 for an image of the reusable cup)** Hanna said, *"It looks and feels like our existing paper cup, maintaining the brand attachment and Starbucks experience. For folks who were previously comfortable with our single-serve cups, this gives them an option for a reusable cup. And, because it's only a buck, if you forget it at home or in your car or office, you can buy another one."*[33] Starbucks offered a 10 cent discount to customers each time they brought in these reusable cups for refills.[34] And every time the customer brought in the reusable cups the baristas offered to rinse or clean the tumbler with boiling water before refilling them.

In November 2012, Starbucks had tested these reusable cups in 600 Northwest Pacific stores before rolling them out in the company's cafés across the US and Canada. According to Starbucks, during this period the use of reusable cups rose by 26% compared to the same month in 2011.[35] According to a YouGov Omnibus survey conducted on January 4-6, 2013, 28% of the respondents reported that they had already purchased or planned to purchase one of the new reusable cups **(See Exhibit VII for the results of the survey)**.

In April 2013, Starbucks introduced the reusable cup in the UK, priced at £1. In the UK, customers got 25p off on their Starbucks drink every time they used the reusable cup. Ian Cranna, vice-president of UK marketing for Starbucks, said, *"We know that our customers really care about saving money and doing their bit for the environment; between 2008 and 2012, the number of people using a Starbucks reusable tumbler increased by 235% and our new reusable cup is a low-cost, high-impact way to help make a difference on reducing waste."*[36] The company made the reusable cups available in select stores across the nation but proposed to gradually roll them out at other places as well.

In July 2013, Hanna said that the reusable cups were popular with its customers. *"After the introduction of the $1 reusable cup, we saw a significant bump in the purchase of reusables overall. But, more importantly, we also saw a marked increase in the number of people bringing their mugs back into the store to be reused,"*[37] said Hanna.

Looking Ahead

While many applauded Starbucks' various initiatives to address the issue with its cups, there were also many skeptics. Since 1985, Starbucks had offered a 10 cent discount to customers who brought their own reusable cups. However, the company was thwarted by the customers' force of habit. Starbucks claimed that, in 2012, 35.8 million beverages were served in reusable cups brought in by customers. *"Unfortunately, we've hovered around 2 percent since the inception of the incentive program. The numbers dance around that a bit, but that's been where we've been locked in for years. What we discovered is that the discount* we offer is great, but it's only a driver for a limited number of consumers,"[38] said Hanna. The company attempted other tactics too. In December 2012, it put a US$30 reusable cup on sale and offered free coffee for the whole month of January 2013. But this failed to persuade customers to use the cup. *"We've done some pretty extensive customer research on enhancing the 10-cent discount that we've always provided our customers for using reusable mugs. Simply upping the discount has no impact on customer behavior,"*[39] Hanna added.

Starbucks' critics contended that the corporate sustainability of the company was akin to greenwashing as the billions of paper cups the company claimed to be recyclable were actually not recyclable by most recycling plants. They also questioned whether the company would be successful with its US$1 reusable cups as people often forgot to carry them into the store. Other retailers had also struggled with this problem as even eco-minded consumers found it hard to remember to carry reusable cups and bags with them.[40] Mary Beth Quirk at The Consumerist, said, *"Once the initial novelty wears off, will anyone actually remember to bring the dang things in? After all, how many times have you arrived at the grocery store only to realize you've forgotten your environmentally-friendly shopping bags?"*[41]

Then, there were the various challenges related to recycling the cups. While acknowledging the importance of the problems posed by the cups, Hanna felt that Starbucks was not getting due credit for the good work the company was doing on the sustainability front as the customers and the media tended to focus only on the cups issue. He felt that these stakeholders did not even understand the complexities involved in sorting out the issue and Starbucks' long-term commitment to solving the problem. Hanna said, *"This is a problem of perception, in that people think that if they can already recycle something in their Blue Bin, why shouldn't they be able to do it at Starbucks? Why can't they do it in the public space? The perception is that recycling should be a very easy solve, and we just need to do it, just put the bins in front of our stores and solve this. Why is it going to take five years to solve this issue?"*[42]

Exhibit 15.1. Starbucks' Financial Summary

As of and For the Fiscal Year Ended	Sep 30, 2012	Oct 2, 2011	Oct 3, 2010	Sep 27, 2009	Sep 28, 2008
Results of Operations	In millions except per share data				
Net revenues:					
Company-operated stores	$ 10,534.5	$ 9,632.4	$ 8,963.5	$ 8,180.1	$ 8,771.9
Licensed stores	1,210.3	1,007.5	875.2	795.0	779.0
CPG, foodservice, and other	1,554.7	1,060.5	868.7	799.5	832.1
Total net revenues	**$ 13,299.5**	**$ 11,700.4**	**$ 10,707.4**	**$ 9,774.6**	**$ 10,383.0**
Operating income	$ 1,997.4	$ 1,728.5	$ 1,419.4	$ 562.0	$ 503.9
Net earnings including non-controlling interests	1,384.7	1,248.0	948.3	391.5	311.7
Net earnings (loss) attributable to non-controlling interests	0.9	2.3	2.7	0.7	(3.8)
Net earnings attributable to Starbucks	1,383.8	1,245.7	945.6	390.8	315.5
EPS — diluted	1.79	1.62	1.24	0.52	0.43
Cash dividends declared per share	0.72	0.56	0.36	—	—
Net cash provided by operating activities	1,750.3	1,612.4	1,704.9	1,389.0	1,258.7
Capital expenditures (additions to property, plant, and equipment)	856.2	531.9	440.7	445.6	984.5
Balance Sheet					
Total assets	$ 8,219.2	$ 7,360.4	$ 6,385.9	$ 5,576.8	$ 5,672.6
Short-term borrowings	—	—	—	—	713.0
Long-term debt (including current portion)	549.6	549.5	549.4	549.5	550.3
Shareholders' equity	5,109.0	4,384.9	3,674.7	3,045.7	2,490.9

Note: Starbucks' fiscal year ends on the Sunday closest to September 30. The fiscal year ended on October 3, 2010, included 53 weeks with the 53rd week falling in the fourth fiscal quarter.

Source: "Starbucks Corporation Fiscal 2012 Annual Report", http://investor.starbucks.com, 2012

Exhibit 15.2. Break-up of Company-operated and Licensed Stores

	Company-operated	Licensed
Americas	7,857	5,046
EMEA	882	987
CAP	666	2,628
Total	**9,405**	**8,661**

Data as of September 30, 2012.
Adapted from Starbucks Corporation Fiscal 2012 Annual Report.

Exhibit 15.3. Starbucks' Growth in Company-operated Stores

	Sep 27, 2009	Oct 3, 2010	Oct 2, 2011	Sep 30, 2012
Americas				
US	6,764	6,707	6,705	6,866
Canada	775	799	836	878
Chile	-	-	35	41
Brazil	-	-	28	53
Puerto Rico	-	-	19	19
Total Americas	-	-	7,623	7,857
EMEA				
UK	666	601	600	593
Germany	144	142	150	157
France	-	-	62	67
Switzerland	-	-	46	50
Austria	-	-	12	12
Netherlands	-	-	2	3
Total EMEA	-	-	872	882
CAP				
China	191	220	278	408
Thailand	131	133	141	155
Singapore	-	-	72	80
Australia	-	-	21	23
Total CAP	-	-	512	666
Total company-operated	**8,905**	**8,833**	**9,007**	**9,405**

Adapted from Starbucks' Annual Reports, 2010, 2011, and 2012

Exhibit 15.4. Country-wise Breakup of Starbucks Licensed Stores

Americas		EMEA		CAP	
US	4,262	Turkey	171	Japan	965
Mexico	356	UK	168	South Korea	467
Canada	303	UAE	99	China	292
Other	125	Spain	78	Taiwan	271
		Kuwait	65	Philippines	201
		Saudi Arabia	64	Malaysia	134
		Russia	60	Indonesia	133
		Greece	42	Hong Kong	131
		Other	240	New Zealand	34
Total	**5,046**	**Total**	**987**	**Total**	**2,628**

*Data as of September 30, 2012.

Source: Starbucks Corporation Fiscal 2012 Annual Report.

Exhibit 15.5. Starbuck's Global Responsibility Goals and Progress

Goals	2011	2012
Ensure 100% of our coffee is ethically sourced by 2015		
Total coffee purchase (lbs, millions)	428	545
Total ethically sourced coffee purchase (lbs, millions)	367	509
Invest in farmers and their communities by increasing farmer loans to $20 million by 2015		
Farmer loans (US$, millions)	14.7	15.9
Build all new, company-owned stores to achieve LEED® certification		
Percentage of new company-owned stores built to achieve LEED® certification.	75	69
Reduce energy consumption by 25% in our company-owned stores by 2015		
Average electricity use per square foot/store/month in US and Canada company-owned stores (in KWH)	6.29	6.36
Percentage change to the 2008 baseline of 6.8 KHW.	7.5	6.5
Reduce water consumption by 25% in our company-owned stores by 2015		

Exhibit 15.5. (continued)

Goals	2011	2012
Average water use per square foot/store/month in US and Canada company-owned stores (in gal)	20.11	20.08
Percentage change to the 2008 baseline of 24.35 gal.	17.6	17.7
Purchase renewable energy equivalent to 100% of the electricity used in our global company-owned stores by 2015		
Purchase renewable energy (in million KWH)	558	586
Percentage of total	50.4	51
Develop comprehensive recycling solutions for our paper and plastic cups by 2012		
As of 2013, operational and being expanded in US, Canada, UK, and Germany; still working toward materials and/or infrastructure solutions for other remaining company-owned markets.		
Implement front-of-store recycling in our company-owned stores by 2015		
Percent of store locations with front-of-store recycling (US and Canada company-owned stores.)	18	24
Serve 5% of beverages made in our stores in personal tumblers by 2015		
Beverages served in personal tumblers in US, Canada, UK, Ireland, and Germany company-owned stores (millions)	34.2	35.8
Percentage of total	1.5	1.5
Improve farmers' access to carbon markets, helping them generate additional income while protecting the environment.		
Mobilize our partners (employees) and customers to contribute 1 million hours of community service per year by 2015		
Total service hours (includes all global markets regardless of ownership)	442,353	613,214
Engage a total of 50,000 young people to innovate and take action in their communities by 2015		
Total number of people engaged	50,050	54,848

Adapted from Starbucks Global Responsibility Report – Goals and Progress 2012

Exhibit 15.6. Image of the US$1 Reusable Cup

Source: www.wehatetowaste.com

Exhibit 15.7. Summary of the Results of the YouGov Omnibus Survey Conducted on January 4-6, 2013

- 2% of the respondents said they had already bought one of the new reusable cups.
- 7% said they will "definitely" buy one but haven't done so yet.
- 19% said they will "probably" buy one.
- 12% of non-Starbucks customers said they intend to purchase one of the cups.
- 70% said they believe it's a good idea.
- 40% said they believe it's a "very good" idea.
- 57% said they probably or definitely won't purchase one of cups.
- 66% had not yet heard of the cup offer.
- 38% said that Starbucks should be applauded for its attempts to go green.
- 23% said they think Starbucks is "generally an ethical company."
- 13% said they wondered if the reusable cup is a "publicity stunt."

*Results were weighted to be representative of U.S. adults 18 and older, with a margin of error of +/- 3.6% Adapted from Karlene Lukovitz, "Survey: Solid Uptake For Starbucks' Reusable Cup," www.mediapost.com, January 10, 2013

Notes

1. Anne Marie Mohan, "Starbucks Approaches Recycling Goal With Systems-Based Approach," www.greenerpackage.com, August 1, 2011
2. Siobhan Wagner, "10Q: Starbucks 2015 Environment Goals are a Cup Half Full," www.bloomberg.com, February 28, 2012
3. Jason Hagey, "Starbucks Makes the Business Case for Sustainability-- And Whipped Cream," www.awb.org, May 7, 2013
4. Lisa McTigue Pierce, "Packaging is the Gateway to a Deeper Conversation about Sustainability," www.packagingdigest.com, July 15, 2013
5. Bart King, "Starbucks Introduces $1 Reusable Coffee Cup," www.sustainablebrands.com, January 3, 2013
6. "Starbucks Disposable Cups Deemed "Unrecyclable" by Major Recycling Companies," www.justmeans.com, July 2, 2010
7. Anne Marie Mohan, "Starbucks Approaches Recycling Goal With Systems-Based Approach," www.greenerpackage.com, August 1, 2011
8. "Starbucks Corporation Fiscal 2012 Annual Report," http://investor.starbucks.com, 2012
9. "Starbucks Introduces Innovative Cross-Channel, Multi-Brand Loyalty Program and Announces Global Social Impact Initiatives at Annual Meeting of Shareholders," http://investor.starbucks.com, March. 20, 2013
10. Brenda Timm, "Sustainability: A Success Strategy at Starbucks," www.greenatworkmag.com, July/August 2005
11. Charley Cameron, "Interview: 8 Questions with Starbucks Global Responsibility VP Ben Packard," http://inhabitat.com, August 22, 2013
12. www.printbuyersonline.com/PrintOasis/PrintOasisContent.asp?id=3814
13. Jason Hagey, "Starbucks Makes the Business Case for Sustainability-- And Whipped Cream," www.awb.org, May 7, 2013
14. Siobhan Wagner, "10Q: Starbucks 2015 Environment Goals are a Cup Half Full," www.bloomberg.com, February 28, 2012
15. Anne Marie Mohan, "Starbucks Approaches Recycling Goal With Systems-Based Approach," www.greenerpackage.com, August 1, 2011
16. "Starbucks Disposable Cups Deemed "Unrecyclable" by Major Recycling Companies," www.justmeans.com, July 2, 2010
17. Marty Lariviere, "Why Can't You Recycle a Starbucks Cup?" http://operationsroom.wordpress.com, November 16, 2010
18. Lori brown, "Can Starbucks Find a Way to Recycle 4 Billion Cups?" http://earth911.com, April 26, 2010
19. Alison Neumer Lara, "Starbucks Cup Recycling: What's the Holdup?" http://earth911.com, April 5, 2011
20. Melissa Allison, "Not a Peet's Peep from Starbucks at Annual Meeting," *The Seattle Times*, March 23, 2011
21. "Starbucks: Improving Cups," http://business.edf.org/casestudies/starbucks-improving-cups

22. Brenda Timm, "Sustainability: A Success Strategy at Starbucks," www.greenatworkmag.com, July/August 2005
23. "Starbucks Brings Thought Leaders Together to Develop a Comprehensive Recyclable Cup Solution," http://news.starbucks.com, May 11, 2009
24. "Starbucks Brings Thought Leaders Together to Develop a Comprehensive Recyclable Cup Solution," http://news.starbucks.com, May 11, 2009
25. Scott Seydel, "Starbucks' Cup Summit: Does the Cost of Recycling Runneth Over?" www.greenbiz.com, May 15, 2009
26. Robin Shreeves, "Starbucks Looks to Recycle Billions of Cups," www.mnn.com, May 18, 2009
27. Clara Kuo, "Starbucks Holds Cup Summit, More Talk About Reducing Environmental Impact," www.triplepundit.com, April 24, 2010
28. "Starbucks Hosts Second Cup Summit to Advance Recycling Initiatives," http://news.starbucks.com, April 22, 2010
29. Tilde Herrera, "Starbucks' Third Cup Summit Marks Progress, Challenges in Recycling," www.greenbiz.com, September 9, 2011
30. Alison Neumer Lara, "Starbucks Cup Recycling: What's the Holdup?" http://earth911.com, April 5, 2011
31. Tilde Herrera, "Starbucks' Third Cup Summit Marks Progress, Challenges in Recycling," www.greenbiz.com, September 9, 2011
32. Miranda Farley, "Starbucks Reusable Plastic Cups — Green Marketing or Greenwash?" www.wehatetowaste.com, March 19, 2013
33. Lisa McTigue Pierce, "Packaging is the Gateway to a Deeper Conversation about Sustainability," www.packagingdigest.com, July 15, 2013
34. "Starbucks to Offer Recyclable Cups," www.bizjournals.com, January 3, 2013
35. Karlene Lukovitz, "Survey: Solid Uptake For Starbucks' Reusable Cup," www.mediapost.com, January 10, 2013
36. Rebecca Smithers, "Starbucks Introduces Reusable Cups," www.guardian.co.uk, April 19, 2013
37. Lisa McTigue Pierce, "Packaging is the Gateway to a Deeper Conversation about Sustainability," www.packagingdigest.com, July 15, 2013
38. *Ibid.*
39. Siobhan Wagner, "10Q: Starbucks 2015 Environment Goals are a Cup Half Full," www.bloomberg.com, February 28, 2012.
40. Karlene Lukovitz, "Survey: Solid Uptake For Starbucks' Reusable Cup," www.mediapost.com, January 10, 2013
41. Ashley Lutz, "Starbucks $1 Coffee Cups Might be the New Reusable Grocery Bag," www.businessinsider.com, January 9, 2013
42. Anne Marie Mohan, "Starbucks Approaches Recycling Goal With Systems-Based Approach," www.greenerpackage.com, August 1, 2011

Discussion Questions

1. What is the role played by Starbucks in furthering the cause of sustainability?

2. How has Starbucks tried to address the issue of climate change?

3. Explain the nature of the problem with respect to Starbucks' cups.

4. How has Starbucks tried to solve the aforementioned problem?

5. Explain the nature of activities occurring at the cup summit.

Index

A

aquifers 266

B

Biofuels 84
blue water 261

C

cap and trade 58, 74
Carbon Cost 83
Climate Change 20
Coal 73

E

electrolyte 131
Energy 161

F

Financing 336
Fossil Fuels 102
fuel cell 120

G

Gaia hypothesis 17
Global warming 21
Greenhouse gases 240
Green water 261

K

Kyoto protocol 23

M

modular product 311

N

Natural gas 74

P

Petroleum 69
photosynthesis 240
Photovoltaics 168

R

Recycle 58
Refurbishing 305
Renewable Energy 87
reuse 10

S

solar power 9

W

wind power 184
wind turbines 89

CPSIA information can be obtained
at www.ICGtesting.com
Printed in the USA
BVHW050313120721
611475BV00004B/148